Patterns, Thinking, and Cognition

Howard Margolis

Patterns, Thinking, and Cognition

A Theory of Judgment

The University of Chicago Press
Chicago and London

HOWARD MARGOLIS is senior lecturer in the Committee on Public Policy Studies at the University of Chicago. He is the author of *Selfishness, Altruism, and Rationality,* also published by the University of Chicago Press.

The illustrations on the opening pages of chapters 13 and 14 (which also appear on the front and back covers of the paperback edition) are from *The Anatomy of Judgment* by M. L. Johnson Abercrombie (published in 1960 by Hutchinson Limited), after a puzzle picture in the *American Journal of Psychology* 67 (1954). Permission for cover illustration granted by The University of Illinois Press.

The University of Chicago Press, Chicago 60637
The University of Chicago Press, Ltd., London
 Published 1987
Printed in the United States of America

96 95 94 93 92 91 90 89 88 5 4 3 2

Library of Congress Cataloging-in-Publication Data

Margolis, Howard.
Patterns, thinking, and cognition.

Bibliography: p.
Includes index.
1. Cognition. 2. Thought and thinking.
3. Differentiation (Cognition) 4. Pattern perception.
I. Title.
BF441.M27 1987 153 87–12805
ISBN 0–226–50527–8
ISBN 0–226–50528–6 (pbk.)

To Tom Schelling

Contents

Preface

This book started out as an attempt to apply recent insights from cognitive psychology to a salient puzzle for students of politics, which is that often what people believe is only weakly correlated with a reasonable assessment of the objective logic of a situation. Of course the same thing can be noticed in the context of private behavior. But there it is easy to attribute the difficulty to personal idiosyncrasy (he is stupid, she is stubborn, and so on). But the mismatch between belief and logic is more striking when the issue has been carefully studied and when the effect extends to people who have no great personal stake in the matter. Both points have larger consequences for social than for private judgment. Even when a reasonable consensus has developed among people who have studied the details of a situation, it is often hard to make that consensus politically credible. The puzzle persists even after making full allowance for differences in expertise and for any nontautological definition of conflicting interests. But if logic + interests cannot account for beliefs, apparently we have to add something contingent on the psychology of belief and persuasion. So we get an extended formulation, allowing for logic + interests + cognition. By itself, that extended formula then merely adds a black box containing whatever it is about human brains that makes it impossible to reliably derive beliefs from logic + interests. The study is an attempt to learn something of what is in that box.

Obviously the study depends heavily on the large prior literature. Some material very prominently in mind when I began the study was the work on illusions and quirks of judgment by Peter Wason and various coworkers in England and by Daniel Kahneman & Amos Tversky here. It seemed to me from the outset that these scholars must be right in their intuition that oddities of judgment that are revealed by special contexts (illusions of judgment, parallel to the familiar illusions of vision) should yield deep clues to how cognition works. Then Gombrich's *Art and Illusion* and in a more detailed way Kuhn's *Structure of Scientific Revolutions* seemed to offer a larger sense of how judgment works that might be used to tease something out of the clues provided by illusions of judgment. The book grew out of a sense that if an account could be worked out that fitted all this material together in a reasonable way, it would have something consequential to say about the political puzzle.

Given this start, I now have to explain why (as you will see) the focus for the empirical applications is almost entirely on the history of science, and barely at all on the explicitly political issues that prompted the study. I reach a clearly political context only in the very last chapter, and then only as a continuation of the account of the Copernican revolution that makes up most of the second half of the book. The reason the book is so much about the history of science is that I early discovered that it is impossible to talk about any familiar political issue with the sort of clinical detachment that we need to deal with the components of judgment that elude description in terms of logic + interests. We can all see that there is such a component to our adversaries' judgment, but we can't see it in ourselves. Even to suggest that possibility is taken as insulting. So it doesn't work to use current political issues as the occasion for cool analysis of the limited role of rational norms in accounting for our political judgments.

But having begun to think about controversies in the history of science—which as much as political controversies are cases of social choice—I realized there is a great additional advantage to a focus on science. For motivation in politics is intrinsically convoluted. But in science it is very rare that an actor has a motivation that seriously conflicts with his desire not to look like a fool to the next generation of graduate students. We get cleaner cases—cases more amenable to a reasonably straightforward analysis—than we find in a political issue of more than trivial interest. When we reach the account of Galileo and the pope in the concluding chapter, readers interested in politics will at last feel at home. But by then the new analytical points are concerned especially with political judgment. The basic cognitive arguments are worked out in the earlier chapters.

Since the study touches on a very wide range of material, I accumulated an exceptional number of debts to people who have shared their expertise with me and from audiences from half-a-dozen disciplines that have allowed me to try out pieces of the argument. The list that follows is certainly incomplete, but it tries to include those who provided comments on a chapter or more of the manuscript: Wilbur Applebaum, John Bonner, Paul & Patricia Churchland, Stillman Drake, Howard Gardner, Owen Gingerich, Ami Glaser, Howard Gruber, Rusell Hardin, Gilbert Harman, Robin Hogarth, David Hull, Jean Lave, George Loewenstein, Robert May, Robert Merton, George Miller, Jamel Ragep, Robert Richards, Marc Roberts, Tom Schelling, Jonathan Schull, R. S. Westfall, and Robert Westman. I owe particular thanks to Westman, who was my initial guide to the Copernican literature and the history of science community as well as a valuable commenter on draft chapters. And of course I owe thanks to many others who offered useful comments or information at meetings or seminars or in response to some particular query.

I was also fortunate to receive just enough financial help to make it possible to carry out the study—which in my case meant a great deal of such help, since while writing the book I lacked a normal academic appointment. The beginnings of the study were done when I was still a research associate at MIT, finishing up an earlier book (Margolis 1982). Harvey Sapolsky, Gene Skolnikoff, George Rathjens, and Jack Ruina were all helpful, as was Daniel Rubinger of the Transportations System Center. The beginnings of the study consisted of papers with a focus on cognitive questions in the context of various public policy issues. I was able to turn full attention to the project during a year as visiting scholar at the Russell Sage Foundation, where Marshall Robinson and Peter de Janosi arranged for support that was generous in every way, and continued that support to supplement the stipend I received from the Institute for Advanced Study, Princeton the following year. Carl Kaysen was helpful in obtaining both appointments. In addition to hospitality and sustenance, the Institute provided the stimulation of a group of fellow-visitors concerned with cognitive questions. I was then able to continue the work while holding a visiting appointment at the University of California at Irvine, where the economics group was particularly hospitable, the word-processing center particularly patient, and the cognitive science group particularly generous in providing me with both advice and access to their subject pool. Finally, I have been able to finish the project at the University of Chicago, a most congenial place for the sort of interdisciplinary work presented here. Albert Hirschman, Julie Margolis, and Russell Hardin were critical for the Institute, Irvine, and Chicago appointments, respectively.

Naturally, this would have been extremely difficult without a very patient family, and in particular a very patient wife. And although I like to think that somehow or other I would have gotten to do this work even without the good fortune of enlisting the support of one exceptional individual, the fact is that I don't see how. So I owe an extraordinary debt to Tom Schelling, who supported publication of my first book, encouraged and counseled me on this book, and encouraged others to support the work. I have offered to do something useful for him, such as mowing the lawn, but he always declines. So he must settle for the dedication.

Introduction

[Bohr] never trusted a purely formal or mathematical argument. "No, no," he would say, "You are not thinking; you are just being logical."

O. R. Frisch, *What Little I Remember,* 1979

This study gives an account of thinking and judgment in which—to lay cards immediately on the table—everything is reduced to pattern-recognition. At least for the tactical purpose of seeing how far we can get with this sort of argument, there is nothing to be said for mincing words, leaving loopholes, making excuses. Further, I will take pattern-recognition as the starting point for the analysis, not as something which is itself subject to analysis. Like atoms for a chemist circa 1900 or genes for a geneticist circa 1950, I treat the central notion as something that is not (here) subject to analysis or even to direct observation. The chemist and geneticist were able to build theories with many observable consequences decades before they could observe an atom or a gene, or say anything at all about what an atom or a gene might really be like. I will try to show that a similarly blunt exploitation of the notion of pattern-recognition can also produce a useful theory.

So I will be working out an analysis of cognition in which everything is reduced to what I will call "*P*-cognition"—to sequences of the sort suggested by the spiraling figure below (see fig. 1). At the top of each cycle

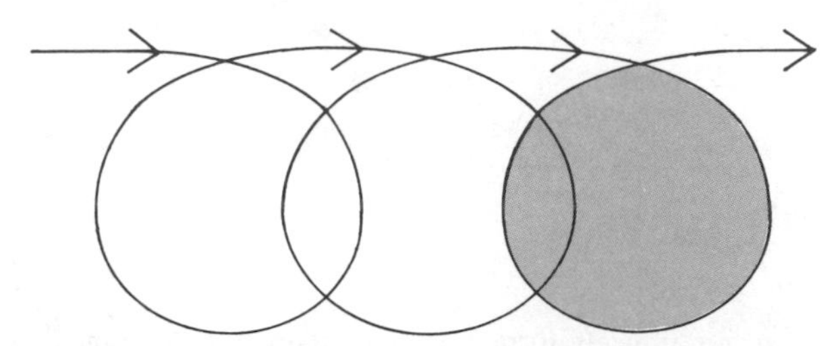

Figure 1. *P*-cognitive spirals. Each spiral represents a cognitive cycle. The arrow at the top of each spiral indicates the prompting of some pattern, contingent on the immediate context, the experience of this individual, and the priming/inhibiting effects of recently prompted patterns. Only a small fraction of these prompted patterns could be expected to come to conscious attention.

a pattern (the arrangement of features in a room, phonemes in a word, for examples) is prompted by cues in the context. That pattern itself then becomes part of the environment which cues the next pattern. Sometimes a pattern, or a feature or subpattern within that pattern, is externally expressed (something is said or done); sometimes (not the same thing) a feature or pattern comes to conscious attention. An externally expressed pattern is more likely to be (or eventually come to be) conscious. Our friends, or enemies, might notice it, even if we don't. Whether conscious or not, though, cues prompt a pattern which, interacting with the rest of the context (hence the spiraling rather than box-arrow-box form of the figure), provides an amended set of cues, which prompts the next pattern. This can go on in several dimensions at once, as when a person plays the piano and carries on a conversation at the same time.

Subjectively (but probably not so at the level of neurophysiology) the patterns are of two contrasting sorts. Sometimes they seem to concern *context* and sometimes they seem to concern *action*. The first is static, like our sense of a familiar face or room. The second involves an unfolding or playing-out, as in a movement or turn of phrase or line of argument. But the distinction between context patterns and action patterns plays little role in the study, and I am skeptical that it has any deep significance. In any case, I will almost never have occasion to explicitly distinguish between context and action patterns, since in context which is relevant is usually evident.

Habits of mind, which play a central role in the argument, should be understood as applying equally to the fluent, automatic prompting of patterns of either sort from the available cues, including among the cues of course whatever patterns (context or action) have been recently prompted.

Further essential complications will develop as the argument unfolds. But the spiral of figure 1 provides a basic metaphor for the cues-to-patterns-to-cues-to-patterns process.

2

That pattern-recognition is central to thinking is a familiar idea. Everyone who writes on these matters discusses—in some way or another—patterns, frames, schemata, attunements, and so on, which are somehow recognized and then provide a key input to whatever happens next. And I am not alone in making the tuning of patterns of response to patterns of experience not just an aspect of cognition but the central notion (see, for example, Dreyfus & Dreyfus 1986; Lakoff 1986; Edelman 1985; Rumelhart et al. 1986, and Holland et al. 1986 for diverse ways in which this sense of cognition can be pursued). For the most elaborate computerized character recognizer cannot yet match a bright six-year-old in recognizing letters of the alphabet from many different fonts. No computer, in fact, can yet match a well-trained pigeon in distinguishing pictures that contain some person from pictures that don't (Herrnstein et al. 1976; Abu-Mustafa & Psaltis 1987).

So it is essentially universal to concede an essential role to the cuing of patterns and patterned responses; but precise articulation of just what is happening when a pattern is recognized is an unsolved problem. Because so little can be said, the dominant tendency (until very recently, at least) has been to move as quickly as possible from pattern-recognition to some algorithmic, rule-following procedure which can be articulated in a discussion and perhaps instantiated on a computer program. Pylyshyn's recent book (1984: 196–97) (to mention one example of several available) comes very close to claiming that anything that can't be handled in the style of propositional logic—such as learning, affect, and intuition—can be shuffled into a box labeled "functional architecture" and defined to be not part of cognition.

Obviously, I risk erring in the other direction. Unlike the many students of cognition for whom a viable cognitive theory must be implementable as a computer program, I have disowned trying to build a theory of that sort. Pattern-recognition is all there is to cognition, on the account I will give; but no one can say much yet about what the brain is doing when it recognizes a pattern, and I will have essentially nothing to say about that. Then how does the study actually say anything concrete? Without an explicit theory of pattern-recognition, we can hardly expect to deduce what patterns will be perceived in a given context. Yet even without such a theory, given an observation that a certain response has been prompted in a certain context, we may be able to think of (guess, intuit) a pattern that would include that response. We can then suggest what cues appear critical to prompting that pattern. We can then test such a suggestion by observations that alter the cues in some specified way.

There is a good deal of that in the study, increasingly so as we move more directly to applications of the argument in the later chapters. This procedure is applied in a literal way to a sampling of well-known cognitive anomalies (Chap. 8), and it is approximated in later chapters in a new analysis of some much-studied aspects of the Copernican revolution. For the historical episodes, of course, we cannot rerun the history with some cues altered as we can replicate the experiments with some cues altered. But the passage of time does something like that for us. The cues and patterns available to Copernicus were different from those available to Aristarchus, Tycho, Kepler, Galileo; and I will try to show how, in terms of the argument of the study, very plausible and perhaps even convincing new accounts emerge of several of the most striking puzzles connected with the Copernican revolution.

3

Another salient question is how the theory handles the empirical fact that human beings, while certainly faulty as step-by-step followers of logical or algorithmic rules, are sometimes able to approximate that sort of performance. We do, after all, follow, and with even more difficulty create, step-by-step sequences like computer programs, mathematical proofs, instructions for assembling bicycles, and so on. But within the argument developed here, that must be done in terms of cues-to-patterns sequences. Rule-following processes, including logic, must be reduced to pattern-recognition, not the reverse. The brain, on the account here, is not properly characterized as illogical or irrational. But it is certainly misleading to call it logical or rational. It is, rather, a-logical and a-rational. What I will try to show is how such a brain can produce behavior and arguments and theories that are themselves logical, where it is always the product of cognition that is (sometimes) logical, never the process itself.

If cognition turns out to be reducible to computation (and in some sense I suppose that must be possible), on the argument here the steps will not be of a sort we could recognize as looking like the steps in formal logic or in mathematics.

In his *Language of Thought,* Fodor (1975: 28–29) remarks that he takes it for granted that mental operations often have a form that (you can see by looking at his list of properties) looks remarkably like what you will find in the opening pages of a text on decision theory. The brain chooses an act by manipulating sets of possible states of the world and their associated subjective probabilities and utilities. But if the argument here is sound, what goes on in the brain is nothing like that. It would not work that way even in the very special case of a decision-theorist trying to demonstrate how to think that way.

4

P-cognition amounts to a kind of gestalt reversal in the roles of algorithmic (rule-following) processes versus pattern-recognition. Perhaps a way to give some quick insight into the character of this claimed reversal is to point to a feature of the analysis of experiments that is common to leading partisans on both sides of the continuing debate over human rationality but which is not permitted under the argument I will develop. Across the lively debate over "rationality," we find judgment treated (often implicitly, but sometimes explicitly, as in Henle 1962 or in Tversky & Kahneman 1981, to mention two particularly influential papers) as a two-stage process. In the first stage the problem is interpreted (Henle) or framed (Tversky & Kahneman). In the second stage, a judgment is produced, given the interpretation or framing.

Characteristically, the first (framing) stage does not involve an explicit process. Rather, the person is treated as somehow recognizing or imputing a particular way of seeing the problem, in a way that is not itself ordinarily subjected to analysis. Instead, it is only at the second (judgment) stage that we get an analysis, and that comes in terms of a comparison with some normative standard of rational judgment—almost always favorably for Henle and her allies, often unfavorable for Kahneman & Tversky and their allies. For both sides, that second stage is treated as step-by-step articulatable (even when the articulation is not consciously followed by the subjects). For Henle and her allies the process is something close to standard logic; for Kahneman & Tversky and their allies it relies on rules of thumb, heuristics, biases, articulatable not in terms of logic but by procedures that often vary from standard normative accounts (prospect theory, decision-by-aspects).

So one side sees anomalous judgments essentially as consistent with logic, though sometimes logic applied to an interpretation of the problem not intended by the setter of the problem; the other side sees such situations in terms of rule-of-thumb procedures cued by the framing of a situation, often in a context for which the procedure is logically inappropriate (Rumelhart 1981). We then get arguments about whether an anomaly is due to something odd at the framing stage or something odd in the judgment stage. Proponents who see themselves as defenders of human rationality naturally prefer the former. For then it becomes easier to argue that the oddity of framing is only different from, not inferior to, what the experimenter intended; or that it was the impoverished or otherwise peculiar context of an experiment, not anything inherent in the way brains work, that accounts for the anomaly.

In the present study, however, no such division of responses into a framing stage and a judgment stage would arise. In the way that I hope will become increasingly clear as the study develops, we have always the spiraling process introduced at the very start and worked out in detail in the balance of the study. This does not fit neatly on either side of the rationality controversy. In terms of *P*-cognition, an anomalous response will almost always in fact be a reasonably logical response to another question (as Henle has claimed), and in particular to a question that means something in the life experience of the individual giving the response. But the other question will often turn out to be a logically irrelevant or absurd interpretation of the context that actually prompted the response.

5

There may be no piece of the *P*-cognitive argument that cannot be found elsewhere. But if there is a fatal flaw in the study, it is not that it too much merely restates what everyone knows about cognition, or what someone else, or a few other people together, have said. That could not explain the common features reported on such much-discussed material as Wason's 4-card test and the Kahneman & Tversky taxi problem. Nor can it explain the new analysis that emerges of so exhaustively a studied piece of history as the Copernican revolution. If I am led to more coherent accounts of these much-discussed experiments and historical episodes, there is a reasonable presumption that I am being helped by a stronger theory than has been available before.

But an unconventional analysis is almost inevitably first seen (above all, first seen by some of its most expert first readers) in a way that lets each bit either look familiar (look like something already known), or confused or wrong. Hence the pervasiveness of the "where new, not true; where true, not new" response to novelty. Naturally, I cannot realistically feel sure that this study will be among the fortunate ones that survive that phase. But that it will have to go through that phase is probably the most reliable empirical regularity in the sociology of science. As with my previous book, I will follow a distinguished tradition and claim only that the most promising feature of the work is the way that it brings together within a common analysis a wide range of empirical material, which many people would have denied could be so treated.

The argument proceeds as follows. Chapter 1 reviews various well-known illusions of judgment and the controversy over their promise as a source of insight into how cognition works. The material is very familiar, and many readers can skim it quickly as stage setting. This is followed by a set of Darwinian arguments which underpin many things later in the study (Chaps. 2 and 3). While the evolutionary argument includes many familiar ideas, it is more detailed than I have seen elsewhere, hence, I think it is fair to say, more open to both embarrassment and support by empirical work.

In Chapter 4, I develop a particular style of talking about the surface phenomena of cognition, in a way that is tied to the evolutionary argument and leading to a more detailed articulation of the notion of *P*-cognition. Chapter 5 then is concerned with the relation between logic and *P*-cognition and with various subsidiary issues tied to that. Chapters 6 and 7 deal with learning: the first with the sort of nonintentional learning that humans share with other animals, the second with the uniquely human sort of learning that involves consciously trying to learn. So I depart from usual practice by allowing consciousness to explicitly enter the argument. I should think that would be a relief to anyone open to William James's remark that it is rather absurd to suppose that the most astonishing thing we know of in the universe is a mere artifact, playing no essential role in how our brains work.

The balance of the study is concerned with empirical applications of the argument. I report some experiments in Chapter 8, in connection with new analyses I will be giving of cognitive anomalies demonstrated by the well-known work of P. C. Wason and Kahneman & Tversky. These anomalies will be analyzed in Chapter 8 through a purely static analysis, in the sense that we need not consider how changes in the individual's cognitive repertoire might affect the response. In the balance of the study, I extend the analysis to the dynamic context, in which change in the cognitive repertoire is the heart of the matter. For these cases of scientific theory-change we then cannot hope to understand what is happening without taking account of how repertoires—the available patterns and their relations to cues—change (or resist change) over time.

Chapter 9 gives a *P*-cognitive interpretation of Kuhn's paradigm shifts, which are seen as continuous with routine instances of the appearance and spread of novel ideas. Chapter 10 gives some preliminary applications to illustrate the "uphill/consolidation/downhill" schema worked out in Chapter 9, mainly in the context of Darwin's discovery. Chapter 11 sketches the background for the Copernican discovery. It starts with a bit of technical detail on the Ptolemaic system that many readers will want to skim quickly and refer to as needed later in the argument. But the balance of the chapter must be read with reasonable care by anyone who wants to understand the new account of the Copernican discovery in Chapter 12. Chapter 13 continues the Copernican story into the essentially social context of the contagion of a radical innovation. This crucial stage, I will try to show, was in fact reached around 1590, with the appearance of the Tychonic alternative to Copernicus' system, so that (on this account) the cognitively crucial stage of the Copernican contagion occurs some two decades before the telescope finally revealed the first empirical novelties supporting Copernicus. Finally, Chapter 14 makes the transition to fully political judgment by way of a new account (turning on the interaction of political and cognitive factors) of the crisis provoked by Galileo's *Dialog*.

I have tried to build a single coherent argument. As you read, there will be many references to points already made, for example, [5.3] refers to Chapter 5, section 3. So points early in the discussion may be clarified by seeing how the material is put to use later on; on the other hand, points later in the argument will often be clarified by reviewing the details introduced earlier. I have tried to arrange the Index to facilitate these moves, including many citations to passages which develop the intended sense of frequently-used terms, such as "entrenchment," "reasoning-why," "habits of mind."

One

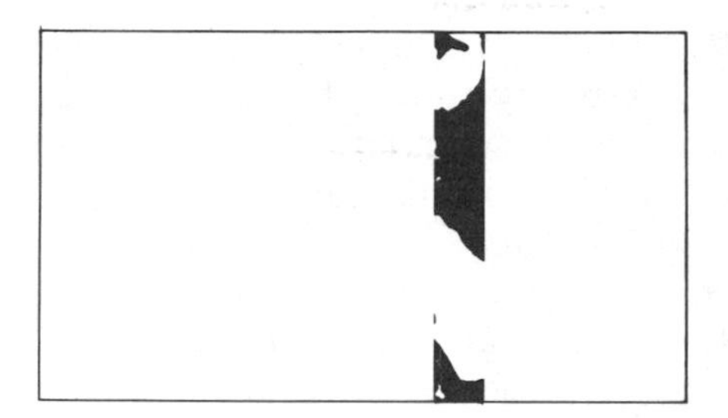

Illusions

We start with a discussion of illusions, meaning cases in which perceptions or judgments are inconsistent with what is really there, or with what is really logically implied. On any account of cognition consistent with the observed competence of human beings, judgment must ordinarily be reasonable, as perception must ordinarily be veridical. But when these conditions conspicuously fail we have a special sort of opportunity to explore how judgment must operate. This survey of (mostly) well-known examples provides a convenient way to set the stage for the later argument.

To avoid immediate questions of who is to say "what is really there," we can limit discussion to cases in which the provision of further information (a tape recording or a measurement, for examples) will ordinarily satisfy the very person we suspect has been the victim of an illusion that, by his own standards, he has indeed been the victim of an illusion.

If illusions in this strong sense can be demonstrated, then it becomes plausible to suppose that what we perceive as our considered judgment could indeed sometimes turn on things we never considered, and perhaps even on things we would not consider it reasonable to consider. Finding that illusions occur does not necessarily imply that they are empirically important either as a direct effect on lives, or as a diagnostic tool for probing cognition. But the first step in making the case that they can be important in both ways is to show that they actually exist.

A particularly familiar illustration is provided by the way we "see" the upper line in figure 1.1 as longer than the lower line.

A book like Frisby's *Seeing* (1979) will give a more adequate sense of the range of visual illusions. These startlingly nonveridical perceptions yield clues to what the brain is doing when it transforms the light that reaches our eyes into the images that reach our minds. And it is at least reasonable to suppose that the same will hold for illusions of judgment.

For perceptual illusions, the effect will often persist even after we have recognized an illusion, but not always. In figure 1.2, in contrast to the arrow illusion, the illusion is subject to correction. If we are given only a quick glance at figure 1.2, it is unlikely that we will notice anything wrong. In fact, figure 1.2 (from Frisby's book) is an actual announcement that had been inadvertently designed with an oddity that escaped detection until after it was printed. I leave it to the reader to find the oddity. After you have noticed it, the illusion completely disappears, illustrating the point that while some illusions (such as the arrows) persist even after we know they are illusions, others (such as the poster) are completely corrected once we have been alerted.

If someone offers you odds of 10:1 that the bottom line in figure 1.1 is as long as the top line, you will not take the bet even though your eyes tell you that the top line is clearly longer. You know your perception is wrong. On the other hand, we can easily imagine someone who has never seen the illusion confidently betting that the top line is longer. After all, unless you refuse to believe your own eyes, it is perfectly clear that the top line is longer.

So we will be interested in a double set of issues: one dealing with the circumstances under which illusions occur; the other dealing with the circumstances under which an illusion, should it occur, can be corrected. Further, you have now seen that there are two ways in which an illusion might be overcome. Sometimes it is possible to come to see the thing correctly (we can correct our intuitive response). Other times we cannot cor-

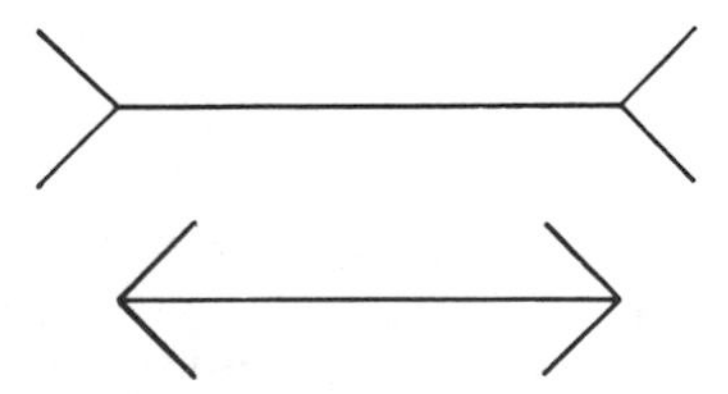

Figure 1.1. Muller-Lyer. The best known of visual illusions.

Figure 1.2. A natural illusion. The poster was distributed before anyone noticed its error, which turns on a visual diphthong. From John P. Frisby, *Seeing: Illusion, Brain and Mind* (London: Roxby and Lindsey Press, 1979). With permission.

rect our perception (in the context of judgment, correct our intuition), but as with the arrow illusion, the context—certainly including, but not necessarily limited to the logic of the situation—is such that we can come to have a second intuition which effectively dominates the first. The upper arrow still looks emphatically longer, but we don't believe it. I will eventually provide an account of how that works.

1.2

Now turn to illusions of judgment, parallel to the illusions of perception. In figure 1.1, we are satisfied that we are indeed facing what can be properly labeled an illusion because we can check to be sure that the part of each pattern at issue (the horizontal line) is identical in the upper and lower part of the figure. The only difference between upper and lower horizontal lines concerns the arrows on the ends, which do not change the lengths at all. Hence if changing the direction of the arrows changes the way we see the length, we know that (so to speak) our brain is misleading our mind. Further, we can satisfy ourselves that it is not a matter of some subtle or mysterious process whereby the arrows actually do change the lengths, for we can measure the lines with the arrows on them, or even simultaneously with our direct perception of the top line as longer.

The most familiar analogues in judgment to such perceptual illusions involve probabilities, such as the well-known birthday illusion. If we are going to ask the next 30 people coming down the street for their birthdays, would you want to bet for or against the following proposition?

A. At least two of the 30 people have the same birthday.

Someone who is not familiar with this puzzle will find it intuitively obvious that the more favorable bet is "no." But the odds are greater than 2:1 in favor of the "yes" response. In fact, it is more likely than not that the birthday coincidence will have already occurred by the time you check the twenty-third person. But even experts in probability theory, for whom the calculation is trivial and familiar, find some such results counterintuitive (Feller 1968:67 ff.).

Provided that we use subjects who are not familiar with A, even the following very much simplified version of the puzzle will often elicit wrong answers:

Shuffle a pack of cards and lay three face down on the table. Then consider the propositions:

B. At least two of the three cards are the same suit.
C. All three cards are different suits.

If we give people proposition B, in general they will judge it as probably false. And if we give proposition C, people will also judge that proposition as probably false.

But if you examine the propositions, you will see that betting against B is the same thing as betting for C. B and not-C (or C and not-B) differ only in ways that have no logical significance, just as in figure 1.1 the lines differ only in ways that have no significance for the length of the lines. As it happens, the chances favor B (hence are against C) by about 3:2.

If, as is ordinarily the case for an unalerted person, both B and C seem intuitively unlikely, then the person is suffering an illusion of judgment, since in fact B can be unlikely (in the sense of less than an even money bet) only if C is likely, and vice versa. The person has an intuitive sense of the probabilities involved that cannot possibly be reliable, since for one proposition or the other (as it happens, for B) the intuition *must* be wrong.

It is crucial here to notice that, as with visual illusions, the typical response is not that we are uncertain, hence being uncertain we sometimes guess wrong. Rather, we have a definite perception that the more probable outcome is less probable. We do not make what seems subjectively a guess. What seems like the best answer is perceived just as clearly as that the upper line in figure 1.1 looks clearly longer than the lower line. And that clear perception is just as wrong on this matter of probability as on the earlier visual illusion.

Even this simple example is sufficient to demonstrate that clearly perceived intuitions need not be consistent (or else our intuition would not mislead us about B). Rather, intuition can be inconsistent even across two contexts as closely related (in fact, two contexts that are logically identical in a perfectly simple way) as those of propositions B and C.

But the human species would hardly have thrived if our brains routinely deceived us. So we should expect to find that our intuitive sense of a situation is usually sound (as with C), even though the existence of illusions shows that it is not always sound. Slips in the way we perceive external objects, or in the way we perceive probabilities (and, more generally, in the way we judge logical relations) therefore invite interpretation as equivalent to rules of thumb, which on the whole have led to good results but in certain situations lead to errors. A Darwinian claim that illusions must be (or must have been) on balance adaptive gives us a point of departure for an account of illusions. But that by itself does not tell us anything substantive about illusions. To go further, we want to be able to say something about what characterizes situations in which a misleading rule of thumb is likely to be invoked, and what characterizes those in which it is not.

1.3

However, as soon as the possible adaptiveness of illusions is suggested, it becomes natural to look for quite a wide range of such effects. For on the Darwinian argument there is no reason to suppose that illusions of judgment will only arise in the context of probabilities. To start, here is a mathematical puzzle involving no question of probabilities. Wittgenstein used to ask his students:

> Imagine a smooth globe as big as the earth. It is 25,000 miles around, and our plan is to stretch a string around the equator so that it just fits. However, we happen to use a piece of string which is just one yard too long. It is 25,000 miles plus 36 inches. When we stretch it around, we are left with a little pinch which sticks up 18 inches. Now suppose we restretch the string so that instead of sticking out 18 inches at one point, it stands out from the globe evenly all around. The question is, how far out from the surface of the globe would the string stand?

An alternate version of this puzzle changes the globe from about the size of the earth (25,000 miles around) to more like the size of the sun (25 million miles around). Again the string is just a yard too long, and we get the 18-inch blip. Again we restretch the string. Now how far out does it stand?

The answers are: for the earth-size globe, the string stands out six inches all around; and for the sun-like globe, with a circumference many times as big, it also stands out six inches all around.

Furthermore, despite the usual intuition that the distance must be microscopic for the earth, and must be much less for the sun, it takes barely more than a line of algebra to show that is wrong.

Let the original circumference be C and the original radius R. And let r be the amount that the restretched string stands out from the globe. Since $C = 2\pi r$, if we increase C by 36 inches we must have:

$$C + 36'' = 2\pi(R + r) = 2\pi R + 2\pi r.$$

But since $C = 2\pi R$, and $2\pi \approx 6$, we get:

$$36'' \approx 6r; \text{ or } r \approx 6''.$$

But virtually no one sees that right away, and virtually everyone is surprised at how trivial it is to calculate the correct but unexpected answer. We are then surprised again to see that the answer does not depend at all on how big the globe is. Not everyone is vulnerable to this illusion (many mathematicians are not).[1] But a person who happens to be immune to the globe illusion can always be fooled by some other no less trivial once it is explained.

As with the probability and perceptual illusions, it is essential to notice that we do not just fail to get the right answer as easily as the trivial math involved would suggest. Nor is it just a matter of being uncertain and guessing wrong. Rather, we have a very definite intuition about what is right, and the intuition is wrong.

However, it could still be objected that although Wittgenstein's puzzle is not one involving probability, it is one involving mathematical ideas. So perhaps it is essentially only in abstract or mathematical contexts that outright illusions of judgment occur. The adaptiveness argument, however, makes that seem likely to be wrong, and indeed it is wrong. The best-known example is a task devised by the English psychologist P. C. Wason (1969). Here is one version of that "selection" task:[2]

Four cards have been taken from two different packs, one with red backs, the other with blue backs. On the face of the cards, we naturally have either a red card (heart or diamond) or a black card (club or spade). The proposition this time is:

D. If the back is red, the card is a least a 6.

The subject has to tell us which of the following cards need to be turned over to be sure D is true or false. Try it yourself.

i. Red back ii. Blue back iii. 9 of hearts iv. 3 of diamonds

When this test is given to unalerted subjects (people who are not reading a discussion of illusions) it turns out that in addition to card i, which essentially everyone sees right away must be turned over, the most common response is to say that card iii should be turned over. Few people see any reason to turn over one or both of the other two cards.

A variant of this test (fig. 1.3) uses envelopes which are either sealed or unsealed, and with either a 20- or a 10-cent stamp. Looking at the upper set of envelopes in figure 1.3, i (sealed) and ii (unsealed) are face down, so you cannot see what stamp they carry; envelopes iii and iv are

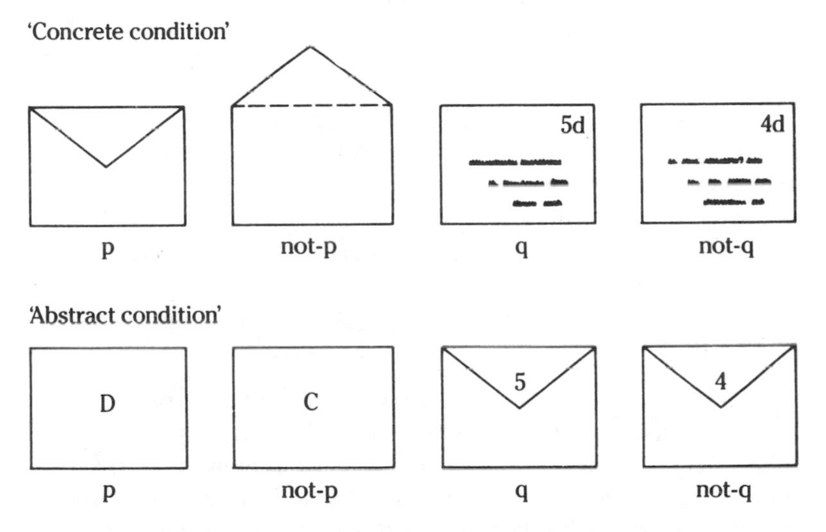

Figure 1.3. Experienced-base vs. arbitrary versions of Wason's problem using treatments of envelopes. The lower version (arbitrary rule) exhibited the usual difficulty. The top version, however, was easily solved by subjects old enough to have experience with a postal rule that requires a higher-value stamp if the letter is sealed. From P. N. Johnson-Laird & P. C. Wason, *Thinking: Readings in Cognitive Science* (New York: Cambridge University Press, 1977). With permission.

face up, so you can see their stamps but not whether they are sealed or unsealed.

Imagine yourself to be a postal clerk sorting these envelopes. The proposition here is:

E. If an envelope is sealed, it has a 20-cent stamp on the front.

The question is, Which envelopes must be turned over to be sure E is true or false?

On this proposition, Wason found that nearly everyone answers that envelopes i and iv must be turned over. An important qualification here is that it was later realized that this changed response was conditional on familiarity with a rule of this sort. For example, it worked with Americans over a certain age but not with younger people unfamiliar with a postal rule that required a larger stamp for sealed than unsealed mail (Wason 1983). Together, versions D and E give a pair of typical responses (i and iii to D, i and iv to E) which are internally inconsistent. For as with the lengths of the lines in figure 1.1, or as with propositions B and C in the three-card puzzle, propositions D and E are logically analogous. In each case, the pairs of test propositions commonly evoke conflicting responses even from people who could be expected to be especially good at logical tasks. Consequently, there seems to be something logically odd in the way the brain processes information.

In either of the two Wason situations, to know which cards (in addition to i) must be turned over we need to see (or at least our brains need to function as if we saw) that:

F. "*X* implies *Y*" DOES NOT NECESSARILY IMPLY "not-*X* implies not-*Y*." And,

G. "*X* implies *Y*" NECESSARILY DOES IMPLY "not-*Y* implies not-*X*."

If you can see why F is true, then you can see why iii does not need to be turned over in either version of Wason's test. If you can see why G holds, then you can see why item iv as well as item i does need to be turned over. But from Wason's experiments and their variants, we can discover that it is possible for the same person to perform as if he understands F and G in one context but doesn't understand them in another, logically indistinguishable, context. As with B and C, and as with the arrows, we get divergent intuitions on logically identical stimuli. Logically sophisticated subjects (for example, scientists and academics who have no doubt been exposed more than a few times to the formal arguments of F and G) perform almost as poorly as less sophisticated subjects (Kern et al. 1983).

However, choosing iii in response to proposition D, but not in response to E, is only slightly interesting. For in many contexts a statement that "*X* implies *Y*" is tacitly intended to convey the wider claim "*X* implies *Y*, and vice versa." Nothing more, then, need be involved than a usage of language that leads us to interpret the rule in its tacitly broader sense, unless additional contextual information warns us not to.

But that only deepens the puzzle about why people fail to see that iv must be turned over in response to D. For the same sort of common extension of meaning that makes it unsurprising that people tend to read "*X* implies *Y*" as "*X* implies *Y*, and vice versa" extends "*X* implies *Y*" to "*X* implies *Y*, so if not-*Y* you can assume not-*X*," as in, "If I get my work done, I'll go to the movie (so if I'm not at the movie, you can suppose I didn't get my work done)."

Hence we cannot explain failing to see "if the back is red, the face is red (so if the face is black, the back can't be red)" by appealing to ordinary habits of language. For now it seems that the logical inference should be reinforced by perfectly ordinary language usage, but we get the answer wrong anyway in response to proposition D.

So although there are only questions of logic here (not probability, or anything else mathematical), we get responses like those in perceptual or in the mathematical illusions. With these purely logical, nonmathematical, nonprobabilistic tests, it is still the case that changing the "decoration," so to speak, turns an easy judgment into an illusion of judgment which fools almost everyone.

1.4

The existence of illusions can hardly be explained by supposing that there is something especially subtle and difficult about the inferences involved. On the contrary, the logic involved in Wason's four-card test is commonplace. In a textbook on logic, propositions F and G ordinarily will be the second and third propositions derived from "if *A* then *B*." The only more elementary one is: "So if *A*, therefore *B*," which is so genuinely intuitive that what the student has trouble seeing is how the proposition is different from its premise.

Further, in some tacit or "as if" sense, propositions F and G are commonly available to any normal human being; and that is so whether or not the person even is aware that such a thing as formal logic exists. If you think about it, you will be able to list equivalents of the envelope version of the puzzles from your everyday experience: for example, what Kahneman & Tversky call "quality control schema." For these situations your intuition gives you the right answer, usually quickly and with no sense of having gone through any logical steps.

So the difficulty we are seeing in the context of a purely logical issue is really more striking than the more familiar probability illusions. With the birthday puzzle, or even the much simpler three-card puzzle [1.2], it is curious that we can get a strong intuitive sense of what the right answer must be, where the right answer turns out to be wrong. But many people would not know how to calculate the probability even for the three-card problem, and very few people would be able to efficiently handle the birthday calculation.

But with the Wason problem the logic is very simple. In fact, if human beings are to be considered competent in logic at all, it is hard to see how logical steps as elementary as F and G could be taken to be beyond routine human competence. Yet what these much repeated experiments show is that it is easy to contrive situations in which human beings fail tests of this capability.[3]

1.5

Nevertheless, in everyday life we do not notice this striking fallability of judgment, and indeed if human beings were ordinarily fooled by such simple problems, it is hardly conceivable that modern civilizations could run. We live in a world in which we depend continually on the ability of human beings to handle much more subtle judgments than are asked for in Wason's tests. So the Wason tests, simple as they are, raise a substantial puzzle for anyone who wants to understand human cognition: somehow we are both competent to handle straightforward logic, and also incompetent, apparently depending on what might have been thought to be trivial details of the test situation.

In fact, the most essential aspect of the puzzle does not depend on a presumption that people have difficulty in consciously grasping the underlying abstract logical relations (F and G). As it happens, people do have exactly that difficulty. Few readers aside from those who routinely have occasion for using formal logic (because they teach it, for example) are likely to have "gotten" F and G at first glance, without thinking it over a bit to be sure the propositions work out. But that is not at all the important point of the Wason material.

Rather the important point is this. If the two pairs of tests each involve questions that are exact logical equivalents, as they do, then why should we find one easy and the other hard?

It is very tempting to say that the answer has something to do with the fact that the second question is closer to a familiar, real-life sort of question. Yet playing cards (which I have used here in place of the abstract symbols in the original Wason test) are familiar concrete objects, and it turns out to be as easy to elicit errors using playing cards as it is with the more abstract symbols Wason first used. However, we ordinarily have no reason to use relations between the face and back of playing cards. So "hands-on experience," narrowly construed, does not prepare us to deal with the problem. But familiarity with the logical relations in a context involving something close to the specific concrete materials used in the test makes the problem easy.

Yet if it takes this *strong* sense of hands-on experience to make even sophisticated subjects fluent, then in what sense is a notion like "logical" really any more applicable to this situation than to tying your shoelaces or to a dog lifting his leg or to any other habitual or instinctive act that efficiently performs its function? If it is only under conditions of strong familiarity that we get fluent performance, then what is happening looks more like habit than reasoning. Eventually [8.3], we will return to the Wason problem and see how the *P*-cognitive analysis leads to a new interpretation of what is going on that is quite different from what both Wason and his various supporters and critics have proposed.

1.6

Turn now to another aspect of the same test material. Wason has said that the most striking thing about the test is the way that subjects typically persist in defending their neglect of card iv even after the cards are turned over and they can see that card iv falsifies the rule (Wason 1977: 122 ff.).

Given a judgment, there is a tendency to hold on to any reason that might justify it; and that tendency to stubbornness extends to far more transparent situations—such as the Wason problem after the cards have been turned over—than we could have supposed on a theory of judgment as applied logic. We can see something of that peculiarity of human judgment at work in the tenaciousness with which writers (on the one hand) who view themselves as defenders of human rationality construct reasons why subjects are really being perfectly reasonable in responding in ways that the experimenters (on the other hand) insist can only be reasonably seen as illogical. Both sides will agree that someone is being stubborn.

To sum up, then, there are well-known illusions of judgment (parallel to illusions of perception) that occur with respect to material that is logically very elementary. That seems to show that however the brain handles judgmental inferences it is apparently not by the literal application of some equivalent of formal logic. Propositions F and G are such elementary logical inferences, yet present such difficulties in simple tests, that if the brain worked by some form or equivalent of formal logic it could hardly have the capability of judging anything at all. It would be hard to contrive real world tests that would not require logical steps of at least the level of complexity sufficient to handle propositions F and G.

An alternative interpretation (Henle 1962) is that the judgment is logical, but the person has answered (logically) a question different from the one posed by the experimenter. Sometimes there is enough ambiguity in the framing of the question, or strong enough ordinary life connotations to the language, that this essentially settles the question of what is happening. (There is actually nothing illogical in the response, or at least nothing that, given the circumstances, reasonably raises any deep question of the role of logic in judgment.) But for many matters (the globe and Wason problems here, for examples) the alternative interpretation necessary to make the usual response logical is either not there at all, or there only under some interpretation so tendentious as to transfer the puzzle about the logic of the response undiminished to an earlier stage.

But then, as Nisbett & Ross (1980) ask: "If we're so dumb, then how did we get to the moon?" What we need is a way to understand cognition that lets us see how human beings could do all the things we do but nevertheless have the logical frailty that the experimental work reveals. Somehow this evidence of logical frailty must fit with what we also know about human competence. Apparently the brain works in a way that produces judgments that very often coincide with the judgments that a strictly logical process would produce. But (given the many exceptions) the actual process cannot be a strictly logical process, and perhaps not a logical process at all.

Other puzzles also arise out of the work on illusions of judgment which deepen the mystery but which may also serve to help in its resolution. If it is reasonable to suspect that human beings somehow process judgments without relying on some tacit equivalent of formal logic, a further puzzle grows out of the very clear disposition we have to see ourselves as making judgments in an explicitly logical manner. The contexts in which we say we are acting intuitively are typically contexts in which we think it would be difficult or impossible to give a specifically logical reason for a choice. Ordinarily, though, we do not answer such a "why" question in terms of intuition but in terms of some sort of (possibly faulty) logical reasoning. Therefore, the puzzle is, If our actual reasoning is not logical in form, then why are we so inclined to believe that there is a logical basis for our judgment, and whence comes the great disparity between our modest facility in handling logic and our great facility in concocting at least superficially plausible reasons-why for our judgments?

In yet another version of the Wason experiment (Johnson-Laird & Wason 1977:155), college students were given what purported to be the correct answer to the problem posed by proposition D. But in fact only some subjects were given the correct answer. The rest were given one of the common wrong answers. Given the (purportedly) correct answer, the subjects were asked to give the reason that answer was correct. They were also asked to rate their confidence in their reasons.

It turned out that whether the answer a subject was given was right or wrong had no significant effect on either the ability to devise reasons or on the confidence rating for those reasons. Subjects giving illogical reasons which seemed to justify the (unknown to them) illogical responses were just as confident of their reasoning as subjects giving the right reasons for the right responses.

Experiments such as this have led Wason and many other psychologists to doubt that a person's conscious reasons for his judgments need have anything to do with his judgments. They have in mind a stronger claim than the sort made by Freud, Nietzsche, and various earlier commenters on the unreliability of a person's own account of his behavior and beliefs. For what was a special sort of situation for earlier writers comes to be seen as the ordinary situation. Given the judgments (themselves produced by the nonconscious cognitive machinery in the brain, sometimes correctly, sometimes not so), human beings produce rationales they believe account for their judgments. But the rationales (on this argument) are only ex post rationalizations.

I will say more about that later. For the moment, however, I only want to point out that the puzzle of these ex post reasons-why gives us yet another class of illusion to ponder. For subjectively, we have no sense that our answer to reasons-why questions may have such a problematic connection with an account of how our brain actually produces judgments. We have no sense that what we think is the reason-why may have nothing to do with the actual reason-why. Furthermore, as I mentioned earlier, we have a striking contrast between the clumsiness in handling simple logical implications that is revealed by the illusions of judgment and the quickness we reveal in concocting superficially plausible reasons-why even when the situation is one that logically makes that task very difficult (because our answer in fact defies logic).

1.7

A striking illustration of reasons-following-choices is the hypnosis experiment in which a subject is told to perform some odd act (for example, jump up and wave his arms around) when given a posthypnotic cue (for example, the hypnotist touches his ear). The subject wakes up. After a while the hypnotist touches his ear, and the subject jumps up and waves his arms. He is asked why, and he gives an answer that seems perfectly reasonable to him. But everyone who has been watching knows that the reason can only be an ex post rationalization.

Further, there is a good deal of evidence that the rationale not only is merely a rationalization but was constructed literally after the fact (that is, it was indeed constructed literally after the action had begun). For even the subjectively certain belief that a decision has preceded an act of conscious volition is sometimes and perhaps even generally only an illusion.

Obviously, this applies only to the proximate volition: if I decide to go to the movie tomorrow, and tomorrow I indeed go, then the decision to go did not follow my going. But the issue here is whether my conscious sense that I have decided to go indeed was the decision, not a consequence of an internal process that already made the decision. Would an ideal observer have been able to say at some point: "The decision to go has been made, so we can predict that soon he will make a conscious choice about going, which will coincide with the tacit choice that has already been made." The answer I will come to on this is that conscious judgments probably do always follow, never determine, choices; conscious reasons nevertheless play an essential role, but that role concerns how judgments are sometimes revised, not how they are made in the first instance.

An experiment illustrating the problematic relation between choice and conscious volition asks subjects to decide when they will make some simple movement (something like, wiggle a finger) and note the position of a clock hand when they do that. At the same time, brain currents are monitored, looking for certain "readiness" potentials which invariably precede movements. The result is striking evidence that the readiness potentials precede consciousness of a decision by about a half-second, so that the "decision" follows the physiological signals that the movement is being organized, not the reverse. It appears, then, that the unconscious beginnings of the act cause both the movement and the conscious sense that a decision has been made to move.

The significance of these results is disputed (Libet 1985). On the argument here, though, a certain aspect of Libet's interpretation is particularly appealing, namely, that the function of consciousness concerns reviewing or checking intuitions and volitions, which sometimes prompts a halting of what is underway. But we do not decide to have a volition or intuition, and only a small fraction of what occurs in the brain that we suppose *might* prompt a conscious intuition or volition actually does so.

In a certain sense, this is empirically commonplace. Every reader must have noticed that, in general, a person does not consciously will physical movements, or consciously choose intuitions, beliefs, emotional responses, and so on. Only a tiny fraction of any of this is consciously monitored. So should it even be controversial to suppose that, for the tiny fraction of consciously considered responses, the most reasonable hypothesis is that it is governed by the same nonconscious processes that govern the vast majority of responses that escape explicit attention? An analogy might be with a baseball pitcher. His choices come from his catcher, and he has no access to the process that yields them. But he can notice the choices and sometimes shake off the catcher's signal when it does not look right to him.

1.8

Some readers, perhaps as much impressed as I am by firsthand experience with their own logical failings, will be quite prepared to believe that in all this material we are dealing with a characteristic of the species, not merely quirks of particular subjects or particular experiments. But the heated discussion that this sort of claim prompts among students of cognition reveals how skeptical others can be that anything very deep is going to come out of such work (Cohen et al. 1981). How you see the thing is likely to depend, in the short run anyway, a good deal on whether you share the sense I had in starting this study that human judgment is often inexplicable in the folk psychology terms of logic + interests. The argument here is that choices become explicable in those terms only by using "logic" and "interests" so flexibly that it becomes tautological that any choice can be explained in terms of logic and interests. In that sense, however, the belief that the explanation tells us anything about the empirical world is itself an example of an illusion of judgment.

Nevertheless, on strictly logical grounds the material in this chapter would not require us to regard accounting for illusions as central to a theory of judgment. For no matter how we proceed, we will certainly have to rely on some simplifying assumptions, and the real question is not whether a "logic + interests" formulation is exactly right but whether it is good enough to give us our most promising route to making progress. That could be the case, even though there are obviously some situations in which whatever internal processes are at work operate in a way that cannot be fully accounted for by the interaction of logic and interests.

Therefore, an argument for treating illusions as inessential to a theory of judgment does not necessarily conflict with the strong assertions I made earlier about the implausibility, in the light of the illusions, of supposing that actual processes in the brain are some functional equivalent of logical calculation. Some writers who doubt the significance of illusions do in fact seem to think of cognitive operations as equivalent to, if not identical to, a logical calculus. But that is not a necessary presumption. It is

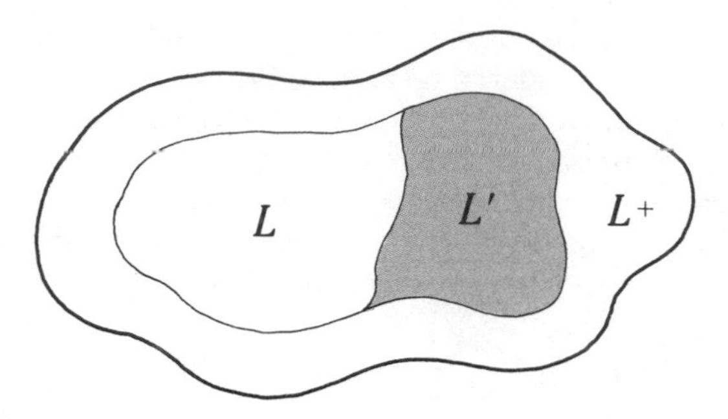

Figure 1.4. A theory might divide behavior into rational and irrational components (*L* and *L*′) with separate accounts of each; or it might seek an account (*L*+) in which judgment which fits comfortably with the label "rational" is produced by cognitive processes indentical to those that also produce judgments we would not want to call rational.

enough for such writers to claim only that for most purposes, and for most of the time, we can analyze cognition "as if" the process were a logical one, as economists ordinarily treat tastes "as if" they were strictly self-interested without necessarily implying that in fact tastes are always or only self-interested.

One way to think about the situation is to suppose that (a) normal human judgment can be essentially accounted for by a theory that treats it as logical, or some functional equivalent of logical (*L* in figure 1.4), with the shaded area (*L*′) covering abnormalities outside the main theory—as in normal versus abnormal psychology. Another way (b) would treat the non-logical side of judgment as an aspect of the normal, by providing a theory (*L*+) that covers both *L* and *L*′. Obviously, this study will try to make plausible a version of (b).

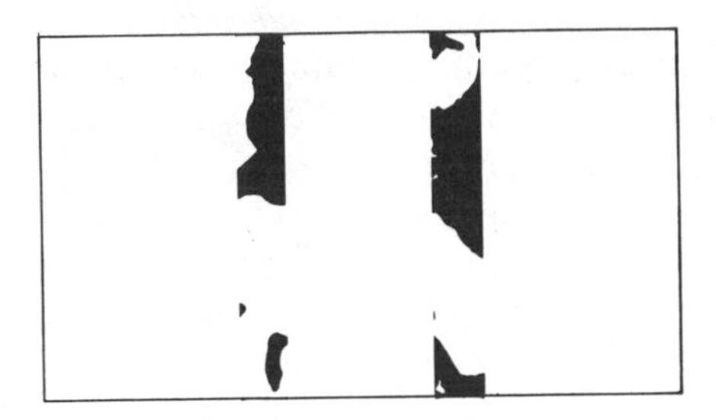

Two Preliminary Arguments

Illusions of judgment are presumably a side effect of generally adaptive cognitive processes. The next two chapters try to work out an evolutionary view of cognition in a way that gets beyond that reasonable but (by itself) not very helpful remark. Accounting for illusions can then be thought of as giving us a checkpoint on the way to a larger goal, since whatever general account we reach of judgment should have properties which begin (at least) to make sense of the particular sorts of illusions we observe.

The analysis starts from an *a priori* argument, which turns on what can be said about any physical system that exhibits what we might think of as intelligent behavior. Unlike the balance of Chapters 2 and 3, it is not Darwinian but would apply equally to computers, people, or anything else exhibiting intelligence. So in the context of the a priori argument, "brain" is a generic term for any physical thing that is good at processing information. If I say something in that context that would not apply to a computer as well as a human brain, or say anything that would apply to a computer but not to a human brain, then I have slipped. By itself, an argument at that level of generality cannot take us very far. But it turns out to have a good deal of force in conjunction with the two arguments that follow, both of which are what William James affectionately called "Darwinizing" arguments.

From a Darwinian viewpoint, everything biological represents the current state of a particular sort of evolutionary pathway. We can think, first, about the properties of any system that is plausibly the product of Darwinian evolution (which I will call the *viable pathway* argument); and then we can think more specifically about the evolution of cognition in terms of a *continuity* argument, since the main result of the viable pathway argument is that a certain restricted sort of continuity can be expected to hold across the evolution of cognition. The a priori and viable pathway arguments of this chapter do not yield a proof that continuity must always hold. But they do yield a presumption in favor of continuity strong enough to justify a serious effort to see how far all of cognition can be accounted for subject to that constraint. As will be seen, the commitment to pattern-recognition emphasized on page 1 is itself a deduction from the continuity argument. The argument of Chapter 3, and indeed of the balance of the study, amounts to an effort to show the fruitfulness of the commitment to continuity and (from continuity) to pattern-recognition.

2.2

A priori, anything which is a sophisticated information processor—any brain—requires a memory. Even to add 2 + 2, a brain cannot be such that it gets only 2 + (what was that other number again?). But the memory could be far from perfect. We need not require equally prompt access to all items in memory, or equally reliable access, or equally durable storage. Failings and peculiarities of memory are such commonplace experiences that no one will doubt that it is possible to be subject to such failings and still be capable of exhibiting what we would accept as intelligent behavior.

But what applies to memory applies to any other feature of a brain. Anything constrained by physical laws will face limits on volume, energy consumption, operating speed, reliability, and so on. A necessary consequence is that nothing can be done with absolute perfection. A corollary of more practical interest is that there will be tradeoffs imposed with respect to what combinations of speed, flexibility, and reliability are simultaneously possible.

In the case of human beings, the most obvious of the physical constraints requires that everything our brain does for us must be handled by an object small enough to fit inside the top of a human head. Within that constraint, we could not expect even approximately unlimited memory, or unlimited computational capability, or unlimited capability for handling input or output flows, or unlimited anything else. Other things equal, the inevitable tradeoffs will mean that more memory will imply less computational capability, more memory of one type (say short term) implies less memory of other kinds (long term), and so on.

However (going beyond the a priori argument for a moment), if in addition to facing physical constraints the system faces competition for survival, then it also will be subject to pressures favoring an efficient balance among its tradeoff possibilities. The systems we see (the systems that have survived) will therefore be those that have somehow chosen relatively well in balancing weaknesses in one area against strengths in others. But "relatively" is an important qualification. For although competition is at the heart of Darwinian evolution, generating strong pressure for efficiency, the search for efficiency itself will be powerfully shaped by the Darwinian pathway constraints that will be discussed in a moment.

To sum up: (1) Any brain subject to the laws of physics cannot possibly be perfect but must be vulnerable to errors and limitations of various kinds; (2) if the brain, in addition, faces competitive pressures, it can be expected to exhibit an efficient response to its problems; (3) subject, however, to constraints on the search for efficiency imposed by the condition that all Darwinian outcomes must be (obviously) the product of a viable evolutionary pathway.

2.3

Though we can contrive cases where the physical constraints are utterly trivial, they can never be absolutely avoided. For example, even a computer programmed to play tic-tac-toe will be subject to tradeoffs and cannot be built with absolute perfection. (There will always be some probability, however minute, that it errs.) For less trivial brains absolute perfection could hardly be approached. So we want to be alert to various properties which, if achieved by a system, would contribute to its ability to perform well despite flaws. For example, it takes resources (at a minimum, time) for a brain to work out a response to a situation. Therefore, it would surely be undesirable to work out the response to every situation from scratch. Rather, we should look for standard responses to various frequently encountered situations (subroutines in computer programming). On the negative side, we could not expect the brain to perform perfectly in identifying situations in which a particular subroutine would be appropriate. In fact, given the inherent imperfection of any physical device, we know that it will not perform perfectly. Consequently, there surely will be some tradeoff between what statisticians call errors of the first kind (hesitating too long in seeking the best response) and errors of the second kind (jumping too soon).

The inevitability of that particular sort of tradeoff (errors of the first kind vs. errors of the second kind) yields the main cognitive building blocks that will be used in the balance of the study. Information processing of a complex sort will exhibit certain basic features tuned by this tension between jumping-too-soon and hesitating-too-long. There will be advantages sometimes favoring one tendency or the other, and we can expect to see (in a complex brain) variants in response to different sorts of contingency. But the balance can always be analyzed into interactions among four subprocesses that I will call "jumping" (reaching a response), "checking" (taking a closer look if something sufficiently adverse is noticed after a jump), "priming" (becoming predisposed to make certain jumps), and "inhibiting" (becoming indisposed to make certain jumps). With no capability for inhibiting, a brain would have no defense against excesses of checking; jumping can go quicker with the help of priming, but quickly jumping would be dangerous if there were no capability at all for checking. And so on. It is hard to see how an account of any complex information-processing could fail to include (under some set of labels) these four features. On the other hand, it is not clear what else is really needed to deal with the hesitating-too-long versus jumping-too-soon tradeoffs. At least in this study, I have no occasion to use anything else. In one way or another, all contingent properties of human cognition are traced to interactions between the a priori balance of errors pressures (yielding the jumping/checking/priming/inhibiting set) and the continuity principle that will emerge from the viable pathway argument taken up next.

2.4

The cognitive balance that emerges from jumping/checking/priming/inhibiting interactions will yield a range of plasticity of response. It is easy to see that a brain which will face some predictable challenges from its environment would profit from a capability to handle these situations in some routine, preplanned way. But unless the brain faces an environment in which all the challenges it could handle are completely foreseeable, it would also profit from a capability to learn. As has very often been pointed out, only if the brain can learn from experience could the preplanning efficiencies be extended beyond contingencies so completely predictable that responses become genetically engrained.

In a sufficiently large and complex brain, then, we could look for more than one kind of routine response (and presumably mixtures of the various kinds). Some would deal with standard aspects of the environment, and these could be preprogrammed from the outset (instincts). Others would be shaped by experience and training to respond to less standard aspects of the environment (habits). Finally, we would also look for the kind of response that we associate most immediately with the word "intelligence": there might be routines with a capability to construct a response to novel situations (judgment).

The output of a brain that had achieved all of these classes of capability would then consist of some blending of instinct, habit, and judgment, all subject to errors and limitations, but on the whole sufficient to make the brain capable of survival in the environment in which it operates. There is a natural hierarchy in the three modes (instinct, habit, judgment). Habits must be built out of instincts; judgment must somehow derive from instincts and habits. So one of the useful ways to categorize brains would be to distinguish among those that work on instinct alone, those that use instinct + habit, and those that use instinct + habit + judgment. (For electronic brains: an ordinary hand-calculator vs. a programmable calculator vs. a chess-playing machine.)

A good deal more could be said, a priori, about what we could look for in a reasonably efficient and sophisticated brain. For example, we would expect to find some redundancy in critical components so that damage will not always be fatal. But as much as will play a role in this study derives from the notions already sketched.

2.5

So far we have been considering necessary features of anything that functions as a brain, whatever the process that produced the brain. But the brain we are interested in evolved by Darwinian selection and therefore must represent the current state of a viable Darwinian pathway. This naturally has consequences, imposing constraints beyond those which would apply to any brain. In particular, there are constraints on a brain that has been produced by Darwinian evolution which would make such a brain operate in ways that would never be seen in a designed brain, such as a computer. Thus we will be led to an account of cognition which differs in some important ways from the sort of account that (implicitly, and sometimes explicitly) takes the digital computer as a model of how human cognition can be supposed to work. Some preliminary points are best made in a relatively uncontroversial context, so I want to start this "viable pathway" argument with an illustration about wings, not brains. There is little in what I will say that will be not familiar from the work of such writers as Gould (1980) and Monod (1972).

The capability for true flight has evolved in three groups of living animals (insects, bats, and birds). The structure is distinct for each group. A person requires no training in biology to be able to tell a bat's wing from a bird's from an insect's. On the other hand, we all can easily see that a bat's wings and a bird's have very much more in common than either has with an insect's.

But all this holds only for the everyday level of analysis suggested by typical pictures of birds, bats, and insects. If we make the comparison *either* in terms of a much broader *or* in terms of a much narrower perspective, the situation is different. At the very broad level, no one has difficulty recognizing that certain appendages of insects, bats, and birds are examples of a kind of thing called a wing, which we readily class with other appendages that appear on certain kinds of seeds (maples, for example), man-made objects (airplanes), fish (flying fish), and so on. On the other hand, in a very much narrower perspective, at the level of how things work at the finest level of detail (inside cells), all the living things are very much alike but differ radically from the airplane. At the level of cells, it takes training to tell a human being from a mushroom, and a fortiori, it is hard to tell a cell from an insect's wing from a cell in a bat's wing.

Such points have a simple evolutionary explanation. For birds and bats are both vertebrates, sharing (sufficiently far back) a common vertebrate ancestor which was neither a bird nor a mammal. The pathways by which bats' and birds' wings could evolve were constrained by the architecture and chemistry of that common ancestor. On the other hand, the airplane wing shares none of this evolutionary history; and while insects also share an evolutionary history with birds and bats, divergence occurs at a point very remote compared to the divergence of the ancestors of birds and bats.

These simple points have large consequences for the evolution of wings (and by very close analogy for the evolution also of brains). At the finest level of detail (cells), vertebrate wings have much in common with insect wings, though no more than they do with other parts of an insect. At a more familiar level of detail, however, vertebrate wings have little more in common with the insect wing than they do with an airplane wing. At the grossest level, finally, we find that all wings, after all, are much alike. We can extend the discussion to the nonbiological wings fabricated by human beings, to the gliding wings of seeds and fish, and so on, to find that various gross design constraints impose similarities across any physical thing that is able to function as a wing. If this book had been about flying rather than about thinking, we could have started this chapter with an a priori discussion of the properties of wings.

On the other hand, though certain common features are imposed on any physical thing which can *function* in certain ways (for example, as a wing or as a brain), no comparable commonality is imposed on the detailed *structure* which supports that function. If at a very fine level of detail, we find that all biological wings (from those on seeds, bugs, bats, and so on) are again very similar, then apparently there must be some conservative principle at work that tends to preserve structures across a vast range of functions. The basic structure of cells in the wings of birds and insects are much alike because the basic biochemical organization of the living cell has changed remarkably little in the last billion years or so. Even the most recently evolved and highly sophisticated of cells, for example, the neurons in a brain, are special cases of a general class: living cells. Their relation to things like muscle cells is clear from numerous fine details.

So we are led, as is usual in discussions of this sort, to emphasize the distinction between function and structure. There are constraints on functional units (wings or brains or whatever) that derive from what they need to do to be what they are, and by how they can do it consistent with the laws that govern physical things. These constraints apply whether the things in which we are interested were evolved or were designed (apply to computers as well as to brains, to airplane wings as well as to butterfly wings).

But for any given function there will be an arbitrarily large number of particular structures that could perform the function. Designs are never unique, though some details may be essentially unavoidable. (If you want to tile a floor with identical tiles, there are only a few shapes that will fit together as required.) The general absence of design uniqueness would hold even if the function were specified in great detail. But if the function is only a loosely conjectured one (as is crudely the case in a Darwinian context, where the things that will evolve are essentially invented by the process itself as it runs along), then the range of detailed structures that could support a certain sort of function is arbitrarily large. This point applies whatever the function: for example, to eating and flying, and also to the function that happens to interest us here, the processing of information.

The reverse is also true (particular structures can be used for many functions), though it may be harder to see why. Suppose we have a structure that is (in geological time) plastic, in the sense that it could be stretched or compressed in various ways, merged with related structures, and so on. Then a given basic structure could be transformed to support a great variety of functions without changing so drastically that we no longer recognize it as a variant of the basic structure. Stephen Jay Gould's essays are often a celebration of this marvel. A conceptual argument that bears directly on this issue is provided by Herbert Simon's "Architecture of Complexity" (1969).

2.6

Consistent only with physical possibilities, there is virtually no limit on how far evolution might move from a given starting point. That the creatures we see in the world must represent the latest step in a continuous evolutionary pathway does not of itself tell us why it should be so easy to notice common details of macroscopic structure in (for example) the wings of bats and birds. After all, a butterfly emerges from what was once a caterpillar by a process which also is constrained to a continuous pathway.

However, chance now intervenes in a way that effectively guarantees that the evolutionary outcomes we actually observe will be a highly biased sample of all those that are physically feasible. That is because chance at the level of individual variation leads not to unpredictable randomness but to virtual necessity with respect to certain important features at the level of species evolution (Monod 1972). Among possibilities that are not foreclosed in any absolute sense, enormous differences arise in the relative probability that various feasible possibilities would ever appear on a Darwinian pathway.

A caterpillar can turn into a butterfly, because this occurs during the lifetime of a single individual. That individual gathers resources during the caterpillar stage that enable it to survive during a pupal stage where the transformation takes place. Aside from parental care, that cannot occur along an evolutionary pathway, where (with only minor qualifications) an equivalent transformation would require that each stage be capable of surviving as a life form independent of what went before or might come after, with the consequence that Darwinian pathways are powerfully biased toward those which conserve structure across a wide range of innovative functions.

What we see in the world is irreducibly subject to chance. But if we toss a coin a thousand times, although it is only a matter of chance what sequence of coin tosses we get, some sorts of outcomes are vastly more probable than others. The class of outcomes with about 500 heads in 1,000 tosses is astronomically larger than the class with anything like 0% or 100% heads. So what we expect to see is a sequence with roughly 50% heads, even though any particular sequence with about 50% heads is no more likely than the one particular sequence that has 100% heads. The conservative principle that works to preserve biological structure across a wide range of functions requires nothing more than that commonplace statistical insight.

Any particular evolutionary development preempts a vast range of alternatives, since the niches it fills are no longer available for exploration by other possible lines. It would be absurd to doubt that among the vast number of preempted possibilities are many that could (given the opportunity to compete) outperform the one that actually exists. The sorts of outcomes we are likely to see are those that can occur in many different ways, as the sorts of outcomes we will almost always observe in a coin-tossing experiment are those that can occur in many different ways.

A Darwinian process, therefore, would not yield unqualified efficiency. The appearance of novel functions will necessarily be constrained by the two fundamental Darwinian properties: (1) Darwinian variation is blind, (2) Darwinian selection is nearsighted. Interacting with the probability argument, (1) and (2) essentially guarantee strong continuity of structure across diversity of function.

2.7

The force behind this principle of continuity of structure across diversity of function is often illustrated in terms of a contrast between hill-jumping and hill-climbing. Solely on the a priori argument that began the chapter we expect to see brains that are imperfect; but on that argument alone we would not expect brains that exhibit clumsy or inessential flaws. For example, a priori, it is a puzzle that human beings who have decided that some behavior is stupid frequently find it impossible to stop behaving that way. We have all had occasion to notice to our regret that the human brain does not seem to work in a way that reliably makes it possible to do what you want to do (Schelling 1980). Or, another example, there seems to be no essential reason why human beings should be so stubbornly vulnerable to many sorts of illusions, even after the individual already knows all about the illusion. Indeed, it is not obvious why a human brain should be grossly inferior in carrying out logical operations to a calculator chip the size of a fingernail.

In general, judged simply by a priori arguments (by the inevitability of tradeoffs among rival capabilities), the way our brains operate seems quite odd. If Alphonse *X* were present at the creation, he would have had something to say.

However, once we think about the brain as necessarily the latest stage of some Darwinian pathway, these puzzles appear differently. We must expect that the architecture and functioning of the brain would be more nearly analogous to the structure of an old city like Paris or Rome than to a contemporary, designed-from-scratch machine like a spacecraft or a computer. And even that metaphor understates the problem, since it is possible to *plan* changes in the design even of something as complex and hard-to-radically-change as a city in a sense that has no counterpart in evolutionary terms. Under Darwinian evolution, novelties emerge from a process of blind variation, favored (or not) by near-term advantage. There is nothing in the process that is more than vaguely analogous to planning ahead.

So while the analogy with an old city may begin to give some sense of the problem of evolutionary novelty, it is only a beginning. What keeps the Darwinian problem from being completely hopeless is that geographic isolation may occasionally give a variant family of a species temporary respite from its closest natural competitors (that is, from its own conspecifics). But the respite could hardly ever be more than temporary, since, even in the case of the most profound geographic isolation, the individuals who reveal novelty presumably will eventually compete against those who reveal reversion toward the older form.

Hence the problem of how radically novel capabilities could emerge from blind variation and nearsighted selection is a fundamental one, as Darwin himself was keenly aware. The basic Darwinian argument makes it easy to imagine evolutionary pathways that lead to local optima (fine tuning, hill-climbing), but hard to see how evolutionary pathways would get to something more than a narrowly local optimum.

Yet quite spectacular hill-jumping has played a tremendously important role in the history of life on earth. There have been repeated episodes in which a minor life form (for example, the mammals) virtually exploded—a very slow explosion, of course, in terms of time scales like centuries—into a tremendous diversity of forms and range of habitats.[1] In particular, radically unsettling things (major climate change, for example) happen often enough to be important in opening the door to hill-jumping. What happens is something like the following.

Consider how an entity might make its way along an evolutionary pathway—or more accurately, make a pathway, since it is only by hindsight that we can talk as if the pathway were there before evolution produced it. If the entity is only climbing a hill, then we can imagine things varying slightly from generation to generation; natural selection will favor variations that fit the entity more closely to its niche. It gradually climbs as close to the top of the hill as its genetic potential permits, and there it remains stable. Sooner or later, changes will occur which alter or eliminate the niche we are talking about. The creature may or may not be able to adapt to that; and when it cannot, the reason is likely to be tied to the very intricacy of its fine tuning to the earlier situation. Or new species have come into the local scene, so that an occupied niche comes into being that dominates the earlier niche. Over time, new species evolve, and nearly all old species become extinct.

But if that were the whole story, we would expect many species to leave daughter species behind after they themselves have become extinct. In the short run, that seems to happen. Nevertheless, the millions of species that exist today are the descendants of a microscopic fraction of the species that existed in earlier times. Apparently, some species develop a new kind of capacity, or some exceptional capability to evolve rapidly to create new niches or seize old niches. The species we see in the world are mainly the descendents of these rare opportunistic species, who somehow, at least for a time, have a capability for hill-jumping rather than only hill-climbing. Every bird alive today is almost certainly descended from a particular species of reptile out of thousands that existed 250 million years ago, and the same for every species of mammal.

The starting point for seeing how that could occur is to note that we are dealing with vast numbers of species over vast extents of time. Therefore, steps that are exceedingly improbable in terms of what we would see if we looked at any single family for a few generations may play a large role in the grand story. Even if you are a passionate bridge player, you are most unlikely to ever get, or ever see, a hand of 13 hearts; but given the total number of bridge hands played every year all over the world, you can be morally certain that such hands will be dealt.

Hence, what counts is not absolute probabilities but relative probabilities. Very rare events can be essential. Nevertheless, whatever process is at work will economize on rare events. For we have a process in which *getting there best* competes with *getting there first*. The winners are almost certain to be reasonably good performers who can get started quickly (here, with relatively few rare strokes of luck), not the best of all possible performers. We can allow for unlikely events, but not plausibly for many more unlikely events than seem minimally required to get to the sorts of outcomes we observe.

Within this general framework, there are two main ways that the capability for hill-jumping (macroevolution) could appear. But since the continuity argument here holds under either version of macroevolution, the relative importance of each is not an issue here.[2] On either account (preadaptation or hopeful monster), the principle of continuity of structure could be expected to hold. For in either context, successfully colonizing a new hill is ordinarily beyond reach. For every such venture that succeeds, there will be a vast number that fail. The successful cases (those whose descendents we see) are going to be drawn essentially from the sort of early steps that will occur most often; and these can hardly help being forms which involve no very radical disruption in established structures.

The reason is that almost any biological structure—a limb, an eye, even a single cell—is extremely complex and finely tuned to its function. Consequently, while a mutation that jumbled the structure might in principle provide the basis for a new function, the mixed-up structure probably would not work at all. Fine tuning would favor atrophy of this worthless structure.

In contrast to the problem of building from scratch a structure that will support a new function, or of reconstructing the pieces of a jumbled structure into something new, consider a structure that is recognizably the older structure but (on the preadaptation argument) facing strong and novel selection pressures, or (in the hopeful monster version) distorted in some way, or duplicated, or abnormally fast or slow in developing. All of these could have some minimal ability to do *something* different from what the creature could do before, and occasionally this something different would be at least marginally useful. Far more quickly than in the case of starting from scratch or from a jumbled structure, they could evolve into doing something well.

So thinking about the intrinsic difficulty of how a Darwinian process manages to do anything beyond fine tuning emphasizes the importance of the notion of structural continuity. The most plausible paths to radical functional novelties go by way of doing radically new things without radical departure from the structures that are already on hand. Evolution most plausibly builds on what it has, not by tearing down or ignoring what it has. If we observe the analogue of a screwdriver, we expect to find that it evolved from something like a knife, not from a hammer, and certainly not from a chain. The more complex the structure that supports the function, the stronger are the grounds for expecting that it will have been built on an older structure and will therefore work something like the older structure from which it was derived. *But since the brain is the most intricate of all biological structures, all the arguments just given become especially forceful with respect to cognitive functions.*

2.8

To sum up the viable pathway argument: very clearly for wings and many other things we already understand, hence (we can surmise) probably also with respect to even the more intricate and much less well understood cognitive processes, evolution is creative with respect to functions in ways that remain very conservative with respect to structure. And the reason why that conservative principle is so powerful turns on the way in which pure chance leads to virtual necessity. Structure tends to be conserved, even while function (in evolutionary time) tends to be marvelously creative, not because that leads to the best of all possible outcomes but because a process that proceeds by nearsighted selection among almost random variations has much better chances to produce new functions out of old structures than to produce new functions *and* new structures. But then those conservatively produced innovations preempt further (and no doubt even better) versions of functions, which could in principle have evolved absent the need to compete with the species that got there first.

The evolution of a bird's wing was constrained by what could be managed without radically changing the structure the protobird had to work with, which was a reptile's claw. A bat, to carry on this metaphorical way of speaking, faced the same problem. The bird and the bat did it differently. But it doesn't take an expert eye to see that both these types of wings are adaptations of a structure that once could have served (and which much independent evidence assures us once did serve) as a reptile's claw.

Analogously, we have good reason—indeed because of the sheer intricacy of the underlying structures we have especially good reason—to consider how human cognitive capabilities might have evolved in a way that is consistent with the principle of continuity of structure as the basis of novelty of function.

In a strong sense the balance of the study consists of analyzing the interactions of the principle of continuity with the a priori features I earlier labeled "jumping, checking, priming, and inhibiting." All the a priori features emerge (in the development of the continuity argument in Chap. 3) before we reach human cognition. For the balance of the study most of the argument will consist in trying to show how these basic features, interacting with the principle of continuity, can produce the characteristically human features of cognitive performance by playing their role on a larger and larger scale.

2.9

Since vision is both far older and in various ways easier to study than thinking, continuity suggests we might start by considering what we know about vision for clues to what we might learn about thinking.

A key notion for the analysis of vision (and from continuity, we are led to anticipate, a possible key to how we think) is the interaction of patterns and cues. And every reader will be familiar with some versions of this notion applied to thinking, such as Minsky's "frames," Schank's "schemas," Kahnemann & Tversky's "framing," and even Kuhn's "paradigms," Talcott Parsons's "templates," Clifford Geertz's cultural patterns. I have already stressed that in the Introduction. In back of all these is the gestalt psychologists' emphasis on the way that we seem to perceive individual parts only in the context of some (often implicit or imputed rather than overtly present) whole. Figure 2.1 will serve to illustrate this key notion.

Even if you are familiar with this grid, and therefore know what comes next, you may see nothing in it right now but a random-looking pattern of shaded blocks. However, if you set the figure across the room (if you wear glasses, it may be enough to take the glasses off and hold the figure at arm's length), then you will easily recognize Abraham Lincoln. You ought to satisfy yourself at the outset that there is nothing ambiguous about these perceptions. When you look at the figure in focus, you see the

Figure 2.1. Seen clearly, the figure is an arbitrary grid; seen out of focus it is a familiar image. Lincoln grid by Leon D. Harmon, by permission of the Estate of Leon D. Harmon.

blocks, and for a first-time viewer you see them with no hint that you might be looking at a human face. But after the figure is placed out of focus you can see—unmistakably, unambiguously—not only something that looks like a human face but one particular face, that of Abraham Lincoln.

So when you are deprived of some of the information in the picture by blurring it, you have no trouble seeing much more in the picture than you originally could see.

Here is what seems to be involved in this transformation. First, the strong pattern formed by the blocks is broken up when the picture goes out of focus. (The celebrated work of Hubel and Weisel on vision in cats demonstrates the existence of elaborately refined edge-detecting neural mechanisms, so that the salience of well-marked patterns of edges is not itself puzzling.) Blurring the picture requires that your brain find some other pattern if it is to make sense of what is seen. Somehow, your brain has engaged in a process that amounts to seeking out some pattern in its repertoire that seems to fit—that satisfices, in Simon's (1955) useful term—the set of cues. Having found a pattern that looks like it fits, your brain seems to impose that pattern on what is objectively no more than an ambiguous suggestion of a real image of Lincoln.

In Darwinian terms, it makes sense that the brain has a bias favoring seeing something rather than nothing, so that it tends to jump to a pattern that makes sense of a situation. Hence, even if there is no pattern objectively there, it tries to impute one. To emphasize that aspect of pattern-recognition, I called the basic building block of *P*-cognition "jumping"—the cuing of a pattern on partial (nonrigorous) processing of cues. When we see a pattern in a situation, we are characteristically jumping, in the fairly literal sense that we see "beyond the information given" (Bruner 1957), or at least beyond the information processed. For the Lincoln grid in focus, the strong pattern of lines dominates the perception. But when the lines are blurred, the brain tunes in on a pattern (in this case, a familiar face) that happens to be roughly consistent with the degraded set of cues that remain.

The perceiver has no conscious control over the situation. Ordinarily we have no way even to become aware that any such process is going on. It is only in odd situations, such as the contrived situation here where you are shown the grid and then see Lincoln emerge as the grid goes out of focus, that you can see what must be happening. It is an important subsidiary point—though not one I will take up in detail until later in the study—that even after you have seen the demonstration there is still a genuine bit of surprise when the image pops out. Similarly, even after hearing Haydn's "surprise" symphony many times, a listener is still surprised by the sudden chord; even after you have seen *Hamlet* many times, even after a wrestling fan has seen many matches, there is still suspense about what is going to happen next.

Further, even after you know that whatever it takes to cue either the grid or the Lincoln image is there, you still cannot see Lincoln until after the figure is blurred, and then you can see only the Lincoln. In other words, and you will see that I make a good deal of this as the argument proceeds, your mind does not work in such a way that you can see many competing patterns in a given situation. Your brain can do more than one thing at a time—even more than one conscious thing. For example, you can drive a car while you listen to the radio or carry on a conversation. But you cannot comfortably listen to two different pieces of music at the same time. When the set of cues exactly overlaps (as with an Escher drawing, or a gestalt figure, or the Lincoln grid under discussion here) there is a strict exclusion. Even after you know that there is more than one pattern available, while you are seeing one pattern, other patterns, including another pattern that you know is there (since you have just seen it a moment before), will be excluded.

2.10

Although the Lincoln grid happens to have been made by manipulating a real image of Lincoln, there is nothing that requires an actual image of Lincoln to produce the effect. The figure was made by laying a grid over a picture of Lincoln, and then scanning with a photoelectric device to get the average darkness of each block in the grid (Harmon 1973). So all the original edges and shading were obliterated, and a new set of (very strong) cues was created by the abrupt shift in tone as we cross from one block to another, but derived in a way tied to the Lincoln image.

Suppose, however, a person were asked to paint a picture without knowing that we are interested in contriving the Lincoln grid. By an iterative process we could get some other example of the infinity of pictures—other faces, scenery, abstract patterns, or whatever—which when scanned and reduced to a grid would yield exactly the pattern of the figure. We could do that without telling the person the result we are aiming for. He or she would just be asked to add a shadow or a deeper color to some parts of the picture, lighten others, and so on, always in a way that keeps the picture a good representation of the tree or building or still life or whatever the person had spontaneously chosen.

So far as I know, the experiment has not been tried; but we would hardly be surprised if it turned out that even the person who made the picture for us found it difficult to see his own picture, rather than a picture of Lincoln, when shown (out of focus) the grid made from his picture. For anyone else, the blurred grid would "obviously" be a picture of Lincoln, though the picture used to make the pattern need now bear no overt resemblance at all to Lincoln.

Once you notice that starting from a picture of Lincoln, or anything that looks like Lincoln, is not essential, you can see that what we are perceiving cannot possibly be somehow just a logical inference from the cues in the picture. For an infinitely diverse variety of other pictures would produce an identical set of cues. From the continuity argument, we would expect that similar effects can be found in connection with other functions of the brain, and that is the case. Many different sorts of brain functions (control of body movements, memory, nonvisual perception such as hearing, and so on) all seem to work in the way suggested by the Lincoln demonstration. We apparently have stored in our brains a large number of patterns, and at least a large part of cognition—on the argument I will try to make plausible, all of cognition—consists of being cued (not consciously, of course) to whatever pattern we first find that satisfies the situation (jumping). We then see (or hear, or feel, or remember, etc.) details that suit that pattern which may have no external correlates at all. The effectiveness of sleight-of-hand magic, the difficulty of catching typographical errors, and many other familiar phenomena can illustrate the point. As with the Lincoln grid, only a carefully devised demonstration is likely to convince us that we are seeing details that are not actually there. But on the continuity argument, the process that produces the Lincoln effect ought to be fundamentally similar to other cognitive processes. In particular, these pattern-dominated effects are not plausibly something that only occurs in the special circumstances where it happens to be fairly easy to demonstrate what must be going on.

This pattern-seeking, pattern-dominated cognition ordinarily works very well, since ordinarily what a creature encounters involves familiar situations to which its responses have been well-tuned. The "filled-in" details are ordinarily correct. The conclusion jumped to will ordinarily work. Even when details are incorrect, or sometimes even if the overall pattern is incorrect, that is either unimportant (we pass on to other things), or further input jars us into looking closer (checking), which may result in correcting the error. When there is a correction, it usually comes so quickly that we scarcely notice the misperception. The "double take" makes easy sense in terms of a Darwinian balance between hesitating-too-long versus jumping-too-soon.

We might then begin to see (and notice how hard it is to discuss thinking without using the language of seeing) how the sort of cognitive process being reviewed here might begin to account for the illusions discussed in Chapter 1. But on the very argument I have been sketching, the only way to get a firm grip on the idea of patterned cognition is by working through the details in a variety of contexts (by practice in going through the motions). The next cut I will take consists of working through a reconstruction of how the evolution of cognition might be accounted for, subject to the *a priori* and *continuity* arguments of this chapter.

Three

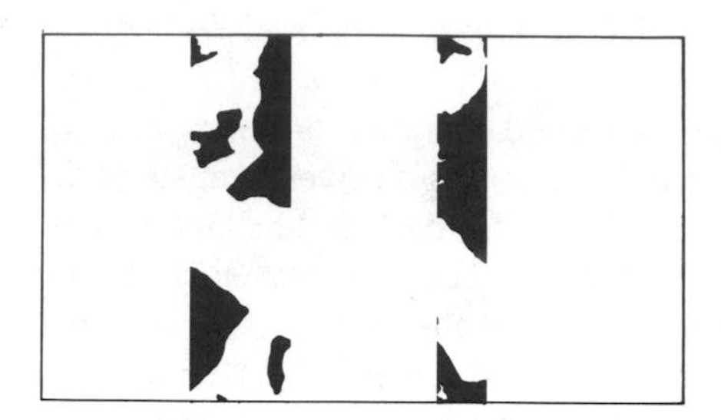

A Cognitive Ladder

A virtuoso violinist will have a most intricate set of physical habits governing the use of his fingers against the strings of his instrument. His performance allows no time for explicit control; so we tend to say, and even more emphatically, the musician will tend to say, that his hand knows what to do. But if we imagine that the hand of the violinist were surgically exchanged for the hand of his identical twin, who had never learned to play the violin, we do not expect the hand to any longer know what to do. The physical location of the habit is in the neurophysiology of the virtuoso's brain, not in his hand. If he decides to change his technique slightly, and masters the new bit of technique, the essential thing that has changed is in his brain, not in his hand. The physical character of a motor habit is therefore not essentially different—at least we have no reason to suppose it is essentially different—from the physical character of a habit of mind (a habitual pattern of inference or way of seeing things). Hence I want to argue that the problem of changing one is not fundamentally different from the problem of changing the other. In an obvious way physical habits engage a more extensive neurophysiology than a habit of mind. But in other ways—for example, interactions with other people—it is the habit of mind that may be harder to change. On the continuity argument we expect that patterned responses will include patterns of judgment as well as patterns of perception and physical movement. What I will try to sketch out is how a continuous Darwinian path might run from the simplest forms of pattern-recognition through finely controlled habitual muscular patterns (as in the violinist's hand) to the most sophisticated forms of reasoning.

We ought to be able to account for the emergence of the a priori features of jumping, checking, priming, and inhibiting [2.3]; and on the continuity argument [2.6] we ought to be able to show how those features can be combined and extended to produce more sophisticated characteristics of cognition, all in a way such that the initial appearance of radical novelty of function never requires radical novelty of structure. Here is a convenient set of steps defining such a "cognitive ladder":

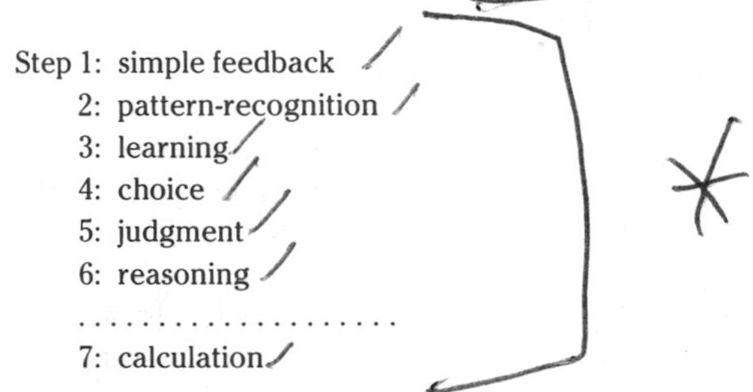

Step 1: simple feedback
2: pattern-recognition
3: learning
4: choice
5: judgment
6: reasoning
. . . .
7: calculation

Of course this could only approximate what must have actually been a gradient, much steeper at some places than others, but not the sharply defined steps of a literal ladder. "Judgment" as used in Chapter 2 is here divided into the three steps of (5) judgment, then (6) reasoning, then (7) calculation. The categories form a cognitive ladder in the simple sense that each step takes us to a more sophisticated level of information processing. The account we want then must be such that each step of the cognitive ladder must require only some refinement or extension of what is already available at the previous step, never a basic restructuring of how things work at the previous steps.

The final step from reasoning to calculation is different from the earlier ones, since (I will argue) it turns on cultural, not genetic, evolution. So for that step, the Darwinian continuity requirement is trivially met, though the argument about how that occurs is not itself trivial.

3.2

Life appeared on earth about 3.5 billion years ago, but for an indeterminate but certainly very long time there is no indication of processing of information. A plant grows toward the light but by means of purely local responses. Differential exposure to light increases the rate of various biochemical reactions in some places (the part of the plant nearest the light) relative to others, and the effect is that the plant grows toward the light. Something perhaps more like what we would think of as information processing is the reorienting response by which one-celled organisms make their way up a chemical gradient toward richer sources of nutrition. But there is still no specialized organ that takes inputs of information and determines a response. If we take the appearance of a specialized organ—something that could be regarded as the precursor of a neuron—as the criterion for the appearance of actual information processing, then even simple feedback (so understood) appeared on earth perhaps as recently as a billion years ago, 2.5 billion years after the first evidence of life.

What I have labeled "reasoning," on the other hand (meaning the explicit use of language as an aid to making judgments), conceivably might date from little before historical times. On the most generous plausible reckoning—dating the emergence of reasoning with the appearance of human forms indistinguishable from ours—the genetic basis of our highest cognitive functions probably appeared not much more than 100,000 years ago.

Therefore, giving a reasonably generous allowance (100,000 years) for the appearance of reasoning, relative to something like a billion years for the existence of some primitive form of cognition (some form of processor mediating between monitoring of conditions and responses contingent on those conditions), the ratio between the two is on the order of 1:10,000. So for practically all of the history of life on earth, and even for practically all of the history of life since the appearance of primitive cognition (upward of 99.99%), there was no creature capable of reasoning. We have, in a sense, plenty of time to work with in considering how the cognitive ladder might have been climbed.

On the other hand, if reasoning appears only at the most recent stage of a very long course of evolution, we would hardly expect that a refinement of reasoning (calculation)—the capability for manipulating abstract rules of inference—would lie at the root of everything, or anything, else. On the argument here, that puts things exactly backward. In contrast, a leading student of perception (Rock 1985) has argued that even perception is based on something like nonconscious processes that incorporate what I have been calling calculation, so that a fortiori reasoning and conscious calculation could hardly fail to be based on wired-in processing of propositions in the brain. An argument like Rock's, or like Fodor's (1975) "language of thought" argument, or like Piaget's formal operations, would be difficult to make compatible with the position I will urge. But I will make the best case I can to show how an argument of the sort I will urge can be made plausible, and even attractive in Darwinian terms, and then can be developed (in the later chapters) into an account of the details of cognition with some analytical and empirical bite, as claimed in the Introduction.

Under this view, even the most abstract sorts of reasoning (formal logic and mathematics) must be reducible to pattern-recognition. The skill human beings exhibit in such matters would be akin to the skills we exhibit (for example) in playing the violin. It is not a natural, inborn, or routinely developed activity of the mind. Even for someone trained to do it, formal reasoning only occurs when the circumstances are favorable for it. Cognition (on this view) is intrinsically "a-logical" in a sense I will try to make increasingly clear as the study proceeds.

Step 1 (simple feedback)

This first step never disappears. Your own body relies on it for many things, such as the knee jerk and other familiar reflexes.

A moth flies toward a light by simple feedback. The moth's right eye is linked to the nerves governing its left wing, and its left eye is linked to its right wing. Neurons in the eye pulse faster according to the brightness of the light reaching that eye, and the beating of the wing is governed by the pulse rate arriving from the neurons in the eye. The result of this linkage is that the moth turns toward the light. If it overshoots the light, or if an air current turns it awry, it turns back to the light. This is the simplest kind of information processing, strictly at the level of such mechanical devices as the governor on a steam engine or the thermostat that controls the furnace in your house.

Deep questions (about consciousness and the like) do not even begin to arise at this level. Nor do reasonable questions arise about whether the creature is "computing" its direction of flight. Of course it is true that we could treat this process in strictly computational terms (simulate what the moth is doing on a computer). But we have no more reason to think of the moth as willing or wanting or choosing or calculating or intending to fly toward the light than to use similar terms with respect to the thermostat's management of the temperature in a house. If we do use such terms it is only in the metaphorical sense that we could speak of a rock "wanting" to minimize its potential energy (so if we let it go, it falls).

Step 2 (pattern-recognition)

For the simple feedbacks of Step 1, there is just one input, direct from the environment via the moth's eye, and one output, going to the wing muscles. Something we are willing to think of calling pattern-recognition (Step 2) emerges from a generalization and extension of that, where we begin to see complex networks of many neurons. Most of them are connected at both ends to other neurons. Only a tiny fraction are connected (at one end of the process) to the outside world, and (at the other end) to action devices inside the body, such as muscles.

The elaboration of the scheme from one-input/one-output to complex networks would not violate cognitive continuity. There is no qualitative discontinuity in sending an output to another neuron instead of sending it to a muscle. Over time—and a very long period of time is available—externally prompted signals might come increasingly to be used to alert other neurons (not to directly control an action device, such as a muscle), and the system eventually produces patterned outputs mediated by cues generated internally. Given a long, long time (given the hundreds of milions of years that are available for this stage), we could get from the stage of one-input, one-output via primitive neurons to the stage of pattern-recognition of a highly elaborate sort, such as is in fact seen in even very primitive multicelled animals. For almost all multicelled animals, however humble, can do at least a few things (recognize and give patterned responses to reproductive cues, for example) with a sort of competence that still seems hard to match on a computer. And by the time we reach the pattern-recognition competence of (say) a pigeon, we are beyond what a computer (as of 1987) could do.

From the fossil record it appears that the stage of pattern-recognition was reached over 500 million years ago. By that time life had progressed far beyond the cognitive step of simple feedback to the step of sophisticated pattern-recognition and patterning of responses. Neurons had evolved a long way toward the elaborate fine structures found throughout the animal world today. In a contemporary brain (say in a cow's brain, or in an octopus brain) there will usually be very large numbers of inputs to a single neuron which carry electrical impulses to the cell body through the dendrites, or through synapses directly on the cell body. The arrangement is such that the neuron gets intensity information from each input through the pulse rate; and the synapse connection can be either of a type that treats the input through that synapse as positive or a type that treats it as negative (excitory or inhibitory). There is also a complicated, still only partially understood, neural chemistry that modulates the electrical activity.

But a simple description is that each neuron adds up its inputs—some negative, some positive—each varying in intensity (pulse rate), and this determines the signal the neuron itself will send to other points through its output axon. Contingent on the signs and intensities of its inputs, the neuron may be stimulated to fire new pulses or not. The neuron's outputs then are carried along the axon, which in turn can divide into many—as with the input connections, often thousands—of fibers, culminating in connections with other neurons or in self-stimulating or self-inhibiting feedback onto itself, or as output signals to the organs of the body.

Suppose, then, that the action to be governed was more complicated than how rapidly to beat a wing, but that it involved coordination of various parts of the body, and coordination of various (not just one) inputs from the outside world, or from the creature's internal store of responses. Instead of just one input and one output, we can get an enormously complex array. A quite typical neuron in a mammalian brain will have hundreds or thousands of input connections and hundreds or thousands of output connections. There will be billions of neurons (in the human brain—no one is yet sure—at least tens of billions), arranged in complexes of levels and stages and specialized subunits, with inputs from internal circuits as well as external senses, and with rich feedbacks from level to level. So from simple feedback, further evolution over an immense period of time has led to enormously complex schemes for responding to various patterns of inputs with various patterns of outputs. As I have mentioned, we can allow 500 million years for the evolution of cognition from the simplest one-input, one-output feedback loop to neural networks of great complexity. There is no reason to suppose that some radical discontinuity was required to get from the simple situation to the astonishingly complex one. Once some processing capability has developed, we have the basic scheme in place which by extension could come to respond to very complex patterns of cues and to trigger complex patterns of responses. We have the makings of a brain.

The beginnings of priming and inhibiting [2.3] would occur even at the Step 1 level of simple feedback, in the form of short-term habituation and sensitization to discreet cues. These can be seen in creatures with very simple brains, such as slugs. Habituation (coming to temporarily ignore stimuli) and sensitization (coming to temporarily respond with special sensitivity) are easily demonstrated with such creatures, and versions of these temporary inhibiting and priming effects can be found in all creatures with more elaborate brains.

Habituation is an adaptation to the risk of a misleading cue (so that the creature will not cycle endlessly in response to some feature that often indicates danger or opportunity but is only persistent noise in this context). Similarly, sensitization adapts to situations where a valid cue is likely to be repeated. If there is a predator in the area, then there would be an advantage to being especially sensitive to cues that so indicate once other cues pointing that way have been detected; if food usually is found near other food, it will pay to be especially alert for cues indicating food once some has been found. So these much-studied mechanisms for positive and negative feedback (sensitization and habituation) would provide the evolutionary basis for the more elaborate forms of priming and inhibiting which become important at later steps. Elaboration of these features would almost automatically (that is, it is hard to imagine how it could be avoided) proceed along with the enrichment of neuronal networks and mechanisms that define Step 2.

Step 3 (learning)

So far, however, all responses are completely stereotyped: learning plays no role in shaping the patterned responses. The creature involved (and for many sorts of behavior—"autonomic" behavior—such as circulating blood, the creature could be you) would have no capability for learning, or it may have the capability for learning and even (as with you) some capability for calculation, but not with respect to the functions under study. When you sneeze, you will shut your eyes. That extra reflex protects your eyes from damage. It is a good idea to shut your eyes if you sneeze, and logical to shut your eyes if you want to protect your eyes while sneezing. But you do not do it because you had that good idea; and even extreme partisans of a view of the brain as a logic device are not likely to suppose that, even unconsciously, your practice of shutting your eyes while sneezing was the product of a logical analysis of the problem.

Similarly, many facial expressions common to humans everywhere (smiling when pleased, and many subtler things) are certainly innate patterned responses, and in many cases involve response to innately recognized patterns (other people's smiles, for example). Many such features appear to be the same across all human cultures and, indeed, similar to signals found in other primates. Other facial expresssions and body language vary from culture to culture. But once absorbed, they are ordinarily triggered by appropriate complexes of cues with no conscious effort or will. Indeed to a large extent we are not even consciously aware that the expression is being expressed, and it is often very hard to keep expressions from coming even if you want to suppress them, though with practice it can sometimes be done.

We want to consider how cognition might get (consistent with Darwinian continuity) from patterned responses that are innate to those that are learned (get from the second step on the cognitive ladder to the third). The simplest account might go this way. Across a wide range of creatures, and for a variety of functions in those creatures, we find the following sort of scheme. Certain complex connections must be made, either while the creature is developing to maturity or beyond maturity, in the process of repairing an injury. In part, there are mechanisms that guide the process in a deterministic way. But especially when the situation grows complex there is an element of trial and error. Tentative, groping connections are made, and those that get used with good effect are reinforced, and those that aren't decay.

So far this grope-and-reinforce process has nothing to do with learning in the usual sense. It is simply an elaboration of the basic development scheme that gets a creature from being one cell (a fertilized egg) to being whatever it is going to be. As more and more complex creatures evolved, more and more selective advantage arose for variations in the development scheme that would reduce the risks of errors in building very complex structures. Later, presumably, these mechanisms got extended to handling repairs after development.

But given this grope-and-reinforce capability, which plays an important role in allowing the creature's brain to develop in the first place, we have a basis for learning. For the same scheme of trial connections followed by either reinforcement (when they work, hence get used) or decay (when they don't) provides the basic step needed for learning. As with the more fundamental step of the evolution of complex pattern recognition from simple feedback, no breach of cognitive continuity is required. In an adult brain, no new brain cells are produced, but there is always a great deal of activity (elaboration and decay of synaptic networks) going on which, if reinforced by particular patterns of experience, constitutes the neurophysiological basis of learning.

Nothing like a complete understanding of the neurophysiology exists today. But from what is known, there is no reason to postulate some radical shift from the structures that are available to creatures who have now reached the stage of pattern-recognition to allow learning to appear. There is no need to doubt continuity. In physiological terms, learning looks like the sort of process I have been describing, not like some process essentially different from what happens in development or in repair.

Step 4 (choice)

So far, the creature never decided between A and B. It only has a repertoire of (partly innate, partly learned) patterned responses, and a particular response is triggered or not. Absent the right set of cues, the creature either routinely is doing something (breathing, say), or routinely not doing it (chewing, say). Given the right cues the behavior is suppressed in one case (the animal holds its breath) or invoked (the animal chews). Choice, in the sense used here, enters when a creature sometimes has more than one response to a situation. The creature is now capable of a two-stage process, in which *first* a context for a particular choice (say, fight-or-run) is recognized, and *then* attention is focused on that context, leading to choice between (or perhaps among) patterns in the repertoire that might be invoked in that sort of context.

Priming and inhibiting would already have a role before we reach the level of choice (Step 4). A creature could learn (or could have genetically engrained) a special sensitivity or insensitivity to certain cues in certain contexts. If *X* is frequently accompanied by *Y*, it will help to be particularly prone to jump to seeing *Y* when *X* is present, and the contrary if *X* is almost never accompanied by *Y*. But the second of these devices (inhibiting) takes on a greater salience, and is likely to be substantially refined when we reach the choice step on the ladder. In particular, at this step an extended risk of hesitation or vacillation—of self-defeating indecision—arises that could not easily occur until the creature had reached the possibility of choice, hence the possibility of extended indecision or ineffectual cycling among its possible choices.

Getting from learning to choice itself poses no special puzzle. If responses are available (either instincts or learned habits) which are triggered by different sets of cues, cases would arise in which the two responses would deal with closely similar situations. Refinement of this process would then lead to the two-stage capability for discrimination between subtly different sets of cues.

But there will be many situations in which it is not clear which choice would be best for a creature (fight or run; continue a search or abandon it; accept a mate or not). Prolonged hesitation will be costly. And once some choice has been made, it will almost always be best to get on with that response, not remain open to vacillation between it and some alternative. Darwinian selection could be expected to strongly favor inhibition of rival patterns of response once a commitment has been made, so that the animal may fight or it may run but it will not try to do both at once or vacillate between the two.

However, checking could be expected to appear soon after the stronger role of inhibition, responding to balance-of-errors pressure [2.3]. For checking amounts to being cued (by something that doesn't fit) to look again before locking in the choice. As will be seen, at Step 5 checking takes on a more profound character.

The neurophysiological bases for priming and inhibiting are much simpler than for jumping or for checking. Since jumping is the basic capacity for pattern-recognition (the term merely intended to emphasize the "beyond the information given" aspect), giving an account of how jumping works amounts to giving an account of how pattern-recognition works. Checking is intimately related to jumping, turning (in the simplest form just characterized) on noticing something that doesn't fit a currently active pattern. A creature is prompted to look closer, do a double take. So how (in terms of neurophysiology) checking works will depend on how jumping works. For both, in order to say very much we would need to know how pattern-recognition works. But however it works, since we have already allowed for feedback loops of a neuron onto itself, and for elaborate connection among neurons, we can easily imagine that a pattern of response could come to include increased sensitivity to cues associated with certain complementary patterns (priming) or decreased sensitivity to cues associated with incompatible patterns (inhibiting). We can expect that once a brain has evolved to the point where it supports the "choice" step on the cognitive ladder, both priming and inhibiting will be well-developed, and that elaboration of these mechanisms to provide the positive and negative feedback I have been sketching will come under strong selection pressure.

Step 5 (judgment)

We next want to get from learning, even learning with choice, to a cognitive stage we would feel comfortable calling "judgment." For this, we want to see more than a capability to gradually construct new routines (capability to learn by trial and reinforcement), and more than that capability combined with an ability to choose between or among standing routines in a particular situation where either might be appropriate. The new kind of ability we require must allow the creature to somehow construct a new reponse to a challenge, solve a problem, resolve a difficulty, by some process that at least looks like the creature can think (can invent a response that it has never made before). The gestalt pioneer, Koehler, provided an account of such a process apparently coming within the competence of a chimpanzee. He set the following sort of problem for his subject.

Inside a cage there is some food, but it is hung from the ceiling out of reach of the chimp. Several boxes are also left in the cage. No one box, though, is big enough to let the chimp reach the food. Will the chimp "see" that by stacking one box on top of another he can solve his problem? It turns out, Koehler reported, that chimps are sometimes capable of doing that, in a way analogous to the following sort of performance:

After a good deal of unsuccessful trial, the chimp wearies or is distracted; but eventually his attention is caught again by the challenge, which is still on view, and which he now studies with what looks like intense concentration. Then (sometimes) he acts, going directly to the job of piling two boxes under the food and getting the food. He has "seen" (an "Aha! reaction," Koehler called it) how to do the job. Although Koehler's original report dates from circa 1915, this sort of account seems commonly accepted by ethologists. A simpler illustration concerns what is likely to happen when a chicken, and later a dog, is put on one side of a wide glass fence with food in the middle of the opposite side. Both begin by trying to bump through the glass. But the chicken never gets beyond that, while the dog eventually is likely to pause, apparently mull a bit, and then run around the window to get to the food. The dog has apparently been able to see the nature of its problem in a way that is beyond the chicken. But the argument here is not contingent on whether the reports on such work correctly interpret what has happened as demonstrating that apes or dogs show judgment of this sort. All we really require is that (beyond dispute) human beings show it.

What looks like an earlier stage in the evolution of judgment is exhibited when a cat goes through preparations for a pounce. What the cat looks like it is doing is what we do before throwing a dart, hitting a golf ball, and so on. It is something beyond the choice level of our ladder, even though the cat's behavior (what happens when one of its routines is triggered) is highly stereotyped.

A cat seeing a rat for the first time usually will attack it, and if so it will kill the rat "in the usual way," biting it in the neck in the way that has evolved as effective over a million generations of cats. Nevertheless, we see something we recognize as a sort of judgment when we see the cat going through its preparatory motions. It certainly looks like it is capable of adjusting the fine points of its stalk/attack sequence to suit this particular occasion: as if what we are watching is not a totally stereotyped performance. We see nothing like that in the behavior of any animals lacking a highly developed brain. For example, no two spider webs are identical. A web is always tailored in various ways to fit the particular site. But there is no sign that this involves anything like planning or judgment. Rather, the spider responds to local cues as it proceeds. But it looks like the cat is planning, albeit within a narrow range of possibilities.

Here is how we might account for the sort of protojudgment the cat exhibits, which then extends readily to the "Aha!" sort of judgment that Koehler attributed to his chimpanzee and which we all attribute to ourselves. The way it works grows out of a capability for internal representations that is increasingly valuable for competent behavior as creatures evolve which exhibit increasingly complex patterns of movement and increasingly rich repertoires of behavior. In particular, once we have land animals of sufficient mass to injure themselves by bumping into things, and for whom rapid complex movements in the presence of obstacles are important, strong selective pressure would favor variations that would contribute to (or improve upon) a capability for carrying some sort of dynamic internal representation of things outside the brain: for the particular sort of pattern-recognition that gives a sense of how the animal's own body is arranged in space.

In all higher animals, that awareness of the arrangment of the animal's own body is elaborately developed. A neural loop provides that in addition to the signals to the muscles setting in motion a pattern of response (efferent signals), there are feedbacks (afferent signals) which inform the brain of what is actually happening. This afferent/efferent setup makes no sense unless there is some internal pattern that allows the creature to use the feedback signals to adjust its performance by comparing the pattern of actual movement (from the feedback loop) to the intended pattern of movement. It seems reasonable to suppose that the pattern a creature recognizes—adjusts to—by sensing its own movements is the same pattern the creature uses to organize its movements.

A similar sort of dynamic internal patterning seems required to allow an animal running along a trail to avoid stumbling over irregularities in the terrain, logs that have fallen across the trail, and so on. For as you can notice for yourself if you walk through the woods, by the time a creature's feet reach the points where it might stumble, its eyes are ordinarily focused on things well down the trail. You do not watch your feet, but they know where to go. For a human, it would take refined experiments to be sure that the feet are not being watched (subliminally) through peripheral vision. But four-footed animals could hardly be doing that.

Finally, a higher vertebrate's memory of its surroundings (its territory, neighborhood, and so on) must provide some internal representation that keeps track of the location of important temporary features of the scene. A zebra which has caught sight of a lion does not forget where the lion is when it stops watching the lion for a moment. The lion does not forget where the zebra is.

This dynamic internal patterning of the world outside the brain emerges much earlier than anything we would be tempted to call judgment. Presumably it always would involve some combination of internal and external cues. That is, as with ordinary perception (such as seeing), the updating of the pattern would exploit instinctive and learned patterns of expectation, given changes in the external cues and the passage of time. What is then needed to get to the stage of judgment is increasingly complete internalization of the process that already exists. What we recognize as judgment arises when we have wholly internal (rather than partly internal, partly external) stimulation of this updating, plus some process for trying out various combinations of patterns of behavior already in memory, until a sequence that works can be found. Signals like those ordinarily sent to the muscles to stimulate action, which then return the monitoring afferent signals, are short-circuited, and the brain—alerted to the situation—receives pseudoafferent signals directly from another part of the brain. When sometimes a version of this process is encountered that produces a result that looks right, the creature acts, sending that pattern of stimulation to the muscles. ("Sometimes" only, since of course a capability for judgment does not require an infallible or flawless capability.) When the pattern does not look right—when it looks wrong—further search is prompted. But this capacity to see something about a pattern that has been jumped to as looking wrong, prompting a look-again response, is just what I mean by "checking." At this step, however, checking appears to take a stronger form that seems to be linked (and might actually be linked) in an essential way with consciousness. For what is being looked at is not out in the world but something inside the brain. In still very limited ways, devices (computers cum sensors) have been built that can be described as seeing or otherwise perceiving aspects of their environment; but no computer does anything even loosely describable as seeing what it is seeing.

Therefore judgment (in the specific sense, of course, of this discussion), and the elaborated capacity for checking that seems intrinsic to it, seems to me the most difficult step of the cognitive ladder, even though it occurs before we reach human beings; and the difficulty comes specifically from the difficulty of going further without supposing that by this step a well-formed consciousness has emerged.

Although checking internal patterns seems closely associated with consciousness, it does not automatically accompany consciousness. When a person is in the peculiar state of consciousness involved in dreaming, the propensity to check is turned off. In accounts of dreams, the most persistent feature is that bizzare discontinuities occur that in waking life would certainly jolt us into a double take, but in a dream they don't. Theatrical performances and various other familiar settings also tend to turn off the critical (checking) tendency, as do emergency situations that prompt intense action. So although internal checking seems to require consciousness, the propensity to check—in particular, to check internal patterns (mulling)—varies widely with context.

We cannot even begin to say something satisfying about just what consciousness means (neurologically) or how it might have evolved, by continuity or otherwise; and so we cannot begin to say with some conviction why (if in fact it is so) there is some essential connection between consciousness and the appearance of brains that have the capacity for checking. But we can follow William James in expecting that consciousness must be essential for *something*. It looks vastly too sophisticated to have evolved and been sustained merely by quirk. The only reasonably obvious thing that it does is permit us to detach ourselves a bit from the internal representations and watch our own behavior.

Aside from the role of consciousness, but given the potential for internally guided dynamic patterns discussed a moment ago, the new stage of judgment requires no radical novelty. What it does turns only on still more of the gradual enrichment of the fundamental theme that appears as soon as simple feedback begins to evolve into increasingly sophisticated tuning of patterns to cues. We have already supposed that primitive learning (acquisition of habits as distinct from relying wholly on instinct) proceeds by trial and reinforcement-or-decay. As will be seen when we come to a more detailed account of learning in Chapter 6, I do not mean that the trials are blind variations of prior behavior. On the contrary, I will stress the Lamarkian aspect of learning (meaning here an account that stresses the response of a creature to its environment). But even this sort of variation-and-reinforcement can proceed with capabilities not fundamentally different from the more primitive processes that allow for development and for repair. What is new is that the trials need not be actual trials, which take substantial time and other resources and may involve danger. Very crudely at first, some capability appears for running the trial in the head. Except for the stress on continuity with the simplest precursors of pattern-recognition (simple feedback), this is a very familiar line of argument which can be traced back through Popper (1968) to Darwin himself.

A human being can do that sort of internalized manipulation very well. It requires sending signals to the internal representation, so you can "see" what you would have done if those signals had been sent to your muscles. (Try mentally tying your shoelaces. Do it carefully!)

And this capacity for trying things out in the head, not just in the world, still does not imply anything beyond extension of capabilities shared with any creature which can run without watching its feet. A creature with a very large brain, capable of storing large numbers of complex patterns, and capable of carrying through elaborate sequences of internal representations, with this capability refined and elaborated to a very high degree, would be a creature like you and me. Somehow, as I have stressed, consciousness conspicuously enters the scheme at this point of highly elaborate dynamic internal representations. Correctly or not, most of us find it hard to imagine that an insect is conscious, at least conscious in anything approximating the sense in which humans are conscious. But it is hard to imagine that a dog is not conscious in at least something like the way an infant is conscious.

The increasingly sophisticated toolmaking that has often been suggested as a critical stimulus to the evolution of human intelligence appears completely within the scope of this "judgment" step of the cognitive ladder. It requires increasingly refined precision of small muscle control, appreciation of subtle cues in the material being worked, ability to see in the mind's eye what is likely to happen if cuts or chips are made this way or that: all the sort of thing that a modern craftsman relies heavily upon, but none of it the sort of thing that very plausibly requires logic, calculation, following out of formal rules, or even verbal reasoning of the most informal sort.

Consider two radically different ways in which we might imagine that a chimpanzee solves the box/food problem (Step 5). In the first case, we could imagine the chimp does (or perhaps now, since the process is wholly internalized, we could say "thinks") something like this: he sees the box, sees himself standing atop the box unable to quite reach the food, sees himself standing on the box needing something to get higher. The problem has now been reduced, as the mathematicians say, to a problem that has already been solved. Almost any chimp can get to food by moving one box to a position under the food. Provided the chimp can visualize himself standing on one box and not reaching the food, he has only to imagine a situation he already knows, which is that of putting a box under the food (one already being there) to see how to do the job. He gets the food. If you think of how you go about locating something you have misplaced, or solving many other everyday problems, or how you would solve the chimp's problem if you were in that cage, you will certainly find that much of your thinking is of this character.

The alternative way of reaching the result would involve a process analogous to what is required in the sort of puzzle where you are given clues about various people, various names, various relationships, and so on, and you must deduce what combination of names/relations/activities is consistent with all the clues. Here we would have clues like: boxes can be moved, boxes can be lifted, chimp can jump a short distance above a surface, and so on. We can thus imagine a computer program that sorts through various ways of putting these elementary clues together to reach the conclusion that "chimp must pile one box on top of another to provide a place to stand from which he can reach the food."

But that does not seem a very plausible account of what the chimp is doing or of how a human being would solve the problem. It is the internal simulation, not the verbal reasoning, that seems plausible. Through arguments of that sort I am trying to show why it is unnecessary to insist that what is going on must amount to something that, could we see it, would look like formal reasoning. On the contrary, on the continuity argument that looks theoretically unlikely, and on the evidence provided by experiments with judgment (I will argue in Chap. 8) it looks empirically implausible.

Step 6 (reasoning)

We now want to take up that peculiarly human capability to make responses that use verbal or otherwise overtly symbolic reasoning, short of the wholly abstract form of reasoning that characterizes formal logic and mathematics (the "calculation" step of the cognitive ladder).

As a step on the cognitive ladder, reasoning appears as the specialized form of judgment that is built out of pattern-recognition applied to forms of language. What Koehler's ape seemed competent to do with respect to boxes a human being can do not only with respect to boxes (and concrete things in general) but also with respect to language. Finding Darwinian continuity between (nonverbal) judgment and (verbal) reasoning is trivial on this interpretation: given a creature with language and with competence at judgment, reasoning comes along free. What we need is enough of an account of how language might evolve to flesh out this notion of reasoning as judgment applied to patterns of language (so that, for example, an inference is a pattern of language that looks right—looks like it fits—given the language and other aspects of context that accompany it).

Checking plays an important role in the case of nonverbal judgment, and it takes on an enhanced importance when we reach the step of reasoning. Once the patterns become abstract (when the internal image is the word "chicken" rather than the figure of a chicken); and further yet, when this becomes extended to more abstract notions like "delicious chicken tomorrow," the risks must become very much greater that a pattern will be prompted which does not really fit. Then the value of being able to mull and check probably changes from an incremental value (though as the sophistication of the capability evolves, an incremental value of great power) to something essential for the capability to use abstract patterns to be worth anything at all.

A musician studying a score will always report that he "hears" the music in his head while doing that. And it does not seem very plausible that the deaf Beethoven could have composed without the auditory imagery of hearing the music in his head, or that Mozart could have refined compositions in his head to the point where he could write down a flawless first written draft without "hearing" the music in his head. But there is no reason to suppose a fundamental distinction between auditory imagery of this kind and the auditory imagery involved in hearing ourselves think, or between that and visual imagery. It is an empirical point that there do not seem to be any musicians without a well-developed capacity to hear music in their heads, mathematicians who do not routinely see in their heads equations and diagrams, organic chemists who do not visualize how atoms seem to fit together, craftsmen who do not examine (as in all these examples, in the mind's eye) the objects of their craftsmanship, athletes who do not review their performances by running it over in their heads. In all these examples, what is involved is judgment in the restricted sense in which I have been using that term in the "judgment" Step 5 of the cognitive ladder. None necessarily or even ordinarily involves the use of language, though all crucially involve the propensity for checking (for being stopped by something that does not look right and doing that part over again, and if it still does not look right, for doing over again with some variation). But although these examples have involved judgment only, and not reasoning in the sense of the cognitive ladder, it is apparent that the same points hold for the conscious imagery of language that makes up the internal monologues that all of us carry on in our heads.

Given that humans have access to language, the ability to reason (to recognize, try out, manipulate) auditory images of language (thinking with language) involves no breach of continuity with the prior step of judgment which turns on being able to do those same things with more immediate images of the external world.

The key puzzle, consequently, for the appearance of reasoning then must be to account for the appearance of language itself. For vision, it is easy to conceive the first steps by which simple cues of light came to give useful information. It is similarly easy to imagine how creature-to-creature signals did that in the context of what would eventually become language. But for language, as for vision, how increasingly rich tuning of patterns to cues gave rise to the marvels of vision or of fully developed language is something that remains a mystery.

We can easily see how to go from simple signals ("danger") to more specific signals ("danger, snake") to simple commands ("come here," "follow me") to specific names ("It's Jack" is something that Jack's dog seems to understand).

All this can be found without getting to humans, but not much more. Seeing the family suitcases being packed, many dogs show every sign of being acutely aware of what comes next; and we would not be surprised if we heard of a dog that had apparently learned to understand the word "trip," showing anticipation at the mere verbal cue. But not many of us would take seriously a claim about a dog that could understand "trip tomorrow" (so that he relaxed today and only showed anxiety about securing his place in the car tomorrow).

For humans, however, the process has been extended to cover reference to things not here now, things not physical but properties defining a class of physical things, and so on. Yet given an adequate brain, a creature could be shown a set of concrete things and learn to associate a signal with the pattern it recognizes in that set of things. Since a pigeon can learn to recognize pictures with people in them (distinguishing them from pictures without people), a creature with a more elaborate brain than a pigeon could learn to also attach some signal (some verbal marker) to the class. We would then have verbal signals for properties of classes, though the members of the class would still be concrete. But having gotten that far, the process could be extended further with the concrete referents fading more and more into the background.

What is hardest to grasp—either with respect to language or with respect to other things, such as vision—is how larger-scale patterns are integrated and understood. For vision, we do not understand how images are made; for language, we do not understand syntax and how that works in addition to and as something that interacts with semantics: how well-formed utterances are put together with no apparent effort and understood with no apparent effort, with the subtleties of communication that makes available. But that it is done, we have no doubt.

On the other hand, although we cannot say much about what makes recognizable patterns, it is reasonably straightforward to sketch out an evolutionary scenario for why such a capability, given that it was biologically feasible, would be favored. For a social species, there is obviously an advantage to elaboration of language in ways that enhance its usefulness to describe to companions, with increasing richness and flexibility, what an individual has seen. The first steps of this can be done with little in the way of sophisticated language use. As those who have traveled in countries where they do not know the language will know, a good deal can be done with nothing more than gestures (such as pointing, grimacing, imitating), supplemented by a few words referring to isolated things (not strung together in utterances). But what we can communicate when we begin to know the language (or when we finally find someone who knows ours) is a vast improvement.

For a species where survival is strongly contingent on social cooperation, there would be powerful selection favoring improvement and refinement of this capability to describe, and with strong positive feedback. As the capability for social cooperation expands, utility of language expands; but as language expands, so does the capacity for fruitful cooperation and the value of intelligence to exploit that.

One of the things that could eventually come to be describable, when this capability has been well developed, is the internal imagery that precedes a judgment. Recall the sketch of the sequence that might lead to the chimp's insight. Describing this sequence would then emerge as a way that one individual tries to prompt another—potentially cooperative—to see something in the same way the first person has.

On the other hand, even a human being today (hence, a fortiori, a remote ancestor of contemporary human beings) cannot easily or ordinarily maintain uninterrupted attention on a single problem for more than a few tens of seconds. Yet we work on problems that require vastly more time. The way we do that (as we can observe by watching ourselves) requires periods of mulling to be followed by periods of recapitulation, describing to ourselves what seems to have gone on during the mulling, leading to whatever intermediate result we have reached. This has an obvious function: namely, by rehearsing these interim results (by using them over and over, as repeated use reinforces developmental and repair connections while unused connections decay) we commit them to memory, for the immediate contents of the stream of consciousness are very quickly lost unless rehearsed.

So there is selection favoring increased capacity to describe to others the way we think a judgment can be reached, and there is also selection favoring those individuals able to sustain mental work on a problem by decribing pieces of judgment to themselves (committing to memory, and some other features that I will describe in more detail in Chap. 5). There would be a symbiotic character of these two functions (language as a vehicle of thought and language as a vehicle of communication). But communication seems primary, since an individual would not get far in the process if he had to invent language for himself. A person learns language from interacting with a community of language users which can pass on from person to person and generation to generation the progressive refinements in the instrument. The process can be observed today in the evolution within a very few generations of sophisticated creole languages (Bickerton 1981).

Reviewing: Given language, we can describe to ourselves what seemed to occur during the mulling that led to a judgment, produce a rehearsable version of the reaching-a-judgment process, and commit that to long-term memory by in fact rehearsing it. And the same capacity for describing internal imagery, or reconstructing it, but now "out loud" rather than only as a private rehearsal, serves the social function of persuasion. Language, intelligence, and sociality are mutually symbiotic, and Darwinian selection would reinforce that.

Step 7 (calculation)

Once language evolves, a learned capability for some use of abstract reasoning (logic, mathematics) can appear even though we have found no reason to suppose any innate capability of that kind at all—or indeed any way to account for how it could have arisen in a way consistent with the continuity argument. We recognize patterns of things in the world. But once language exists, a new class of things exists on which that same pattern-recognition can operate. In words we describe all sorts of special cases, the special cases requiring only tuning to experience, not to any general logic. "If lightning, then thunder." "If knife sharp, then cuts better." But we can then recognize patterns of these forms of words, such as "if . . ., then . . .," in a way that prompts another pattern of language (what we call an inference) that the first, by persistent association in the world, comes to cue. This is not yet formal or abstract logic but reasoning in the sense of Step 6. A novel concrete form is handled as something that looks like it fits a recognizable class of concrete forms. A "looks like" insight will not always occur, nor need it always turn out to be sound when it does. But the process, as I have described it to this point, works strictly by "looks like" responses to patterns of language, not by an analysis of its abstract form.

Logic emerges when the blanks in the forms are merely placeholders for anything (forms like, if *X*, then *Y*) rather than concrete things. On this account, verbal reasoning does not depend on logic. On the contrary, logic on the account here is a by-product of the capacity for informal reasoning (pattern-recognition applied to patterns of language). Informal reasoning, then, is not degraded logic; rather, formal logic is a refinement of informal reasoning. This point will be developed further as the study proceeds. It is an important point, and I hope it will come to seem completely obvious. For much writing on the matter certainly sounds like its author has supposed the contrary.

Reviewing again: On the account sketched, reasoning is a specialization of pattern-recognition applied to language, and logic is a further specialization of reasoning characterized by its fully abstract character, using *X*'s and *Y*'s, not specific referents like lightning and thunder. On this account, devices that allow even these abstractions (like *X*'s and *Y*'s) to be given an embodiment that assimilates them to concrete experience should have great value. We see that in the important role that diagrams play in facilitating abstract thought through the use of Venn diagrams in logic, coordinate geometry in mathematics, Feynmann diagrams in physics, Edgeworth boxes in economics, and so on. Formal logic and mathematics are then no more products of biological evolution than Chinese. Rather, the wherewithal for carrying on abstract reasoning is a special kind of language, and like other particular languages a product of cultural not of biological evolution. Hence logic is not something that we can expect to find built into the brain. If we came to understand the brain completely, we would not on this view expect to find some special place that specializes in logical operations. The tendency of both language and formal reasoning to be co-located on the left side of the brain is not a clue to the genetic evolution of logic but a tautology.

On the argument here, therefore, we expect to find competence in informal reasoning in every human society (as we find language in every human society); it is always there because the capacity for it and the propensity to develop it is innate to the species. But calculation (by which I mean to include formal logic and other sorts of abstractly defined, formal inference procedures) is not something in our genes. Therefore while we would expect to find a capacity for calculation in all normal human beings, the form it takes, how far it is developed, and the role it plays will vary across cultures and across contexts within a culture. On that point, following Goody (1977), I will stress the role of literacy when we come to a more extended discussion in Chapter 5.

The balance of the theoretical argument (Chaps. 4–7) seeks to put to work the notions that are now in hand from the arguments of Chapters 2 and 3. What I will be trying to show is how an expanding scale of interactions among jumping, checking, priming, and inhibiting (expanding the scale of interactions of the same basic notions that have already been used so often in the cognitive argument of the present chapter) can yield an account with strong empirical implications.

Four

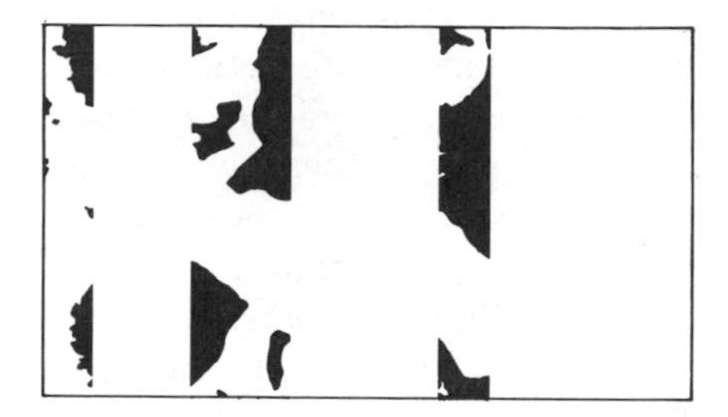

P-Cognition

We do not have direct access to our own mental processes. Still, what we can observe by introspection necessarily has some connection with what is going on.

What appears to happen when we think is akin to tuning a radio. There is a rapid scanning, then a tuning in on some pattern. If the scanning takes a short enough time (for example, because we are already tuned to the right part of the dial), we may have no sense of this step at all; and if the scanning and tuning together take a short enough time, the whole process seems immediate and processless. We just switch on the radio, and there we are. Open the mouth and (in routine situations) words flow out. As in the radio analog, it is distinctly uncomfortable to be just sitting on the dial not clearly tuned in to some station or other. On the other hand, once a pattern is tuned in, the process is sensitive to cues that are "on the right wavelength" and is insensitive to cues that are not (priming and inhibiting).

Here is a simple example of priming. Suppose I ask: "What do you open to go into a room?"; then, "What is the part of an apple you don't eat?"; then, "What is the opposite of less?"; then, "What is an odd number?"; then, "Is four really an odd number?"—you probably experienced an example of priming.

A simple example of inhibiting is provided by this sentence: "Jim Wright, who lives right down the street, to the right of the Kelly's, is a playwright who is studying the Masonic rites, so he will soon have all the rights of qualified members." You have no difficulty understanding that sentence. You do not stumble over the rival meanings of "rite," nor would you do so even if the sentence were spoken (rather than written, so you would have no spelling cues). Rather, you jump to an appropriate meaning in a subjectively instantaneous way. Though a moment later you see another meaning, and a moment earlier were seeing another, all those rival meanings are inhibited when they would get you in trouble if you noticed them.

However, the radio model of tuning runs into trouble as soon as we ask, Who is controlling the dial? Apparently the radio somehow tunes itself. The radio notion works fine as a metaphor but collapses quickly if we take it seriously as an analogy.[1] But there are other images that add a capacity for self-tuning, and these can be pushed a considerable way. I will start with a concrete illustration adapted from Neisser's early text (1967) on cognitive psychology. Similar analyses have been offered by Hebb (1949) and others. I then turn to a description of statistical curve-fitting as a more general analog of the tuning process. The purpose of this exercise is to give some concrete sense of the notion of tuning of patterns to cues. As I stressed in the Introduction, what goes on inside the cycles of *P*-cognition cannot be directly observed. Nor can anything very detailed be said yet about how pattern-recognition works. But the curve-fitting analogy will give a general sense of what might be going on.

Neisser (following Selfridge) sketched how a computer might be programmed to recognize letters of the alphabet. This pattern-recognition device (fig. 4.1) would work as follows. The device is presented with a letter from some arbitrary font. The device transforms the image into a fine-grained pattern of dots, which an eye does also by way of its individual receptor neurons. The machine then analyzes the pattern of dots for cues (the "feature demons" on the left of fig. 4.1). In the example, there would be detectors of rows of dots in various alignments, or for cues such as the closed perimeter in some of the letters. At the next stage (the "cognitive demons" in the middle of the picture), various more aggregated features are looked for: particular combinations of first-order features. And although how this works in a brain is not understood, the existence of a series of stages of processing, with feedback, inputs from memory and so on is not in question.

In the analogue of a computer character-recognizer, we eventually reach a stage in which the characteristics are "*A*-ness," "*B*-ness," and so on. The device now "recognizes" the letter it is shown as whichever of these final stage characterizers fits best.

In contrast to the radio, such a device would have a capability for tuning itself if there were feedback from the outside world that depended on whether it got a letter correctly. By giving different weights to the various cues, and elaborating and adjusting positive and negative feedback among cues, the tuning at various levels could be gradually adjusted toward better and better performance. In the biological world, the feedback that could tune the process itself (make better brains) is natural selection, rewarding favorable variations of the mechanism that governs sensitivity to cues. The accumulating evidence from neurophysiology is consistent with the sort of scheme we have been discussing. Hubel & Wiesel's feature detectors (neurons that respond to edges at various orientations, and so on) look much like the first stage of the figure 4.1 scheme.

Taking account of the immense periods of time in which mechanisms for tuning the perceptual pattern-recognizers of vision, hearing, and so on were shaped by Darwinian selection then yields the structure/function argument of Chapter 2. The very recently evolved processes that govern thinking are not likely to be simpler or easier than those that govern perception; and they are not likely to have evolved other than by being built on those earlier developments. We can reasonably expect well-marked continuity with those older processes, not eruptions of something

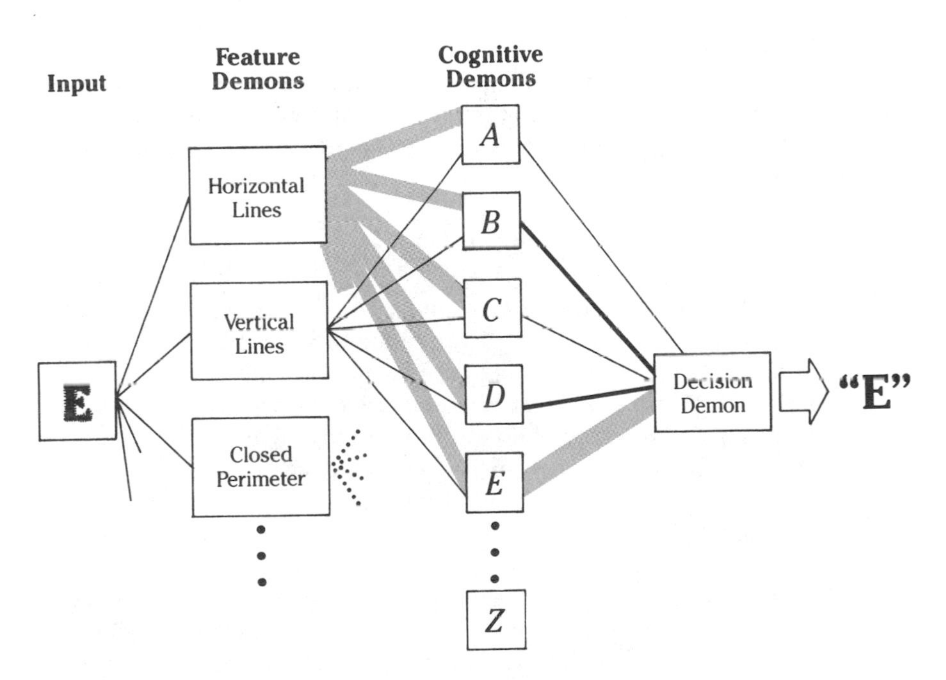

Figure 4.1. Conceptually, a computer (or a brain) might recognize letters by some such scheme as in this one (adapted from Selfridge & Neisser 1960).

radically different. What I will try to do next is to squeeze some insight into how this continuity in pattern recognition might operate from a familiar distinction between two sorts of scientific models.

4.2

Consider the contrast between a mathematicized theory, with clearly stated axioms leading to logically deduced theorems (henceforth, the logical model), versus the sort of empirical curve-fitting that emerges from statistical analysis of data (henceforth, the curve-fitting model).

Given a set of axioms and a set of initial conditions, what the logical model implies about a situation is (in principle) determined. In practice, outside of pure mathematics, this is never quite so. Outside of pure mathematics, and in subtler ways even within pure mathematics, it is almost impossible to construct a fully explicit deductive model. But in principle two competent individuals will expect to reach identical inferences. In the event they differed, they could say (echoing Leibniz's dream of how conflicting judgments ought someday to be resolved), "Let us calculate." They then would be able to agree on what results logically follow from the initial premises. Further, if the side conditions change, but not the axioms, the new results will be consistent with the old ones.

But none of that is true for the curve-fitting model. The model is just a "best fit" to data, sharing the mindless, if you will, character of the computer program described with the aid of figure 4.1. The simplest example is a regression equation:

$$y = a_0 + a_1 f_1(x_1) + a_2 f_2(x_2) \ldots$$

where y is the value being estimated; and a_1, a_2, etc., are weights chosen to give a good fit to the data. The data may be raw inputs, or they may be functions (f_1, f_2, etc.) transforming the raw inputs. And, finally, there are the basic inputs (first-level cues). In the equation, x_1, etc., are the particular inputs included. The string of dots at the end says that more of the same (beyond x_2) might be included, up to x_8 or x_{23}, or whatever.

Figure 4.2 shows an example of a regression for the simplest case, where a single independent variable, x, is used without transformation to estimate a dependent variable, y. The model estimates a value for y given a new sample point (say $x = 10$) as lying on the "least-squares" regression line. That least-squares fit amounts to choosing the single linear fit to the data that is least improbable, given the sample of data. In a very reasonable sense, it is the best fit, given the data, but as in the illustration here, it may not be a very good fit.

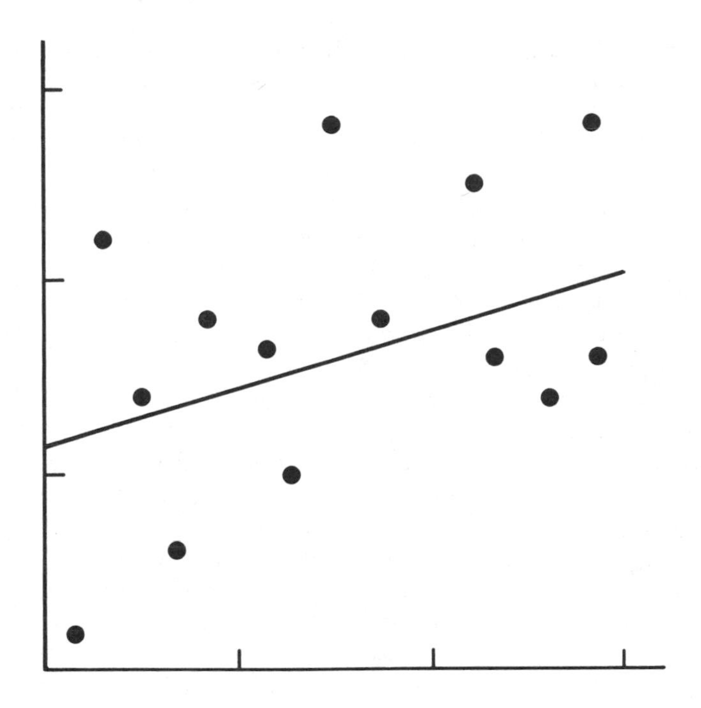

Figure 4.2. Scattergram of data points with correlation $r \approx .4$.

We can imagine the character-recognition program of figure 4.1 as being built up out of an elaborately nested sequence of simple submodels analogous to the regression equation illustrated in figure 4.2. For curve-fitting, the models can be vastly more complicated than the simplest version illustrated by figure 4.2. The term "linear regression" is used to describe these models, but it is misleading since the regression is linear given the transformations but not necessarily linear in the original variables. The functions (f_1, and so on) might convert all the x's to log x's; or convert x_1 to $\log x$, x_2 to $x^{3.2}$; and so on.

There is, then, an enormous range of possibilities for models of this sort. With enough grinding of the computer it is possible to find some model—in fact, many models—that fit any set of data. An elegant demonstration of this point derived the biological clock of the unicorn from a table of random numbers. The potential for tuning that might be possessed by a brain is hardly less impressive. There is an astronomically rich neural network (billions of cells, with trillions of connections) mediated by a complex chemistry. The brain has access to processes qualitatively richer than the on/off switches out of which a digital computer is built.

The relation between curve-fitting or existing computerized pattern-recognition versus the way the brain actually works might be like the relation between a slide rule and a hand calculator. The slide rule and the calculator perform similar functions, but the physical processes are radically different, allowing the analogue device—the slide rule—to be vastly simpler than its digital equivalent. For a fixed level of system complexity, such a device might be able to quickly get roughly right what its digital competitor may after great effort get exactly wrong.[2]

Imagine an attempt to model how the slide rule performed, undertaken by someone familiar with digital computers, but who had never heard of a slide rule. That person is told only what outputs correspond to what inputs, after what time delays. The researcher discovers that the hidden device is extremely quick at some things (for example, computing compound interest), but apparently very clumsy at what might seem much simpler tasks (for example, addition). It would be an extraordinary feat if our imagined observer (given the input/output information) were able to divine the intrinsic simplicity of the way the answers were being produced. For all there is to the slide rule are some scratches on two pieces of wood, tuned to the properties of the number system.

Operating the slide rule involves following some simple rules to produce the calculations, and the rules do not at all look like (although on analysis they must turn out to be roughly equivalent to) the logic of the calculations. What the brain does is vastly more complicated than what the slide rule does, and the organization and operation of the brain is vastly more complex than that of a slide rule. But the analogy sketched makes sense in terms of the general account of the evolution of cognition we have been going through: it is a device that usually gets things roughly right, and there is no reason to suppose that the means by which it does so look at all like the manipulation of propositions in logic or like the algorithms that run contemporary computers. Similarly, we should not expect that the tuning arrangements in the brain will look anything like the structure of a computer chip embodying a regression routine. Nevertheless, as a functional sketch of what seems to go on, the regression routine looks (on the *P*-cognitive argument) much more plausible than a formally logical account.

4.3

Now let me point out some things that characterize the curve-fitting (in contrast to the logical sort of) model and that also appear to apply to the cues and patterns of *P*-cognition.

To start, the variables included in a regression equation (analogous to the cues the brain uses to govern *P*-cognition) have no necessary causal relation to the thing being estimated. Producing a regression model does not require that the modeler first propose a causal hypothesis, then specify a model embodying that hypothesis, and then proceed to test it. Published reports read that way. But the actual process is often quite different, with the model deriving mostly from a good deal of grinding through of one combination after another, trying to find a variant that seems to work respectably well.

Some examples of curve-fitting particularly apt for the analogy with *P*-cognition are "stepwise" regressions and factor analysis. In stepwise regressions, the computer is programmed to sort through more cues (variables) than will actually be used, and the choice of which variables to include is largely the result of the computer picking out the cues (or combinations of cues) that add most efficiently to the accuracy of the fit. Differences in the way various versions of the stepwise process operate, or changes in details of the operation (for example, the order in which certain variables are to considered), can lead to large changes in the model. Once the model is tuned, it neglects variables dropped out, though eventually they might turn out to be the key causal factors.

Factor analysis is a more aggregative process, analogous to the higher-order cuing illustrated by the "character demons" in figure 4.1. The computer searches to find a set of weightings of first-order cues (a set of factors) which seems to sort the data into categories. In both these situations, a causal account is often constructed after the computer has located a good fit. For example, the factors that come out of the machine simply as unlabeled sets of weightings of the cues are then given labels like "intelligence," "sense of humor," "creativity," or whatever else might seem, ex post, to explain what is happening.

The connection between the grinding out of the factors and the ex post imputation is much like the connection between the reaching-a-judgment process (*seeing-that*) and the justifying-a-judgment process (*reasoning-why*) that will play a large role in the argument to follow.

On this sort of model, the parameters will reflect the particular set of data to which the estimates were fitted; and at the same time the set of data that is used (because of the cut-and-try approach to specifying the model) will partly depend on the specification, which itself was shaped by what the data seemed to best fit. If the model-building is part of a sequential process, then choices made at an early stage (about what variables to include, for example) will influence the later developments.

What I have been saying about curve-fitting models is not (of course) a description of what a textbook will tell you about how to do curve-fitting. But it is a fair enough description of what commonly occurs. As the phrasing was intended to reflect, this sort of curve-fitting is a round and round, tuning-in process, not a step-by-step logical construction. The brain, in terms of *P*-cognition, appears similar. As my son has told me, the brain is an adjustor, not a calculator.

Now consider what we could expect from two equally qualified people, but varying in details of personal experience and training, with different samples of data at hand, perhaps slightly different "canned" statistical routines on the computers they happen to be using. Realistically, as anyone familiar with such work will know, we cannot expect that they will produce very similar regression models. It is only to the extent that we think there is a large amount of shared experience, or direct interaction on shared problems, that we would expect substantial consistency across such models.

But the classes of situation in which we could in fact expect a high degree of convergence are important, since without them we could not be the social creatures we are, hence also not capable of such competence as we turn out to have, since much is contingent on social knowledge [5.13]. If we were looking at a set of models fitted to very large samples of qualitatively similar data, and constrained by some strong prior assumptions on functional forms, then there would tend to be a good deal of convergence. In our context, the prior constraints are provided by the genetic endowment; and the large body of similar data is provided in the cases of such things as language and visual imagery by the fact that almost everyone in a given society has very comparable visual experiences, and (at least within a given social class in a society) very nearly equivalent exposure to samples of language. So although there is no reason to expect that in general intuitions should be much the same across individuals, for deeply shared experiences we would expect that to be true.

In particular, for many of the most common matters, such as ordinary language and visual perceptions, we can expect across-persons consistency to hold. But we would expect that consistency because the special conditions (shared experiences, social interactions) apply, not because it is intrinsic to the way cognition works. For many other matters, even such closely allied matters as technical use of language (as distinct from ordinary language), or seeing things in a technical context (through a microscope, for example), we have no reason to suppose the *P*-cognition will so often be the same or closely similar across persons, beyond a narrow community within which the special conditions hold.

In contrast to a curve-fitting model, however, which is explicit, cognition is tacit. Hence, it may not be obvious at first to people who are misunderstanding each other that they are seeing things differently; and even once they notice, each may be simply puzzled, or appalled, that one claims not to see what the other sees as so obvious that it cannot be missed.

4.4

Now suppose that indeed cognition can be considered to operate much like the sort of judgments that would be made by a person with access to a curve-fitting model, but lacking any real familiarity with how the model was developed—as we lack direct insight into what is inside the black box that in this case is our own brain. Then the judgments that the individual would reach—or better, find himself reaching—would have the peculiar properties I have been stressing by using the analogy with curve-fitting. But those properties would all lie "out of sight" of the chooser, as the way we reach our intuitive judgments (the way we reach judgments by *P*-cognition) is indeed out of sight of the chooser.

We could not expect the pattern to reflect with equal weight everything in the experience of the individual. Rather, where we get to may depend a lot on where we start from. Hence, the details of cognition will depend not only on the experiences of the individual but also on the time sequence of those experiences.

Further, I have already mentioned that (as with a curve-fitting model) cognition—even if we could articulate the cues—need not reflect causal relations between cues and the responses. If things work well, we know that there is a correlation; but a good correlation does not guarantee a causal relation between the cues and the events subsequent to the cues. It remains an open question to account for the fit when the model is working well, or to account for what might be going wrong when it is not.

As a set of *P*-cognitions finely tuned to a class of situations takes hold, the individual acquires, to pick up Polanyi's (1958) phrase, tacit knowledge. Without being able to say just how, he knows how to handle situations of this class, including (in a way I will come to) some ability to adjust for the variations that usually mark each new instance of the now-familiar class of situations. But in contrast to the drift of Polanyi's discussion, we have to emphasize that at least sometimes what Polanyi calls tacit "knowledge" is wrong (leads the individual to a bad result). Or, picking up another celebrated insight, we can say that Pascal was right to say that the heart has reasons which reason knows nothing of; but we have to add that those reasons may sometimes be very bad reasons.

In all these things the analogy between curve-fitting and *P*-cognition allows us to point to a process outside the brain, hence examinable in detail, that exhibits the sort of tuning of patterns to cues that characterizes what I am calling *P*-cognition. We know that a process with these key properties is possible, for we can point to one.

But statistical curve-fitting actually takes place outside the head, and in a way for which we can give an explicit step-by-step account. The patterns it yields can be explicitly defined. So although the statistical routines are serving us as a model for *P*-cognition, it is actually using a step-by-step articulatable process as a model for a process that has (for now, certainly) just the contrary character. The analogy works well, though, because few users of statistical curve-fitting have any knowledge of these details. Like *P*-cognition, the action takes place in an out-of-sight process, inside the computer, which functions outside the comprehension and awareness of the typical user.

An instructive reminder of the richness of the patterns at issue is to consider the varied roles that human beings take on as parent, warrior, lover, hunter, and so on. Each person plays many roles, and while we are playing a particular role we hardly are troubled by, and indeed we are ordinarily entirely unconscious of, contradictions between the behavior we exhibit in one role (one pattern of responses) and the behavior we exhibit in another role. We are somehow cued into various roles (as, for example, warrior, lover) by the interaction of internal and external cues, only some of which we are likely to be able to articulate. Frequently, we are left puzzled or annoyed by our own moods and choices.

But if we turn to physical habits, what goes on is more visible. Such habits can be understood as another form of cued, pattern-governed activity, but with more in the way of overt external (observable) features than their purely mental equivalent.

As Schelling (1983) has brought out with special clarity, we all have habits (motor or otherwise) we would just as soon not have. It is hard to give up smoking, or eating or drinking too much, or whatever else we do that we "prefer" we did not do. If we are in a situation that cues these habits, and it is often hard to adequately articulate what those cues are, we will sometimes find ourselves doing things we know we would really rather not be doing. But knowing that we would really rather not be doing them is not enough to stop us. We can (often) develop habits we would like to have where these are not in conflict with entrenched existing habits (that is really what practicing the details of how to play the piano, drive a car, run a computer involves); and we can (sometimes) learn to turn off a habit that we wish we did not have. But neither is really easy, and the second often proves impossible. All this will be taken up in more detail when we come to the evolution of cognitive repertoires within an individual.

Summing up the argument as it now stands:

1. Cognition is built out of sequences of patterned responses to cues.

2. The cued patterns of judgment can be expected to involve important parallels with other cognitive operations, such as those governing physical habits, memory, moods, emotions.

3. These patterns are not necessarily or even ordinarily consciously perceived (recall the walking without watching your feet example [3.5]); nor are they usually articulable if perceived (you can confidently recognize a face without being able to say anything confident at all about why you are so sure).

4. Patterns play a dual role: we recognize patterns in making sense of the world; we are prompted to use patterns to guide activity in the world. The same patterns play both roles, as is perhaps most clear if you think about the case of efferent/afferent signals that guide and report on our physical movements [3.5].

5. Explicit reasoning is based on patterns of language (which in turn is again based on cued patterns) and, through language, tied to sociality.

4.5

We can begin to develop these points by setting up a taxonomy of judgments in terms of their superficial subjective properties. For a particular judgment, either we have a conscious awareness of an extended process or we do not. And if we have a conscious sense of a process, either it is something we are aware of only in a loose sense, or it is something that is articulated in a step-by-step way. This gives the following taxonomy:

1. Completely intuitive judgments, like recognizing a familiar face, of which we have no conscious sense of how the judgment was reached; nor, even after being conscious of the recognized pattern of features (for example, John's face), can we ordinarily say how we recognize the pattern.

2. Judgments in which there is a conscious sense of a sequence of thinking leading up to the judgment, but the sequence is not exactly or reliably reportable, as when we study a blurred image for a while before we "figure out" what it is, though we can rarely say much at all about what the "figuring out" consists of.

3. Judgments that emerge from sequences that have an organized, rehearsed, repeatable, step-by-step character, of which the extreme case is a formal argument (as in mathematics or logic).

In terms of the cognitive ladder worked out in Chapter 3, these categories correspond to pattern-recognition, judgment, and reasoning, where the last two must somehow be built up out of the first.

Subjectively, there are no sharp lines to be drawn between the categories. Rather, judgment shades imperceptibly from (subjectively) processless intuitions to the most rigorous sort of argument. A judgment that can be articulated in a step-by-step way might be reached also by the less-explicit processes, as I can tell you just by rote memory that $8 \times 7 = 56$, but if you insist I can work it out in a step-by-step way also.

For subjectively processless pure intuitions (broadly understood, to include perceptions, judgments, affects), not only are we unaware (by definition) of a specific process but we are ordinarily oblivious of any span of time during which a process might have taken place—as, for example, when catching the meaning of a familiar word or phrase. This instantaneous sense of the thing is misleading, since even what seems subjectively an instantaneous response can be shown to require a measurable span of time. Something is happening during that span of time. Objectively, what "subjectively instantaneous" means is that the duration of the process falls within the roughly 100 milliseconds (0.1 sec.) minimum subjectively perceptable processing interval.

Sometimes, however, a single cognition does take an easily noticeable span of time. In the case of random-dot stereograms, an apparently patternless array of red and green dots, viewed for 20 or 30 seconds through glasses with colored lenses, seems to eventually organize itself into a three-dimensional pattern. In the last instants, the pattern seems to rapidly emerge from the noise; but earlier, despite a definite sense that the brain is working on the problem, there is no sense at all of anything in particular going on, or of progress being made. Frisby (1979) gives some elegant examples. Something, however, *is* going on. If not, then you would be as likely to see the pattern in the first 5-second interval as in the fifth (given that you hadn't seen it yet). But that is not what happens. After the first few intervals, the probability you will now see the thing goes up sharply.

So (from the last example and many others) it is clear that we do not have subjective access to all the patterns which are recognized on the way to a consciously perceived judgment. And of course it would make no Darwinian sense for cognitive processes to be limited by the rate at which conscious sequences of perceptions can be managed. On the other hand, at least some internal steps on the way to a judgment that are usually unconscious can be brought to, or may be prompted to come to, conscious attention in a particular situation. It takes slightly longer for a person to say whether canaries lay eggs than to say whether canaries sing. Apparently singing is linked directly to the sense of what cognitively defines a canary, as is being a bird. But laying eggs seems to be linked to "birds" and not directed to "canary." The extra step (canary—bird—egg vs. canary—sing) takes a bit of time. Yet nothing essential seems to be involved in whether or not the person recalls seeing an image of a bird, or saying the word "bird" to himself, in the interval before saying that canaries lay eggs. The overt imagery (whether visual or auditory) seems essential only in the context of rehearsing and of checking: that is, in relation to the memory process [3.6] or to the "look closer" response, or in otherwise going over a prior sequence, as in looking for a lost glove, or polishing an argument or a poem or a melody.

The penultimate pattern in the canary example may be the pattern cued by the word "bird"—where, notice, even that that one-syllable word is actually a pattern of phonemes which, in turn, is a more complex set of patterns turning on subtle sublinguistic details of the production of those phonemes. Given that penultimate pattern—in a context where the question is, "Does it fly?"—we get a direct intuition of the response, as in a dark but familiar room we get a direct intuition of where to reach for the light switch without necessarily consciously "seeing" the room. But if we have been away a long time, then in fact it is likely that the process will include seeing (in the mind's eye) the room.

Of course, there is nothing that guarantees that we will actually reach a conclusion when we look at or think about something. Sometimes we try to look at something, or try to think about a question, and reach no result. We say we can't recognize the thing, or we are confused. Further, if we do reach a judgment (get an intuition of what is the case), the process does not guarantee that it is a good or effective intuition, or one that will stand up to critical reflection in the event we or someone else is motivated to examine it critically. The point of the illusions reviewed in Chapter 1 was to exhibit situations in which unsound intuitions are predictably elicited. A very important further point is that when several tries prove necessary to reach an intuition—whether the intuition turns out to be sound or unsound—a critical point of the later turns of the spiral may be additional cues (or altered sensitivity to cues) provided by the earlier failures.

4.6

With respect to judgments that emerge from an extended sequence of conscious intuitions, there is now a sense of things developing, most obvious to us when (with our mind's eye, or ear, and if you were a dog, perhaps with your mind's nose) we seem to see how things would go if we did this or that. We have the capability [3.6] to simulate something, often accompanied by a loose internal commentary on what we "see." Other times there is a sense of a sequence of imprecisely discerned patterns. Visual imagery is an important aspect of thinking for most of us, and language (auditory imagery) is impossible to avoid, even if we try (try!).

This surface phenomenology (what it feels like or looks like when we try to notice our own thinking) cannot be taken literally. We cannot inspect a mental image in the way we can inspect an external image. What looks like seeing new detail when we try to look closer should be understood as a new image, not a detail of what was already there; and internal monologue (especially those that occur during mulling rather than rehearsing) is probably much more fragmentary than we can observe. We get discreet sequences of *P*-patterns (the outputs indicated by the spirals of the figure on p. 2), successively evoked from the individual's cognitive repertoire by the changing set of cues available from the most recently prompted pattern, from the fading remnants of closely preceding patterns, and from external sensing. Each turn of the spiral cues some shift in attention, or a recall from memory, often colored by an emotional response, with the altered set of cues then prompting the next *P*-pattern. "Closely spaced" *P*-patterns may be seen as continuous movement simulating an external process we are trying to think about, like the frames of a motion picture or cartoon.

Adding a conscious sense of a process—going from subjectively simple cognitions to sequences of such cognitions—does not yet bring us to explicit reasoning (Step 6 on the cognitive ladder). Sequences leading to a judgment of any complexity probably never proceed in a fully controlled way (even when attempting to follow an argument provided for that purpose). Rarely is that even approximated. But over time, we can rehearse a sequence so that it becomes more or less repeatable. We get, formally introducing now some language already mentioned, a distinction between

> *seeing-that* . . . by which we reach a judgment. This is to be interpreted as recognizing a pattern whether the result comes in one jump (as in recognizing a familiar face or word) *or* comes after a sequence of jumps (as in recalling where we left a book, or in recalling whether canaries lay eggs if a conscious "bird" intuition intervenes before we see that they do) . . . versus
>
> *reasoning-why* . . . by which we describe how we think we reached a judgment, or how we think another person could reach that judgment.

At a minimum (I will say a good deal more as the argument unfolds), reasoning-why always has an additional important cue absent from the seeing-that process which first produced a judgment: namely, it has the conclusion. The seeing-that/reasoning-why distinction is intimately connected with the mulling/rehearsing distinction [3.6]. Since we can hold only a few things at a time in immediate awareness, a judgment of any complexity requires somehow reducing the number of things that must be simultaneously in short-term memory. We can reach interim results, so that the next intuition need only start from an advanced point; or we can bundle together pieces of the situation into recognizable—but not necessarily or characteristically articulable—patterns. The rehearsable reasoning-why that we perceive as the more organized sort of internal monologue can serve both needs. It produces a sense of pattern to the overall situation, so that one can quickly recapitulate or come back to the point where one was having difficulty getting further; and it produces bundled-up patterns of argument that can be treated as single cognitive objects at later stages of thinking through the problem, perhaps right now, but also when we return to the problem at another time. A consequence is that our thinking will be biased toward "bundle-able" or "patternable" sequences, so that what we find plausible and comprehensible is intimately tied to what can be assimilated to patterns already in our repertoire and built out of bundles (smaller patterns) already in our repertoire.

4.7

This bundling-up point suggests something about how, by "seeing" the pattern of the situation, a person judges a piece of music as good or bad, or a piece of scholarship. Somehow we do that, as somehow we recognize faces and grammatical sentences, and do vast numbers of other things that must turn on highly sophisticated recognition and use of patterns. Even for something as elaborate and open to explicit analysis as the work of scholarship, subjectively confident judgments may be based on a quick scan of the material made with little overt effort; and such judgments, for scholarship as for music, are not always unsound. Nor, at the other pole, are carefully considered judgments always sound.

But it does not even occur to someone recognizing a face that he should be able to articulate how he did it; and unless you are a serious student of music, you will not be embarrassed to be unable to say much about how a judgment on the music was reached. But there are other things, as with the work of scholarship, where we expect to be able to articulate why we reach the judgment we do. And when we do articulate reasons for a judgment, we do that with no sense that the reasons-why we give have only a problematical relation to what in some objective sense could be construed as the actual reasons-why (to the cues that in fact were critical to prompting the judgment). There is very ample evidence [1.7] challenging the commonsense view that what we honestly see as the reasons why we believe something are in fact the reasons why (Nisbett & Ross 1980; Wason & Evans 1975). But it remains as hard to see that problem in ourselves as to see the arrows of the Muller-Lyer illusion as equal in length, or to hear ourselves speaking with whatever accent we ordinarily speak with. We can come to know about these things. But we cannot directly see them. They can never become part of the surface phenomenology of judgment.

The situation becomes more complicated when we deal with extended rather than simple judgments, or with repeated judgments. On the account I have given, the original reasoning-why for any step can *only* follow seeing-that, coming in the way I will describe [5.5]. The reasoning-why grows out of rehearsing, but the original insight (by definition) could not. Once on the scene, however, reasoning-why may serve to confirm and reinforce an intuition; the inability to construct explicit reasoning-why that looks right, in a context where a person expects to be able to do that, prompts doubt about the intuition, and perhaps eventually disbelief. So with respect to a repetition of some step of judgment, reasoning-why can play the role it subjectively seems to play. When the repetition takes place across persons (*A* gives *B* a reason for believing *X*, and *B* does believe it), the reasoning-why might easily be essential for prompting the seeing-that. So sometimes there is a reasonable basis for supposing that the reasoning-why a person gives in fact explains the seeing-that he reports. But that is always a question to be explored, never something about which a person can have direct insight. Even under circumstances that might be supposed especially favorable (as in solving a set-piece puzzle), reasoning-why can never be taken at face value as an adequate account of how a judgment was first reached. I will develop that point a good deal further as the argument unfolds. It has many consequences. A crucial topic for us will be the relation between intuitive seeing-that versus analytical reasoning-why. Seeing-that has a straightforward interpretation in terms of *P*-cognition (seeing that a situation "looks like" a pattern we know). But the connection of reasoning-why with the spirals of *P*-cognition is not so obvious. On the argument of the study, it must be somehow reducible to seeing-that. Chapter 5 will give an account of that reduction.

4.8

Associated with the seeing-that/reasoning-why distinction are two other aspects of the surface phenomenology of judgment every reader will be able to recognize from self-observation. The first turns on the role of "checking" in the strong sense developed at the "judgment" step of the cognitive ladder [3.6]. We get a distinction between ordinary cognition (seeing a tree, for example) as contrasted with a reflective mode (seeing that you are seeing a tree). The distinction is like that between what is happening when a football fan is watching a play, seeing what happens versus checking a replay, watching closely to see whether what he thought happened did happen; or between a musician playing a passage, and replaying the passage listening to check for something that seemed to produce a particular (pleasing or displeasing) effect.

When the individual is surprised by a perception or intuition, he is prompted to a "look-closer" response, or "double take." Or if a person does not see something immediately—whether it is a visual thing or an idea—he may be challenged to look closer, which in the context of an idea or judgment usually means to look closer at some argument for the judgment. What *P*-cognition does in the context of cues that prompt checking amounts to seeing if the cognitive response already experienced (or claimed to have been experienced by someone else) will be repeated.

This will commonly involve a second feature, which is a narrowing of the focus of attention and tightened control of scanning. The combination of checking and close control produces a critical, detached, dissociated sense of what is occurring. Checking a judgment requires an object, as visual checking requires something to be looked-at again. Hence, it requires an externalizable description of a seeing-that sequence: though not necessarily actually externalized, for a person can be self-prompted into checking his own internal rehearsal of reasons-why.

As usual, I will stress the continuity between this sort of cognitive work and more remotely evolved operations of the brain. The shifts between jumping and checking and between loose and close control of attention are noticeable things (not like trying to hear your own accent). They are continuous with the shift from tying your shoelaces or driving your car in the usual, largely automatic way to one of concentration in a step-by-step way on just what is being done. Because of the explicit character of mathematical argument, readers who have done mathematical work are likely to have a particularly sharp sense of recognition as soon as attention is drawn to this distinction. Even in the case of a mathematical argument, a person will commonly "run over" the argument and be left with an intuitive sense of how the proof works. But he can also (and will, if he is checking a proof of his own, or reviewing a proof submitted for publication) look closely at the proof, checking the argument step-by-step. Here the question is, does each step of the proof, as it comes along, prompt the response that the author of the proof claims for it?

As some of the examples I have used were intended to show, the "look-closer," double-take, checking response is not necessarily, or even ordinarily, focused on reasoning-why. Nor, even when we are attending to reasoning, are we always actively checking to see if the claimed step is in fact prompted, or always operating with the tightly controlled focus and scanning that is most likely to prompt checking. But there is an easily observed contrast between a tightly controlled situation versus that of relaxed looking. Relaxed looking is characterized by bare consciousness of the flow of intuitive jumps being prompted by the argument. We would not then be narrowly focused on the argument, or trying to concentrate our awareness on just those aspects of the situation that are the explicit focus of reasoning-why. From that contrast (relaxed vs. focused looking) we get the subjective sense of an intuitive, holistic way of thinking versus a critical, analytical way of thinking.

Figure 4.3 illustrates the situation. Jumping versus checking is a binary pair—when looking we either are checking a pattern we just saw or that we know we are expected to see (checking), or we aren't (jumping). An actual sequence would include both. But the probability that checking will be prompted with respect to a spiral of *P*-cognition will vary not only with the cues and patterns involved but also with higher-order features of the situation. In some contexts we are intently alert to the possibility that something is coming along that needs checking, and in other contexts we are relaxed. Whether checking is what is happening at a particular moment is a yes or no question. But the likelihood that a person will check when in a critical frame of mind is very large compared to the likelihood of checking when a person is in a relaxed frame of mind.

In figure 4.3 the horizontal axis represents narrow-to-broad focus, and the vertical axis gives tight versus loose control of scanning. Checking can occur anywhere in the diagram, but it is particularly likely to occur when focus is narrow and scan is under tight control. The control of focus and scanning is continuously variable. But the distribution of cognitive activity is not at all like a hill, with most activity piled up somewhere near the middle of the figure. Rather, we seem to spend much more of our cognitive effort in the corners than would be immediately obvious given the feasibility of being elsewhere.

4.9

Now consider a bit more detail of the seeing-that/reasoning-why distinction. Figure 4.4 gives a schematic view of the situation.

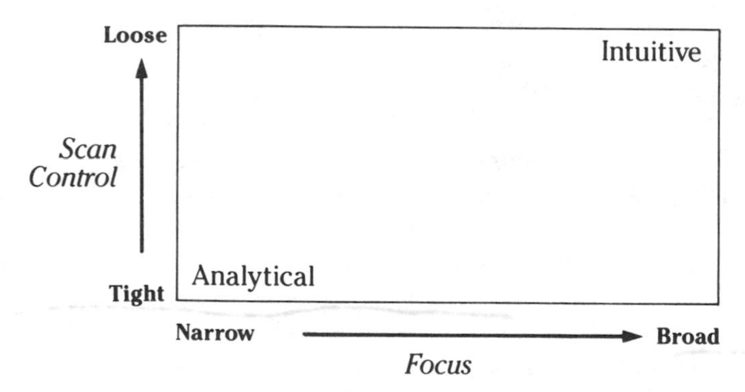

Figure 4.3. Cognitive activity (looking, thinking) will tend to cluster in the lower-left and upper-right corners, producing the familiar sense (for thinking) of a dichotomy between intuitive (holistic, etc.) vs. critical (analytical, etc.) thought styles.

The vertical objects represent the process of reaching a judgment. Moving from left to right in the figure moves us from naked intuition to the most detailed, step-by-step logic. The blobs in the figure refer to consciously noticed steps of *P*-cognition. At one extreme (on the left), we have just a single all-at-once leap from context to judgment. But as we move to the right, we see the blobs (jumps of intuition) getting smaller and smaller. We eventually get segments in which there are no longer any apparent intuitive leaps at all. This gives the apparently solid segments, which are stretches of explicit logic. A dotted stretch of line is a looser (less rigorous) sort of argument, where you must imagine (since it would be inconvenient to crowd it into the diagram) little blobs between the dots and dashes.

Think of the gaps in the vertical structures as parts of the process covered by little jumps of *P*-cognition, where the individual reaching the judgment scarcely notices, or does not notice at all, that he or she has just slipped from one point to another leaving a gap in the reasoning-why. Large intuitive jumps in reaching a judgment become rarer as we move from left to right in the diagram, and the intuitive jumps that lie between the dashes and dots of the vertical lines become ever subtler. As we get nearer the right-hand pole, we get closer and closer to the normative ideal of rigorous step-by-step deductive reasoning. There is no useful boundary along the horizontal axis where we pass from nonrigorous, intuitive thinking to analytical thinking.

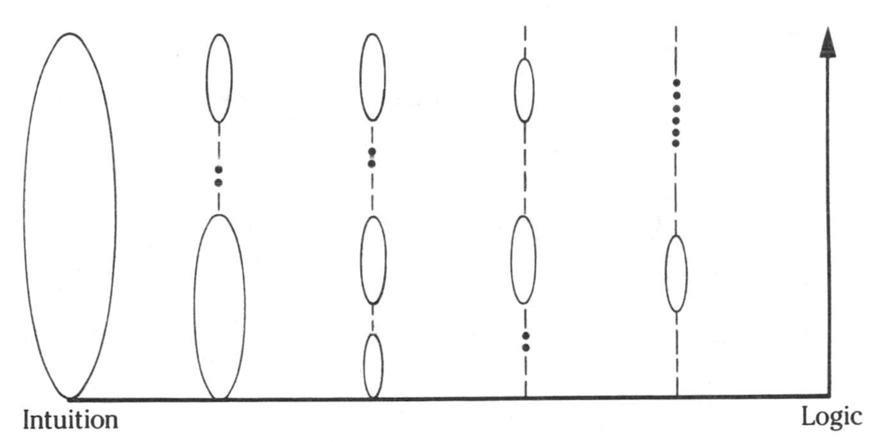

Figure 4.4. The structures move from all-at-once intuitive to step-by-step logical as you move from left to right in the figure; but there is no sharp line to be drawn. Within the figure, each structure can be interpreted in terms of either seeing-that or reasoning-why, as described in the text.

Further, even the solid lines depicting tight logical stretches, when examined closely enough (with the aid of the sort of "cognitive microscope" provided by mathematical logic), turn out to be sequences of *P*-cognitions with the individual steps so small that only, so to speak, under a cognitive microscope can we notice that they are not the seamless arguments they appear to be. Hence, the paradox that some of the most difficult logic and mathematics (associated with names like Bolzano, Peano, Russell & Whitehead) involve precisely those topics we ordinarily take to be so trivial as to require no analysis at all (on the order of how we know that $2 + 2 = 4$).

In sum: as we move up any of the vertical sequences in figure 4.4, a judgment is being reached. In the horizontal dimension, as we move from left to right, we see sequences that we would think of as more analytical, less intuitive. But even the most analytical line of argument must rely ultimately on intuition (axioms, simplifying assumptions, definitions, etc., which appeal eventually to an intuitive conviction of reasonableness). There is no boundary that marks off an intuitive judgment from a judgment based on close reasoning. Both the seeing-that sequence and a reasoning-why description of a sequence will commonly include visual images (diagrams, sketches) as well as words. Even in contexts where that might seem difficult—in oral discussion, or in a text where it is not feasible to use illustrations—gestures in the first case and metaphor in the second function as prompters of imagery.

Since the figure is intended to suggest the actual process of *P*-cognition, all across the spectrum we are looking at instances of "seeing-that." Consider even the far right of the diagram. It would cover the case in which a person is slowly working through a written-out, closely reasoned argument, as in going over a mathematical proof. As drawn, no checking is shown. But we could extend the diagram to note checking also, for example, by horizontal marks on the structures. The critical point is that reasoning-why does not enter somewhere between the poles of pure intuition and pure logic. Rather, it is an attempted *description* of a vertical structure—and not necessarily or ordinarily a very accurate description of any actual instantiation. It is a description of how a person thinks he reached, or thinks someone else might reach, a particular judgment. Any sequence in figure 4.4, however far to the left, could be described (could be converted into reasoning-why) by inserting phrases at suitable places like, "and looking at that, it seems to me . . . "

There is then an essential distinction between reasoning-why accounts of judgment and the actual sequences of *P*-cognitions that they may purport to or are intended to describe. For any actual chain, there can never be an absolute guarantee that at the top of the chain the individual must reach the same judgment if he repeats the same steps, even if he has written them down and is going over them step by step. Anywhere along the line, perhaps at the very last step, the person may find himself seeing something different from what he saw before (seeing "that step may not be right" where he had seen "that step is obviously right" before). For the repetition could never be exactly the same. Things have happened in the interval. Even if the person is trying to go over, step by step, a sequence of reasoning-why he has written down, it is possible that a new intuition will appear at any point (uninvited, so to speak). I may now see something wrong, or think I see something wrong is an argument that had seemed impeccable. For a familiar line of argument (one seen and worked through many times in the past), that is unlikely, but it always could happen. If that happens, checking the step will be prompted and that will usually result in seeing the thing in the usual way after all. But failing to repeat the intuition that had always held in the past is always a possibility.

Therefore a reasoning-why description of the chain cannot itself be a judgmental process. Even if I have only described the chain to myself (it is specifically reasoning-why I not only think would lead me to this judgment, but going over it in my head I find it does), I can't be absolutely sure it will do that again if I go over it again. I may still (having rehearsed it, and certainly if I have written it down) be able to repeat the reasoning-why, including its conclusion. But it cannot be guaranteed to coincide with my judgment.

Summing this up: a chain of reasoning-why is an essentially different thing from a chain of seeing-that, even when the seeing-that is produced by very carefully going over the articulated reasoning-why. The relation between the two is like the relation between an image and the object imaged. Consequently, figure 4.4 has two essentially different interpretations: as a schematic for the spectrum of actual sequences of *P*-cognition occurring in a person's head; *or* as a schematic for the spectrum of descriptions that people give of how (they think) they reached a judgment, or how they think someone else could reach it.

Adversary situations (not excluding self-criticism) push a person's reasoning-why toward the right of the spectrum in figure 4.4 (toward logic), where it becomes increasingly hard for a critic to simply resort to skepticism ("I don't see any reason to reach the conclusion you do"). The critic also pushes the argument in that direction, where it becomes increasingly easy for the proponent to be exposed as relying on contradictory intuitions. Contrariwise, seeing-that tends toward pure intuition (grasping the thing at once, as a whole).

4.10

But a particular thinker does not move from intuition to analysis (from left to right in the spectrum of fig. 4.4) as a judgment crystallizes.

What we remember as a flash of insight, on a matter of any complexity, probably never comes out of the blue. Characteristically, and probably inevitably (for reasons developed in Chapter 6), it follows an extended period of mulling, of coming close to the insight but losing it, of noticing and rehearsing partial or related insights that turn out to tie into the major result. A particularly rich account of this process has been teased out of Darwin's notebooks (Gruber 1981a). From a background of mulling and partial or tentative insight, someone prompted to produce reasoning-why will find much more in hand than a naked intuition. Further, once an insight is firmly established, the subsequent evolution of judgment is not exclusively toward more elaborate reasoning.

The meticulous character of their arguments has let mathematicians provide especially detailed accounts of this process. Poincare's (1914) account of his discovery of Fuschian functions is the best known. After working with a topic, finding bits of argument that look relevant (precursors of lemmas), trying out special cases to get a concrete feel for how things go, a conjecture arises. It is not an undiscussable bolt from the blue. The mathematician can explain to colleagues in a more or less coherent way some reasons for supposing the conjecture plausible.

Over time, he tries to fill in the holes in the argument. Sometimes the effort leads to seeing that the original conjecture cannot be right, though (very often) seeing that prompts a new, related conjecture that turns out to work. Not all conjectures turn out to be provable, or lead to (interesting) alternatives that are. But all theorems first come to articulable notice as conjectures. As mathematicians have often insisted, no one discovers a theorem by deducing it. Quibbles aside, every theorem is born as a conjecture. Like the rest of us, the mathematician's seeing-that precedes reasoning-why. But it is an important point (as any mathematician will confirm) that the mathematician comes to see his conjecture as a theorem at some stage prior to, and indeed usually a good deal prior to, being able to prove it. This point is very important for the analysis of belief. It can, I think, shed some light on the very old and still lively controversy among philosophers about the distinction between knowledge and belief, which plays a central role in the next chapter.

Refining a judgment invariably involves a good deal of the mulling/rehearsing [3.6]. But rehearsal is not just a commitment of the argument to rote memory, for rehearsing has the effect of creating a nonarticulable sense of the pattern of the argument. Once the mathematician is able to give a formal proof, he does not need to memorize it, and he will characteristically report that he has not memorized it. Rather he has a sense of a much stronger intuitive grasp of the theorem than he started with, commonly with a sense of apprehending the whole thing at once. He can then construct or reconstruct a proof, or some piece of the proof, by what subjectively seems like turning over this object in his mind and considering what he sees there, as Poincare reports about his mathematics or Mozart about his music. The mathematician can present a formal proof, but he does not ordinarily do it by rote memory. Rather, he reconstructs the proof as he goes along. What he cannot externalize is a description of the tacit sense of things that is much more important than the proof, since once a proof has been found, it characteristically turns out to be relatively easy to find a variety of proofs. He is aware of the pattern of the theorem, but cannot explicitly describe it. He reaches, in short, both poles of the spectrum in figure 4.4, with a sense of being able to see the whole thing at once, and also an ability to produce a detailed step-by-step argument.

4.11

Before proceeding with the main argument, let me end this chapter by elaborating a bit on the axiomatic versus curve-fitting analogy introduced earlier [4.2], making use now of the seeing-that versus reasoning-why distinction. The curve-fitting process is vulnerable to jumping too soon (seeing a pattern when it is not there). Offsetting that vulnerability (but not with a guarantee that the compensation is always worth the cost) is the promptness with which the process is ordinarily capable of yielding *some* pattern, hence enabling some response. The virtue of a scheme that may jump too soon is that it does not have the failing of hesitating too long.

In contrast, a step-by-step deductive model is open to critical reflection. Its assumptions are intended to be explicit (so even if it is actually not quite so, an issue can be raised about what is missing), and the steps from assumptions to inferences also aim at being fully explicit. But if human cognition were analogous to a deductive model—we will be considering the condition under which something like this holds—then we would be highly vulnerable to hesitating too long (failing to accept a pattern that is actually good enough). A person who sets out to construct a deductive model can work for many years and at the end have no theory to show (Einstein's long years of search for a unified field theory being a famous example). But give a batch of data to someone who knows how to push the right buttons on a computer, and you can have a regression model by lunchtime.

And if you consider the situation of a creature in an evolutionary context, a scheme that gives a crude best fit to the prevailing environment is not such a bad thing. If the environment does not change radically or rapidly, then it may be a good start toward an empirical fit which is very finely tuned indeed; eventually it need not be at all crude. Even if it is not such a good thing in the long run, it may be the best thing available, since evolution has no direct way to reward long-run good effects: survival has to take place in the short run before it can take place in the long run.

Nevertheless, flexibility and ability to modify patterns of *P*- cognition obviously also offer advantages. In the analogy with curve fitting, a user who can only take whatever model fits the available data best is not likely to do as well as a user who can take some account of the causal plausibility of a specification. Even in the short run, a user will do better if he is able to reject a specification that implies a plainly absurd empirical account, yet gives only a marginally better fit to past experience than some causally plausible model. But the problem is not whether some departure from rigid adherence to choosing a mindless "best fit" might be helpful. Rather the problem is one of a tradeoff between the costs of flexibility and the costs of rigidity.

For the statistical analogue, the contrast between seeing-that and reasoning-why comes out starkly when the proponent of a model churned out by a canned statistical routine is pressed by a skeptical audience to give a plausible causal interpretation of the model. He will try to construct the best reasoning-why he can manage. Clearly, the reasoning-why follows the seeing-that and could not have influenced the seeing-that (since the computer was just grinding out its canned routine). However, at the next stage the model-builder will be influenced by the reasoning-why in his judgments about what to do next, what new data to gather, where (given that he can't question everything all the time) he should be especially skeptical about what his information sources tell him, and so on. So while it is very clear in this special case, it should also be clear enough in general that saying that reasoning-why follows seeing-that cannot mean that reasoning-why has no influence on seeing-that. At the next turn of the cycle, it could be crucial.

Five

Knowledge, Belief, Logic

Sometimes a person makes judgments about judgments, perhaps prompted by some salient inconsistency with another judgment, or prompted in some other way to special care or second thoughts.[1] Whether it is your own judgment or someone else's, checking a judgment (looking again to see whether it looks right) will be akin to what occurs when a person looks again at some visual perception, checking to see whether it still looks like the impression he had first gotten, or looks like what someone else has suggested it looks like.

The way we check a judgment, like the way we check a perception, can occur over a small-scale version of the range sketched out in connection with figure 4.3. In particular, we will sometimes do that in an intuitive, mulling way; or we can do that in a way that is narrowly focused, with tightly controlled scanning of the steps of reasoning that might justify the judgment. Specialized to the context of explicit judgments and supporting reasoning-why, the distinction between cognitive activity in the upper right (the intuitive "*I*-gestalt") versus the lower left (the critical "*C*-gestalt") of figure 4.3 becomes a distinction between an intuitive and a critical sort of checking. We can make judgments about judgments (whether our own or someone else's) either:

> . . . Intuitively (the *I*- question), in terms of "Does the *result* look right?" *or*
> . . . Critically (the *C*-question), in terms of "Do the *reasons* look convincing?"

Within the usual account of rational choice, these alternatives—the *I*-question versus the *C*-question—are just two ways of framing the same issue. If the way you respond to a judgment is effectively to process a probability that the judgment follows from the argument, then there is no difference between the *I*-question and the *C*-question. I would confront you with an argument presenting evidence in a way organized to lead to the result. If the response to the *C*-question is that the argument is sound, then the conclusion is sound. If the response is that the argument is probably right with $p = .73$, then the response to the *I*-question must be the same; and if your response to the *I*-question was $p = .73$, that could only be because you had (perhaps only tacitly) considered the *C*-question and judged $p = .73$.

If people operated with the algorithmic sort of judgment characteristic of a computer program, that would be a reasonable way to think about how judgment works. But a little observation (watching what you and your friends say) will assure you that people do not work that way. On the contrary, a human being will often respond "I accept that" to the result, even though he thinks he sees a definite error in the argument. If we ask the *C*-question first, the answer does not leave us sure how the person would then respond to the *I*-question. For we know that sensible people, such as ourselves, can feel quite comfortable saying things like, "That's a lousy argument, but the conclusion sounds OK," or "I don't see what's wrong with that argument, but I can't believe the conclusion." The .73 probabilities just mentioned provide an illustration. Axioms that most of us would find hard not to accept, together with an argument that is logically impeccable, lead to a theorem that says that subjective probabilities (like .73 above) are well-defined (Savage 1954). But intuitively, even among those who understand the argument, not everyone can believe that. In fact, for some years it has been recognized that it is psychologically false (Samuelson 1952). Much of this chapter will be an exploration of the puzzles provided by such conflicts.

5.2

Ordinarily what we see or hear or feel is what is there (meaning, consistent with independent measurements, consistent with what we and other people see when looking at the thing from another angle, and so on). That perception will have this ordinarily veridical character makes Darwinian sense. For the same reason, we would expect that by and large human beings should manage to reach judgments consistent with the logic of a situation, though not necessarily by a process that looks like logic. There are, in fact, many situations in which our perceptions can be tied in a completely explicit way to the logic of the information reaching us. For example, stereoptic cues to depth effects or stereophonic cues to where sound is coming from logically imply that we should perceive certain things (depth effects, sound origin) about what is there. Ordinarily that is the way we do perceive things in the presence of those cues.

Yet it is easy to show cases in which perception overrides logically unambiguous cues. We may see fool-the-eye paintings as convincingly three-dimensional, even when we know perfectly well they are flat. A mask with features painted on the inside will usually be perceived as a normal face. Our expectation that a thing that looks like a face will "face out" to us is so strong that even knowing we are looking at the painted inside of a mask, and despite the logically incontrovertible stereoptic cues, we see the thing as convex instead of concave. There are limits to how far this can be pushed. As the viewer comes closer to the mask (which makes the stereo cues increasingly gross), the illusion is eventually overcome.

An analogous illusion can be observed if you watch an opera on television that is simulcast on the radio. Even if the radio speakers are well away from the TV, after a while you will find that the sound appears to emerge from the lips of the singers on the TV screen, even though you know that is false, and even though that knowledge is being given continuous reinforcement from the cues which ordinarily would locate where the sound is really coming from. Again the illusion can be overcome by making the contrary cues strong enough. But as with the mask effect, it is quite startling how gross the cues have to be before your perception matches the way you already know things are.

All this is quite exactly analogous to the role I want to suggest for logical argument. Sufficiently gross logical disparities will lead to reconfiguration of our judgment, as sufficiently gross perceptual cues will lead to correction of illusions of seeing and hearing. I only want to claim, as I think few people will doubt if they think about their own experience, that it would be naive to suppose that we will believe something whenever the logic of the situation implies it. And it would also be naive to jump to the polar opposite and see logic as having no capability to influence judgment.

In terms of continuity, the Darwinian process which shaped our brains probably could not produce thinking with the well-defined structure of formal reasoning [3.7]. Rather, evolution produced (as it seems almost bound to produce) thinking and language built upon patterns tuned to networks of cues. Even under perfectly ordinary conditions, this will yield performance that will have the a-logical properties (inconsistency, incompleteness) I tried to suggest [4.2] by the curve-fitting analogy. But it is an empirical fact that human beings have the capacity, at least under certain conditions, to do formal logic, construct abstract mathematics, and so on. I have compared that [3.7] to the capacity to play the violin, arguing it was something human beings could learn to do—hence necessarily something compounded out of more elementary capabilities which *are* part of our natural endowment—but not something intrinsic to the way we think or the way our brains work. In this case the elementary capacities amount to pattern recognition applied to forms of language (so that we become "tuned" to various patterns of language ("if . . . , then . . . , " and so on), augmented by the propensity to construct reasoning-why [3.7]. Both would have evolved in the way discussed in connection with mulling and rehearsal [4.10]. Finally, in practical use our capacity for logic can be refined by the cycles of seeing-that/reasoning-why feedback [4.11].

5.3

But the argument as just summarized must need some qualification or amendment or extension. For it would be consistent with the argument so far for human cognitive operations to be rigorously logical to a degree that even the strongest proponents of cognition-as-propositional-logic would not claim. The argument that we are not genetically endowed with logic might be true, yet still be irrelevant to an account of human judgment in empirical contexts, in the following sense. Logic might be nowhere in your genes, but it could still come to be wired into your brain, as the capacity to read English is now somehow wired into your brain. It is empirical evidence (such as the illusions of Chap. 1), not a general argument, that shows that logic isn't in our brains in even that sense (Johnson-Laird 1983). So we need to say something about what apparently is in our brains that lets us (under the right circumstances) do logic and mathematics, while still specifying something about that tuning process, or about the nature of logic, or both, that keeps us from learning logic to the point where we can do it with the fluency with which a person can learn to play the violin or speak a language.

The key to the situation, I want to argue, is that natural language (meaning all spoken human languages, without exception) works in such a way that what a word means depends on context. Recall an earlier example [4.1]. "Jim Wright, who lives right down the street, to the right of the Kelly's, is a playwright who is studying the Masonic rites, so he will soon have all the rights of qualified members." I used that sentence to give a simple illustration of inhibiting, which saves us from the cognitive overload that would result if we were not somehow blocked from seeing any sense of "rite" but the one that looks right at the particular point in the sentence. Formal logic and mathematics do not allow that sort of multiplicity of meanings. On the contrary, exactly what is indispensable is that things be defined in a way that is context-free, which then permits building up structures of language that do not depend on details of context; hence, Bertrand Russell's definition of mathematics as the subject in which we never know what we are talking about. Being context-free, the structures of formal logic are not about anything in particular. But if we believe that certain premises apply in some local context, we can put a built-up structure in place, jumping in one step through what may amount to a very large number of "stored-up" elementary inferences. So logic can be a very powerful tool of thought. But it is not itself thought.

Formal languages are context-free in this sense, even though it is often the case that the same computer command (for example) will mean different things in different computer languages, or a particular term or symbol will mean different things in different parts of mathematics, or even in different parts of the same paper. Meanings can change; but only in a spelled-out way. The intended sense is never something that the person just knows, in a sense that relies on an unarticulated, and in good part unarticulable, sense of what goes on in this context. Further, a formal language will have a special syntax, which is always unlike the syntax of any natural language, though in some Chomskian sense its syntax may be governed in ways shared by natural language.

Refinements of ordinary language can pick up something of the character of formal languages. Given a particular overall context (an operating room, a law court, a game) we can identify terms that are almost always used only in a unique way, though outside this special context they carry other meanings. But the language used in these situations never is really context-free in the free-of-pattern-recognition sense I am stressing as the characteristic that makes a formal language. Language users in those situations must almost continually be able to discriminate (recognize) the special context where a word or phrase is to be understood in its technical sense, despite much else that is going on in the flow of language, which isn't appropriately understood in that way.

5.4

In general, we learn by adapting and tuning patterns we already know to new situations. I will develop that in detail in Chapter 6. But on the argument so far, it should be clear that the study is committed to making that (familiar) sort of learning not just one way we learn, but with only minor qualifications the only way we learn when we are talking about more than fine-tuning something already in the repertoire. If that is so, then we must be led to the following sort of argument about how we learn *reasoning*.

In general, words serve as cues to patterns, where the patterns cued will depend on the totality of verbal and nonverbal cues available in the context [3.7]. But just as individual words have meanings that (like all those "rite's" in [4.1]) are context-dependent, so do patterns of words that trigger inferences (other patterns of words).

Nicholas Rescher (1982) distinguishes between what he calls "use-conditions" of ordinary language and the "truth-conditions" of formal logic. We use language on what amounts to a tacit understanding that with respect to everything we say there is a qualification that adds, "As a practical matter, in this context, I am prepared to talk as if . . . " Foreshadowing an argument to come [5.8], this is what accounts for why we routinely say "I know" in contexts where we know perfectly well that we would not really claim anything more than "I think I know. . . " And although without additional cues (from external context, tone of voice, gesture) you cannot tell whether "that's a likely story" means it's likely or unlikely, in context you are never confused about that. Ordinary language patterns of inference (pragmatics) will not only go beyond what would be warranted in formal logic but we will use many patterns that are mutually inconsistent. Proverbs are a common feature of informal reasoning-why. But for every proverb there is an antiproverb ("Too many cooks spoil the broth" vs. "Two heads are better than one," and so on). Further, informal reasoning-why need not exhibit much of the linearity characteristic of formal argument. There may be pointing to this thing and that, followed by a claimed inference, with no explicit line of argument connecting the features pointed to with the inference drawn. When this proves persuasive, as it often does, the things pointed to have apparently been sufficient to prompt the inference, but not in any explicitly reasoned fashion.

I will use subscripts ($\underline{\qquad}_u$ and $\underline{\qquad}_t$) suggested by Rescher's "use-" versus "truth-conditions" to emphasize the distinction between the reasoning process in the head of a human being versus the formal process that is described in a logic text. The usage is partly akin to Gilbert Harman's notion of an asymmetry between implication (what the language logically implies) and inference (what a competent listener will infer), to the stress of writers like Grice and Quine on the tacit components of language use, and (most generally) to the contrast in linguistics between the formal properties of language and the pragmatics of language. An essential distinction, which applies especially to Grice's use, as I understand it, is that (using the subscripts) the meaning$_u$ of a piece of language is by no means equivalent to the meaning$_t$ as adjusted to take account of various things (Gricean inferences) that the speaker presumably intended but did not spell out. Rather, the meaning$_u$ will often vary from the meaning$_t$ in ways such that both speaker and hearer may find it hard to articulate what

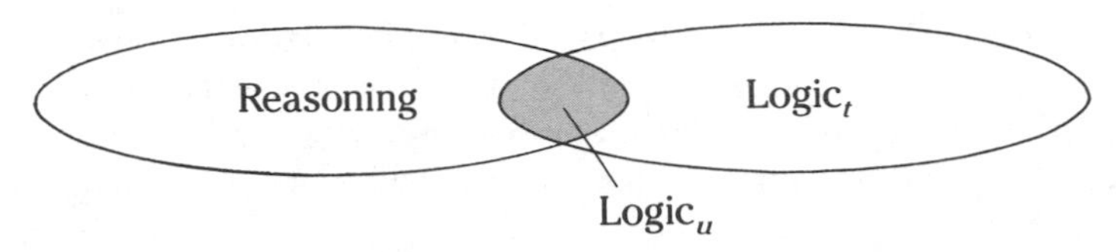

Figure 5.1. Logic$_u$ can be thought of as the zone of overlap between formal logic$_t$ and human reasoning. But strictly speaking, what goes on in the head is never logic$_t$, even in the special case (such as a teacher going through a proof for a class), where an explicit attempt is being made to do logic$_t$.

warrants seeing the language as meaning something different from its meaning$_t$.

In general, then, what a person infers$_u$ in a given context is not necessarily what the language implies$_t$. As a set of abstract rules of inference, *logic$_t$* can be embodied and given an existence independent of any human brain. It can be externalized in Popper's (1972) "World 3." In contrast, as a style of reasoning-why that goes on in peoples' heads—as a process in principle reducible to neurophysiology—*logic$_u$* is continuous with informal reasoning and indeed with unarticulated hunches and guesses. As something out there (something that could be found in a well-preserved book by a visitor from another galaxy a million years from now), formal logic (logic$_t$) has no connection with the pattern-recognition processes I have labeled *P*-cognition. But logic$_u$—what goes on inside an individual head which is *closest* to logic$_t$—is still pattern-recognition, mediated by neurophysiology operating in the manner sketched in Chapter 4. Schematically, the relations are illustrated by figure 5.1.

This patterned thinking includes what happens in the special contexts in which a person thinks very self-consciously, in a step-by-step way, trying to be sure to stick to steps warranted by logic$_t$. But contexts in which that happens must themselves be cued, and usually must be very emphatically cued, and must not be confounded by other cues which prompt choosing quickly, or which prompt not bothering to choose carefully. The zone in figure 5.1 in which *P*-cognition is shown as looking like the operations of formal logic (in which the neural outputs pretty nearly coincide with the logical steps) is the small zone that I described [4.9] in terms of the polar extreme of the spectrum of cognitive sequences. Restating the point of that: it is not a zone of cognition where things are no longer operating by pattern-recognition but a zone in which attention can be narrowly focused, step-by-step, on sequences of intuitions that seem compellingly linked. Even when such intensely focused step-by-step thinking is going on, the fallibility—but also the openness to "fruitful misperceptions"—of logic$_u$ distinguishes it from a literal instantiation of logic$_t$ at the level of neurophysiology. It is only after extensive seeing-that/reasoning-why feedback that we eventually get inferences that are stable against further scrutiny, including scrutiny across individuals. That is what is embodied in World 3 logic$_t$.

In sum: logic$_t$ is a product of human thought but not an example or description of human thought. Operationally, it has different properties from logic$_u$. Logic$_t$ is reliable but never capable of fruitful misperceptions. Logic$_u$ is highly fallible, and even the most brilliant minds stumble, so that it is commonplace that the most remarkable insights are initially justified by logical blunders. The distinction I want to stress between logic$_t$ and logic$_u$ is essentially an old one, dating back at least as far as Frege's distinction beween logic and psychologic.

5.5

There are many versions of logic$_t$, and the way that arises is important for the larger argument. As an example, consider what is implied$_u$ by the form "... *or* ...":

Example	*Neither*	*Either*	*Both*
A. "soup *or* salad"	x	x	
B. "$100 *or* 10 days"		x	
C. "cream *or* sugar"	x	x	x
D. "wash *or* sweep"		x	x

If your waitress says A, you know that what is implied$_u$ is neither or either, but not both. But if the judge says B, he means either, but not neither, and not both. The hostess C means neither, either, or both. A child, told that he must wash the dishes or sweep the floor if he wants to go to the movies, knows that means either or both, but not neither. It is a convention, not something spelled out in the reaches of eternity, that in standard formal logic "or" implies$_t$ "either or both, and not neither."

For an isolated case of this sort, there is nothing essential about what we choose, anymore than there is anything essential about driving on the right versus the left side of the road. We just need to avoid the confusion that must result unless *some* convention coordinates expectations. We have a coordination problem, not a moral dilemma or an intrinsic test of rationality (Schelling 1978). But once some forms of inference are chosen to be part of (a version of) logic$_t$, any others are constrained by the need to avoid inconsistency.

So it turns out (given earlier choices that it would be clumsy though not impossible to avoid) that, if we want to make use of negative numbers in arithmetic (and avoid inconsistency), the product of two negative numbers must be defined to be positive. If we want to have a definite true or false label for every proposition, we have to define "if *p*, then *q*" to be true whenever *p* is false (material implication). Neither is a natural language form (experience in the world does not provide contexts which would make these patterns familiar), and consequently students have great difficulty making sense of them when first encountered. So logic$_t$ (broadly understood, to include other locally context-free languages, such as those used in mathematics and computer programming) comes to look very different from reasoning in any natural language.

Reasoning in natural language, even very careful reasoning by people who happen to be trained in logic$_t$, does not have this "tight" character, hence has no need for things like material implication. Such a form would not appear at all in an account of logic$_u$, with the exception of a person well-trained in formal logic. Even for such a person, material implication would not hold except for contexts recognized as the special context in which logic$_t$ is intended. It would therefore make no sense to anyone who has not been drilled in using it in the context of logic$_t$. For if language use, like all cognition, comes down to pattern-recognition, how could we possibly make sense of a pattern that doesn't look like anything familiar from our experience in the world, or at least from our experience with language? We can understand (and speak) particular utterances we have never experienced before, and that is crucial (Chomsky 1975). But on the argument here, we can only do that to the extent that the novel utterance looks like a pattern of words we are familiar with. A drastically novel use of language requires not just definition but drill to get that novel pattern worked into the cognitive repertoire. ("How would you explain school to a higher intelligence?" says a kid in the movie *E.T.*)

5.6

We have reason to deny that ordinary-language reasoning will look much like even logic$_u$. Nevertheless it must be possible to tease out of ordinary reasoning some crude approximation of logic$_t$. For ordinary reasoning must work from patterns of inference (forms of language that cue other forms) which experience would have supported. Tuning to the way the world works will shape even the most informal or primitive reasoning. But unless there has been opportunity for elaborate training, it would be naive to expect reasoning with the neatness of carefully honed argument. We could expect to find wide use of forms that someone from outside the culture or language community could only with difficulty recognize as embodying widely shareable forms of inference. For it is only a person within the culture who would automatically recognize the inference$_u$ in appropriate contexts and only in appropriate contexts.

Given persistent patterns in the way the world works (reality is not just chaos), then forms of language used in reasoning-why in any culture would become tuned to the world. Those that are reinforced will be just those that have proven effective in often drawing correct inferences in a given sort of context. Within the constraints of what the society can manage (of which I will shortly stress the role of literacy), and in particular in the context of highly familiar situations (providing extensive opportunity for good tuning), we should expect to find a crude protologic, however bizarre the forms might look at first to people outside the society (for whom, in particular, the contextual cuing essential to informal reasoning will be meaningless).

Once we find societies in which a version of explicit reasoning has developed, there is then a model for logic$_u$ that, although still very crude by the standards of logic$_t$ (still very dependent on the tacit recognition of what is appropriate for the local context), will nevertheless *begin* to capture some of logic$_t$'s neatness and sensitivity to contradictions. But until literacy emerges, that cannot go very far, since the whole process is restricted (by definition) within the limits of what unaided memory can handle (Goody 1977). This is not a very narrow restriction for cognition in general. The chunking or bundling-up of information into patterns, and patterns within patterns, and so on to more elaborate nestings allows the information a particular person has in his head to be very large. The social division of labor then allows the totality of bundled-up information available to a society to be very large compared to what any individual can personally know. So in an absolute sense, a human being can know a lot, and a human society can know much more. But as for extended chains of explicit reasoning, the restriction to unaided memory is a severe one.

But with the appearance of writing, reasoning-why can be made into an increasingly unambiguous, inspectable, self-consciously articulated thing, subject to extensive across-persons seeing-that/reasoning-why feedback. Without writing, there cannot be much progress toward creation of generalized structures capable of supporting extended chains of reasoning. Hence the invention of writing is a crucial step, expanding not only how much information can be socially stored but greatly extending the length of chains of reliable reasoning-why which can be feasibly managed by human brains. Even though I can only jump a small distance, if there were some way I could store up my jumps, I could jump an indefinitely long distance. I eventually will make a good deal of an argument that a crucial turning point for human beings was the discovery of technologies (writing) that created the possibility that logical jumps could be stored up for future, repeated use. And in a different way, the invention of printing, I will later argue, provides a further fundamental step in the capabilities of societies to learn new things.

5.7

Now reconsider a puzzle mentioned earlier [5.2]: Why don't we become fluent in the use of formal logic, in the way we can become fluent in playing a musical instrument or skiing, or as we are all fluent in ordinary language use? The very fluency with which we recognize patterns of ordinary language (jumping, with little occasion for checking, to patterns that work) makes us vulnerable to misreading language intended to be understood as logic$_t$. Where we have the best chance to follow the hard-to-follow path$_t$ is where the language itself is so artificial that it continually cues its intended interpretation within logic$_t$. This is not the only reason that logic$_t$ and mathematics are so closely associated with the use of symbols which immediately differentiate them from natural language. But it is a significant reason.

Ordinary language reasoning does not look much like the sort of careful and well-schooled reasoning characteristic of a legal brief or a piece of analytic philosophy; and that legal and philosophical reasoning is far from the disembodied and quite artificial (in the sense that no human being would ever talk that way) patterns of inference of formal logic. I am trying to construct a well-reasoned argument for the view of cognition presented here. If it is successful in attracting attention, it will prompt writers with other views to put forth what they regard as tightly reasoned demonstrations of its inadequacy. But a close parsing of either side of such a debate, as of any reasoning in practical language, will look horrendously sloppy if judged by the standards of rigor customary in mathematics or formal logic.

5.8

Now return to the *C*- versus *I*-questions which introduced this chapter. The propensity to produce reasoning-why is a universal feature of human performance. A corollary to that is the propensity to pay some attention to reasoning-why, and for reasoning to play a role in judgment. That follows as strictly as that the propensity of males to engage in performances intended to attract a mate should be accompanied by a propensity of females to pay attention to those performances. Absent the response, how could natural selection favor the initiative?

It makes sense, therefore, that we should have a propensity to look at judgments from what amounts to two gestalts: in terms of "does-the-result-look-right?" and in terms of "does-the-argument-look-right?" Ordinarily, a person's judgment is what he sees in the relaxed, *intuitive* cognitive mode in the upper-right corner of figure 4.3. But challenged by other people who see things differently, or in anticipation of such a challenge, or faced with sufficiently gross and immediate conflicts between a current judgment and other judgments, a person can be prompted to the concentrated *critical* activity in the lower left of figure 4.3.

The various combinations of response then yield nine possible belief states, allowing for "yes," "no," or "maybe" responses in what I have called [5.1] the *I*-gestalt and the *C*-gestalt. Checking might reaffirm the earlier intuitive impression, so that what was seen at first (or what another person has claimed was there to be seen) is seen as fitting the situation. Or that intuition might, with this closer look, seem wrong. Or the closer look might yield no clear answer. Similarly, a closer look might confirm (prompt belief), contradict, or leave open the claim that reasoning-why leads to the result. These possibilities yield the 3 × 3 "belief matrix" given by figure 5.2.

We would not expect the matrix to be part of a person's conscious sense of how thinking works, any more than we expect (or observe) people to be self-consciously aware of how debatable the connection may be between their own seeing-that and the reasoning-why they produce. Nevertheless, we are not surprised to find linguistic correlates of the intuitive versus critical distinction (phrases like "I've got a hunch" or "It just looks that way to me" vs. "I've got a good reason" or "Logically, it just has to be that way"). Similarly, though we do not expect a person to have in mind anything like an explicit "belief matrix," we do expect to find linguistic correlates of the cells in that matrix. It would be an embarrassment to the argument if I could not point to aspects of the surface phenomenology of judgment that seem to provide observable correlates of this underlying structure.

The labels of the cells are intended to suggest those observable correlates. In the matrix, I^+C^-, for example, means "yes" to the *I*-question, "no" to the *C*-question, yielding the cognitive state I have labeled "paradox." The pair (I^+C^+ and I^+C°) (or its converse, I^-C^- and I^-C°) gives an interpretation to the distinction between knowledge and belief that has been disputed since Plato and remains a leading topic for philosophers. These cells can also be related to a much-discussed point of legal philosophy, turning on the way juries are instructed with respect to the two standards of proof used in Anglo-Saxon law (by a preponderance of the evidence, and beyond a reasonable doubt). Legal scholars have difficulty

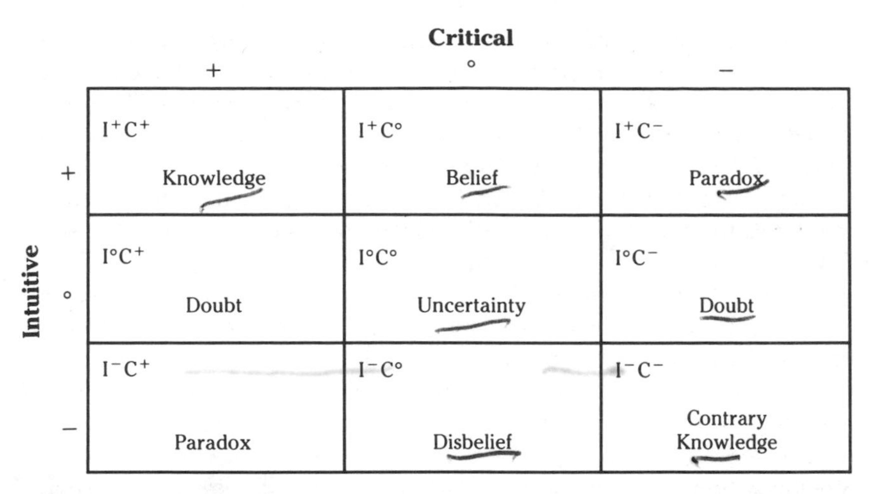

Figure 5.2. The belief matrix illustrates the cognitive states available, contingent on the pair of responses to a cognitive claim in the intuitive *I*-gestalt vs. the critical *C*-gestalt. $I^+ C^\circ$ may also be seen as a puzzle.

agreeing on a coherent account of just what the jurors are being instructed to do. But the jurors seem to have no great difficulty. By and large, they think they have an intuitive understanding of what they are being told. So there is a parallel here with the philosophical disputes about an adequate account of knowledge versus belief.

On the simplest view, there is no special problem about knowledge versus belief beyond the general question of what moves us (or ought to move us) along a spectrum from complete uncertainty or ignorance to complete certainty. We can distinguish various zones along this spectrum—such as "opinion," "belief," "knowledge"—analogous to the shading of a white-to-black spectrum. There would be not only no sharp boundaries, as would also hold for a color spectrum, but also no qualitative distinction between the zones. Aside from things true by definition, if there were things we think we know as infallibly true—things we can say we know—they would define the end point of this spectrum, not some different sort of thing from belief.

Yet it is certainly a common feeling that somehow a qualitative distinction jumps in at that endpoint, though that is certainly ridiculous from a mathematical viewpoint. Further, even philosophers use the word "know" in contexts where the claim to knowledge could hardly be taken to be a serious claim about infallibility. Even within the context of philosophical writings, things that were once seriously thought to be infallibly known (that space is Euclidean, that the sun moves around the earth, and so on) have come to be regarded as false, so that the set of things that are empirical statements but plausibly infallible is not large; on most accounts it is empty. In any case, examples that appear in philosophical discussions commonly involve perfectly ordinary claims on the order of, "I know that John will want a martini when he gets here."

Clearly, there is in all this (on the one hand) some strong intuitive appeal in the claim of a knowledge/belief dichotomy, but (on the other hand) there seems to be no nonmystifying logical basis for such a dichotomy. Knowledge is intuitively seen by some, but logically denied by others, to be a cognitive state which is qualitatively different from belief. In the usual phrase, knowledge is said to be justified true belief, but no one has ever been able to give an account of what that means that settles the issue. For all the ink that has poured on this topic, we are still left with an unresolved tension among (a) the logical view that knowledge is the endpoint of a belief spectrum, (b) the intuitive sense that it is something different, and (c) the empirical observation that we use the word "know" (in the sense of justified true belief) in contexts where we also know we might be wrong. Further, and an important point for argument here, it is not only among philosophers that we find people who sense a difference in kind between knowledge and belief. In fact, the opposite is the more important aspect of the case. The philosophical argument reflects an apparently universal cross-cultural tendency to mark the knowledge/belief distinction in ordinary language. Even very primitive languages mark this distinction.[2]

But on the argument here, there is a straightforward explanation of this famous dispute. A conventional logic of belief treats belief as one-dimensional (as if we chose in the way discussed at the very beginning of this chapter), while on the argument here, a psychologically valid account would have to allow for two dimensions of belief, one based on the *I*-question and the other on the *C*-question. So we have to expect that people could sense (hence the pervasive language distinction) a distinction which defies an account in terms of standard (one-dimensional) accounts of belief.[3] An unfavorable response on the *I*-question but a favorable response on the *C*-question yields a sense of paradox ($I^+ C^-$). Uncertain responses to both questions yield a sense of uncertainty or ignorance ($I^\circ C^\circ$). And continuing in the manner illustrated on the figure, we get the remaining states of doubt ($I^\circ C^-$ or $I^\circ C^+$), and contrary knowledge ($I^- C^-$). Belief ($I^+ C^\circ$) means seeing-that lacking good reasoning-why. So this state will sometimes be sensed as a *puzzle.*

On the view here, then, the enormous philosophical literature on this matter (a bibliography of papers and books since World War II alone would run to hundreds of items) reflects the incompatibility of strong intuitions and clear logic. The conflict cannot be resolved so long as the analysis is confined within the standard one-dimensional notion of judgment (where there is no room for the cognitively distinct beliefs of the *C*-question vs. the *I*-question). On the other hand, in terms of an account that does allow for that duality, the matter is not mysterious at all. A favorable response in the *I*-gestalt that fails to elicit a favorable response in the *C*-gestalt yields a different affective state from the same favorable response in the *I*-gestalt but with a favorable response also in the *C*-gestalt. The first situation yields the affective state, belief. The second, though, yields something more, and "knowing" is our label for that something more. This label in fact meets less than half of the usual philosophical definition, since it involves justified belief, but only subjectively justified, and not necessarily true belief. Subjectively, a person cannot say "I know" when he feels he might be wrong. But it is a commonplace observation that we all in fact say "I know" on many occasions when a little prodding will force a retreat to "I feel I know, though conceivably I might be wrong to feel that way."

Evidence that the belief/knowledge distinction is a real distinction (not just a demarcation along an ignorance-to-confidence continuum) is that, empirically, pairs such as $I^- C^+$ can be observed. Two distinct dimensions are necessary to characterize such cognitive states. The various affective states identified by the belief matrix will play an important role in the account of learning in later chapters.

5.9

Since rigorous standards for reasoning-why are more salient in mathematics than in ordinary human activities, it is instructive to notice what happens in mathematics with respect to the knowledge/belief distinction. What follows reflects my own modest mathematical experience as confirmed by conversations with some mathematicians of much less modest accomplishments. It elaborates on the earlier discussion on the conjecture/theorem distinction [4.10] (see also Hadamard 1945).

A mathematician will ordinarily say he believes his conjecture is true but that he knows his theorem is true (not in a self-conscious, deliberate way, but just as the natural way to talk, as in usage like, "Well, I know from theorem 2 that . . . ").

However, the point at which a mathematician characteristically begins to think of his proposition as a theorem not a conjecture, and to talk of knowing it, or knowing things that follow from it, is *not* when he proves the theorem. Rather, the completed proof of a theorem quite ordinarily follows, and sometimes by quite a bit, the point at which a mathematician begins to say he knows the theorem is correct. In *P*-cognitive terms, that makes sense. For what we would expect would be sufficient for the mathematician to feel he *knows* is only that the proposition looks intuitively right to him, plus that he has some reasoning-why in support of the proposition that looks convincing to himself. Neither requires more formal mathematical rigor than his claim, in another context, to know that John will want a martini when he arrives.

By convention, however, a formal proof meets a much more difficult standard. It has to be an argument built up of such tight steps that even a lout with no mathematical intuition at all, though we require a technically competent lout, would have to admit it is correct. Proofs in this strong sense did not exist prior to the mid-nineteenth century. Euclid is far from rigorous by current standards. And although mathematicians prior to the middle of the nineteenth century certainly felt they knew that the calculus was essentially sound, they also knew they had no rigorous proof of that. They further knew that their intuitions could sometimes lead them astray, which is why the convention grew up of demanding a standard of proof that relied minimally on appeals to intuition. This standard is accepted even though every mathematician will agree that anyone who can only judge a theorem by whether its proof meets a high standard of technical rigor will never amount to much as a mathematician.

In terms of figure 4.4, what is happening is this. The mathematician considers his own informal reasoning-why for believing his proposition is true. He looks at his own view of the situation in the *C*-gestalt, as he is easily stimulated to do by the prospect of making a fool of himself before his peers. If he sees no jump in the argument that looks to him as if it could fail to be OK, he finds himself with a sense of knowing the proposition is correct. The argument may still be far to the left in figure 4.4, based not on anything that even looks like a publishable proof but on such things as computing several simple cases and finding they work out fine, and on a sense that the theorem very prettily extends the pattern of some theorems in a related field. On this sort of basis, the mathematician "knows" it is OK. And it almost always is. For although *P*-cognition by its nature will defy a completely coherent logical account of how it works, an absolutely essential point is that applied to contexts that are familiar (contexts in which good tuning of cues and patterns has had an opportunity to occur) is nevertheless remarkably effective.

In general (not just in mathematics), working out an argument amounts to explicitly filling in (as far as possible) the holes left after reasoning-why has already become subjectively persuasive. At that point, reasoning-why becomes focused on what it takes to make the argument as compelling as can be managed to other people. In that context, the checking response is very prominent. But even in a mathematical context, and even when the work is at the final stages of cleaning up a proof, the reasoning is never just a matter of purely formal logical operations. For while the proof is still being polished, there is always a good deal of trial and error, shoving things around, watching for something that looks like a good pattern. The process is more like putting together a jigsaw puzzle than following detailed directions to sort some unambiguous object. For a theorem of any nontrivial difficulty, the initial formal proof will usually be clumsy, and the original author or others will eventually find the much neater, less roundabout, more highly valued even if no more rigorous kind of proof that mathematicians label "elegant." As I have mentioned before, this trial-and-error aspect, when we come to develop a more detailed account of it in Chapter 7, is not like the blind variation of a Darwinian process. But it is not at all algorithmic. The person doing it cannot explain exactly how he is doing it (any more than he can explain how his stomach is digesting his lunch). Often the original proof is not only inelegant and nonrigorous but formally wrong, as was true in the first published versions of Arrow's "Impossibility Theorem" or Kepler's construction of his planetary laws or Einstein's general relativity argument.

5.10

An important aspect of the relation between seeing-that and reasoning-why is illustrated by an experiment mentioned earlier [1.6]. Wason & Evans (1975) found that their subjects readily provided reasons-why for whatever choice in the Wason selection test they were told was the correct choice. It made no difference in either the subjects' ability to respond or in their confidence that half were given (as the purportedly correct answer) the actually correct answer, and half were given the usual incorrect answer. Whether they were producing logically sound reasons for the correct answer or illogical reasons justifying an incorrect answer, people turned out to be equally facile in coming up with reasons-why, and equally confident that their answers were right. Apparently, Wason & Evans concluded, human beings have a process for coming to judgments, and another process for coming up with reasons justifying the judgment. The two do not (Wason & Evans argued) seem to have much to do with each other. On the argument here, however, while it is essential to distinguish seeing-that from reasoning-why, both must be produced by the spirals of *P*-cognition. There could not be one process that is logical and another fundamentally different process that is illogical, as in Evans (1982). An analogous point will be made in the larger context of discovery versus justification in science [9.3, 13.1].

Further, experiments in reasoning almost always deal with static contexts (no learning), so that very little work casts light on the role of feedback between seeing-that and reasoning-why. In the context of such experiments, reasoning-why could be interpreted as nothing more than rationalization of judgments made in some other way (Evans's dual process). But in a broader context, the reasoning-why can be seen to be an essential component of human competence. Even for an isolated individual, reasoning-why at time t_1 would provide new cues for a related seeing-that judgment at time t_2.

5.11

The following puzzle then arises. Where does the argument come from on which the critical process operates? How does the brain (which on the argument here is intrinsically a-logical) manage to produce reasoning-why that often looks quite logical in form and that is sometimes even rigorously logical in fact? As the Wason & Evans experiment and much other evidence shows, human beings are remarkably facile at producing "reasons-why" even when logically we shouldn't be. Therefore, we want an account of how multiple-step reasoning-why gets produced, and particularly how, as has so often been observed, it often gets produced with striking facility.

Yet nothing more is required for this performance than a further reliance on the same sort of *P*-cognition that produced the original seeing-that. The reason-why response again relies on simple *P*-cognition to prompt some pattern of argument (out of many we would become familiar with as we gain experience in the world) that seems to fit the judgment already intuited.

It may help to think about a more abstract process that offers a parallel to the reasoning-why case, but one in which we are likely to have less ego-involvement. Its very artificiality may help bring out what is going on. What an expert chess player has in his head that an ordinary player does not have—or at least one essential thing he has in his head—is a vast array of somehow-stored tacit chess patterns, aside from the many "book openings" that he has explicitly memorized (Dreyfus & Dreyfus 1986). Consequently, far more readily than an ordinary player, he recognizes characteristic situations that he feels familiar with and has only to adjust marginal details. That is why an expert player can play lightning chess (10 seconds a move) of a quality enormously superior to what an ordinary player can produce given 10 minutes a move. But clearly, evolution did not provide human beings with a specialized capability to deal with chess patterns. So the competence of a chess expert, and indeed of experts of all kinds, must depend on some much less specialized knack for discerning, organizing, storing, recognizing, and calling back from memory a rich store of patterns, shaped and tuned to cues by experience in the world.

Cues & patterns

For chess players this intrinsic capability of the brain is trained especially on chess. But since human beings can acquire a very rich sort of competence in handling the artificial patterns of chess, we can hardly doubt that we are also competent to acquire facility in handling patterns of great functional importance. Then, given the competence demonstrated by our facility in handling patterns in faces, languages, places, and even entirely artificial things like chess and concert music, it is hardly implausible that we would also exhibit facility in handling patterns of argument (reasoning-why). It is then no surprise that we observe ourselves to have some knack not only for dealing with chess and music but also patterns of reasoning-why with a more obvious relation to biological fitness.

Another way of putting this is to say that since we have language, and indeed handle intricate patterns of language (grammar), we can expect this pattern-handling capability to operate also in the realm of patterns of words, and specifically patterns of words that describe relationships we learn from living in the world, where we repeatedly encounter various conjunctions of cues and patterns. We acquire a rich store of verbal patterns and conjunctions of verbal patterns, though unless we have specific training they will not look like formal logic. Even if we do have that training (outside of the special situations in which formal logic is emphatically cued), the patterns of reasoning we use will not look like formal logic, nor will they be rigorous, globally consistent, nonredundant, and so on.

Given these patterns in our head, then, it is easy to resolve the puzzle of how we get very facilely organized multiple-step reasoning-why out of the intrinsically one-step-at-a-time spirals of *P*-cognition. We produce a pattern of argument, as we set in train the pattern of motions that will get us out of a chair. In either case the act releases a chain of connected things which are the components of a pattern for carrying out through this piece of business. In either case, we may observe the performance as a whole, and we also may examine the elicited pattern in a step-by-step way. With respect to judging a piece of reasoning-why, our awareness of the overall performance (as it occurs in the context of whatever else is going on) constitutes what I have been calling the *I*-gestalt. The step-by-step concentration defines what I have been calling the *C*-gestalt, focused on a critical appraisal of the argument.

5.12

Something out of order may cue the intense "look closer" checking of the *C*-gestalt. On the other hand, noticing that a pattern of reasoning-why does not seem to reach the conclusion may only prompt an adjustment, with no more than momentary awareness that a correction has been made, as we sometimes reach for an object and overshoot or undershoot in our reach, and this leads to corrections of the physical pattern. But sometimes we have trouble finding a correction, which may lead to a change in judgment.

As we often practice a physical act until we think we have gotten it right, a novel bit of reasoning-why (a pattern used in a new way, or varied in a new way) will prompt a more self-consciously iterative process of argument, look closer, adjust—an elaboration of the mulling/rehearsing process already discussed [3.6, 4.10]. In a social context where reasoning-why supporting an intuition is likely to be challenged, or is being challenged, particularly important cues to look closer will come from the counterarguments of adversaries, or anticipation of such arguments. We now have an altered context, consisting of the original context, plus the reasoning-why, plus (perhaps) contrary reasoning-why. Absent the last, attention may simply move on, but especially in the presence of contrary reasoning-why another look at the situation is prompted. The judgment could change; the judgment could stay the same or change just enough to blur gross vulnerability to the counterargument. Even if the judgment does not change, we still may get an amended or more explicit line of reasoning-why. This cycle can continue until a very refined argument finally emerges, apparently stable against further checking, providing a match between judgment and subtly reasoned argument that would satisfy the most finicky criterion of human rationality. On other occasions, however, we see nothing better than fluent production of patterns of argument in the service of self-deception. A goal of the account of cognitive evolution in the next several chapters is to better understand the circumstances under which one rather than the other outcome is more likely.

On this account, to repeat a central point, there is no discontinuity between logic and intuition, so long as it is understood that we are talking about human thinking (about logic$_u$ not logic$_t$). Formally organized chains of argument lie at the extreme of the cognitive spectrum (fig. 4.4). It is the most refined form of reasoning-why. But there is no discontinuity between the subtle intuitive jumps of a tightly logical argument versus the larger, less universally shareable, steps of argument we would call intuitive (if we like the result), or by some such label as sloppy or hand waving (if we don't).

Further, subjectively simple intuitions can recognize as well as produce logical bits of argument, or notice when a pattern looks good or looks bad. A "doesn't look right" intuition about the reasoning-why triggers step-by-step critical work. Nevertheless, even when focused on the *C*-question, it seems almost as difficult to discipline ourselves (stay in the *C*-gestalt, attend only to the logical cues) as to attend only to the stereo cues in the earlier illustration of visual and sound illusions [5.2]. Even the most elaborately trained and experienced reasoners are embarrassingly liable to slips. On the other hand, it is only in the artificial context of exercises in formal logic that it is always an error to attend to nonlogical cues. In natural settings, a person who failed to use such cues (a hypothetical personage, since no human being does that) would find ordinary language incomprehensible and the tasks of ordinary life overwhelming. I will have occasion to elaborate on this in the analysis of cognitive anomalies in Chapter 8.

5.13

If we consider the whole argument to this point, we are easily led, as many writers have been led, to see language, reasoning, and culture as things that evolved together, each reinforcing the others. On the account I have given, as on almost any other, reasoning depends on language; but the evolution of language must be contingent on the existence of important things for human beings to talk to each other about. As I mentioned earlier, anyone who has had occasion to visit a country in which he does not know the language must have discovered that a good deal of human discourse can be handled using no more than a grammarless handful of nouns and verbs, supplemented by shrugs, pointings, facial expressions, etc. [3.6]. Language apparently developed because human beings had something more to say to each other. The most striking candidate for the "something more" is that human groups thrived that were able to explicitly plan and coordinate activities tailored to novel contingencies.

Planning (devising responses, divided into roles for individuals, tailored to particular contingencies) profits from language beyond grunts and gestures; and coordination across individuals requires agreement, which in turn creates the problem of what to do when one person's intuition does not coincide with another's. So we can envision an evolutionary sequence in which reasoning-why joins seeing-that (encouraged by the advantage of supplementing seeing-that by reasoning-why), and the *C*-question joins the *I*-question in the response to reasoning-why, both promoted by selection pressures favoring groups that can share information, coordinate behavior, resolve differences of judgment (especially in a way that allows for inventiveness and strategy). Although it is not an issue I want to push right now, this would be synergistic with the embedding of social motivation discussed in Chapter 3 of my earlier book (Margolis 1982). A society cohesive enough to carry out a cooperative plan if it can be chosen will be where increases in the cognitive capacity to design and choose a cooperative plan contribute most to Darwinian fitness.

An initially crude capability for reasoning-why, we can suppose, is refined from this incentive to coordinate across rival intuitions. But once even a primitive reasoning-why supplements seeing-that, then a new and powerful further advantage to reasoning-why emerges. For we then have the competence to refine and improve judgment through the iterations of seeing-that and reasoning-why under the pressure of criticism by others with contrary intuitions. That propensity toward reviewing, and thereby elaborating and refining, reasoning-why is deeply ingrained. Even in the most primitive societies, or the most autocratic ones, institutions or customary practices that serve to elicit reasoning-why on public issues (councils of all sorts) are always found. Evans-Pritchard's (1976) account of witchcraft among the Azande has acquainted many of us with the odd procedure of "trials" in which guilt is decided on the basis of whether a chicken fed poison dies. It is often not noticed that before the poison is administered extended arguments for and against guilt are made, addressed to the poison, but among those who hear the arguments is the person who will administer the poison.

Within a society, the store of responses to situations (the social repertoire) grows with time. Culture, we can say, grows richer and more intricate; and this in turn creates pressure for evolution of ways to store and transmit this expanding repertoire. At the core of this, observation of human societies will tell us, is an innate propensity to internalize the patterns of life in which we find ourselves (most obviously and facilely, but not only, as children). This certainly predates the emergence of the species (Bonner 1980). We know how to do, sometimes we explicitly want to do (feel uncomfortable not doing), but for the most part we just routinely proceed to do all sorts of complicated things (fill social roles, speak our native language, and so on) which we know without ever explicitly learning. Usually, we do all that without so much as noticing that there was anything to be learned. But beyond a stage long passed in human history, the interacting elements of language, resoning-why, sociality, and culture come to depend on techniques for augmenting the capability of individual brains to command the social repertoire, hence permitting expansion of the repertoire. The division of labor provides an essential contribution here [5.6]. But a radical change in constraints comes when patterns of reasoning-why and explicit rules for knowing-how become established in a way that has a permanence that does not depend on the memory of any single individual. Popper's World 3 emerges.

In preliterate societies, the bounds of this process are limited. What can be managed depends on facilitating memorization, which in turn depends on embedding what is remembered onto a familiar framework. A reasonable and widely held hunch (though I will not develop the point here, and it is not essential to the argument to follow) is that the pervasive role of myths and proverbs in human societies, and the very salient role of myth in preliterate societies, has its roots in the use of stories as templates on which to organize patterns of information and patterns of informal reasoning. This suggests a deep continuity between the mythmaking of primitive societies and the scientific theory-building of modern societies. Mathematics itself—the building of extremely intricate structures of reasoning-why which can be applied to many situations in one step—becomes the endpoint of a spectrum with primitive myths about monsters and ghosts near the other end.

A critical turning point comes with the discovery that it is possible to turn abstract things, such as ideas, categories, and patterns of argument, into concrete objects by embodying labels for them (encoding them) in physical artifacts. For with this (that is, with what eventually becomes writing) it becomes possible to radically transcend what any single person can remember and what any single person can critically review in his head. Human beings become capable of creating structures of argument and ideas that are now abstracted from specific stories; and further, these are structures that can be extremely complex, yet be remembered and reviewed and corrected and transmitted to others, as Goody (1977) has stressed. The cognitive limitations of the human brain (with respect to, for example, attention span, memory) are relaxed. In particular, once patterns can be given a concrete embodiment (as symbols on paper, for example), they can be looked at and thought about and become familiar things on their own.

So without making any superhuman demand on the human brain, it becomes possible to gradually build up vastly extended structures of argument and information. In turn, these allow the creation of intricate devices that extend human capabilities in other dimensions, lately including even devices that feed back into this very process by providing human beings with machines for extending their capabilities for reasoning-why. To understand that process, it is naturally important to pay close attention to how these structures of information and reasoning-why themselves evolve. Somehow, at the level of conscious seeing-that (Popper's World 2), but consistent with the processes innate to the physiology of the brain (Popper's World 1), individual human beings acquire competence which depends not only on their physical experience in the world but also on their access to stores of socially available ideas and patterns of reasoning-why (World 3). Individuals also build (or amend and expand) that World 3. Reasoning-why is carried out and stored in a socially shareable way. The Darwinian evolution that shaped the World 1 cognitive machinery (brains) has led to non-Darwinian processes that turn on the way cognitive patterns evolve in individual brains. That in turn influences and is influenced by the state of the essentially social (rather than individual) structures of cognition that define a particular culture.

Figure 5.3 gives a schematic view of some of this. What goes on in an individual brain is bounded by the triangle. Strictly World 1 neuropsychology (no consciousness) yields sequences of *P*-cognitions. At the level of consciousness (in World 2), however, human beings exhibit a capability—tied to the command of language—to use as inputs to further *P*-cognitions, not only perceptions from the external world but perceptions of objects of thought (words, symbols, diagrams, bits of argument, and so on). We are not only able to look at the world but also to look at patterns that originate wholly inside our heads. When articulated into socially shareable reasoning-why, a further level at which cognitive patterns can evolve emerges (World 3) which escapes the limitations of attention span and

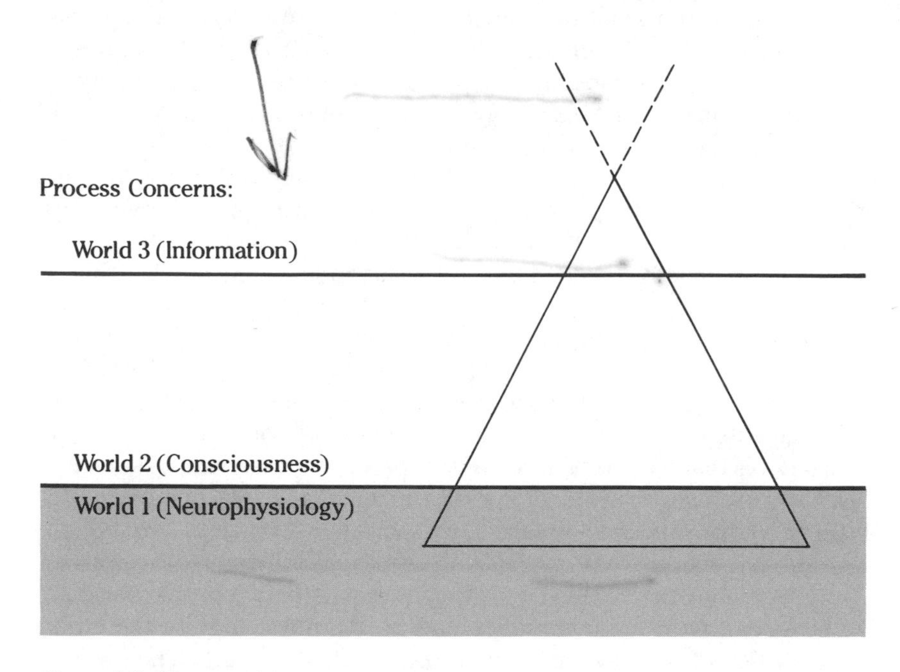

Figure 5.3. The cognitive pyramid provides a heuristic for the relation between what goes on inside a brain and Popper's (1972) "three worlds" schema. The peak of the triangle extends into World 3, since information in the heads of individuals can be transmitted across a community and across generations, most obviously in the case of preliterate societies. Likewise (hence the dotted extension beyond the peak), an individual has access to far more information than is available in his own head at any moment.

memory of individual brains. The whole process then becomes vastly more powerful when technology emerges (writing) which allows these socially shared patterns to be embodied outside *any* brain.

A merely static account of this will be inadequate. For the World 1 patterns for an individual are not fixed; nor are the World 3 patterns available to an individual or a society fixed. Hence conscious intuitions (World 2) also are not fixed. All that is significant for understanding judgment over time by an individual, and it becomes crucial for understanding judgment as it manifests itself across generations or across cultures. We will have occasion later to explore something of that in the concrete context of the Copernican revolution. Even to understand a static context (taking repertoires of cognizable patterns as fixed), it turns out to be essential to see that as a stage in an unfolding process by which both available repertoires of simple *P*-cognitions (subjectively processless seeing-that) and socially shared reasoning-why jointly evolve over time.

In the next several chapters, I will try to sketch that process of cognitive evolution. It is convenient to reinterpret the cognitive pyramid (fig. 5.3) in a way that focuses attention on certain levels of cognitive evolution (levels at which patterns are changing). Spelling out that "levels" taxonomy will be the first business of the next chapter.

Six

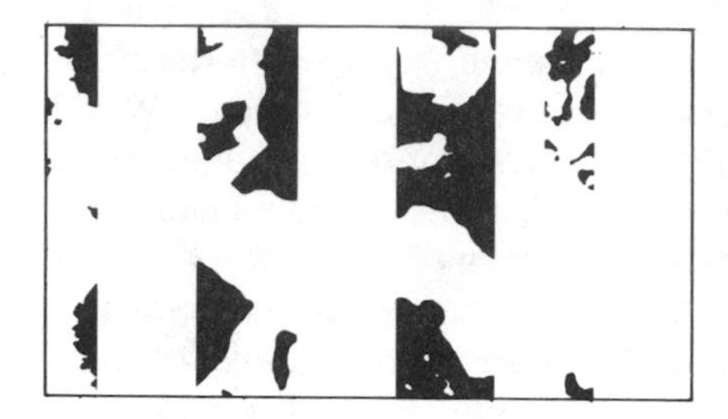

Learning: Level 1

A creature can learn a new thing by copying what someone else is doing, and it can also learn on its own (discovery). However, even copying an immediate demonstration must actually work by copying an internal pattern (a memory of the external), since at the moment the copier is doing a step the demonstrator has already done it. If I say "do this," by the time you're doing it I am not doing it, or at least I am doing a later piece of the pattern.

In general, there can be no sharp line between copying and discovery. You may sometimes be copying what I am doing (presumably by using bits of movement you already know); or you may have recognized the whole pattern as one you already know, or like one you already know. In the latter cases, you are not really copying the outside pattern (via short-term memory) but steering through a now-recognized pattern from your own repertoire in a new context. Or there may be no one cuing that pattern, but simply some (chance or characteristic) feature of the encounter with this context that cues the novel use of a pattern already in the repertoire. I recall learning to hit an overhead in tennis by following a suggestion to pretend that the problem was to hammer a nail about the height of an outstretched arm. It would be harder to remember when we copy an internal pattern in a novel context without outside instruction. For then there need be no conscious sense of trying a pattern in a novel context (the tennis example), but simply doing what looks like it fits the situation. Discovery on this account amounts to something akin to "fruitful misperception." When the pattern that serves as a model for something new is a pattern of language, rather than a pattern of physical movement, we call the pattern an analogy. Over the range of such cases, there will be a continuous series moving with no noticeable break from situations we would certainly call "copying" to ones we would certainly call "discovery." In evolutionary terms, this spectrum must start from the "discovery" pole. So contrary to the usual way of talking about this, copying must derive from discovery. Responding with a pattern in a novel situation that looks like a familiar one is a more fundamental step than the more specialized recognition of a context that prompts copying.

In general, Darwinian evolution would favor neural variants such that, facing a novel challenge, a creature would respond with some pattern cued (even if only weakly cued) by features of the situation. (Recall the Lincoln grid discussion which concluded Chap. 2.) That would rarely do worse and will often do much better than some blind variation of prior behavior. For something close enough to a familiar situation, we would say that the creature knows what to do in this situation. But in a novel situation, if the pattern is of the right sort, we would say the creature has discovered what to do. If the pattern is only crude at first, we would say it is learning what to do; or if the pattern prompted is one of exploring and trial, we say it is searching for what to do.

If I ask you to put a pencil between your toes and write your name, you will be able to do it, though it will take a good deal of practice to do it smoothly. Since this performance can be done immediately, what happens must involve steering through a series of bits of action the newly used muscles already know (move left a bit, then up, and so on) under the guidance of the old pattern stored in memory. In this way, a pattern could be copied even to muscles that are not at all equivalent to those used in the old situation, as you might write with a pencil in your mouth instead of between your toes. The novel behavior will be crude at first, but it gradually becomes more polished, fluent, and automatic. Eventually, we will have a smooth variant of the old pattern specialized to this context. In all these cases the copying could hardly be exact, and variation in the cruder early attempts to use the pattern will be reinforced in a way that yields a variant on the original pattern which better fits the novel context. So there can be a certain fruitfulness (for producing a variant pattern) of that inexactness in copying. But it would be radically misleading to see the inexactness of the copying, rather than the cuing of a pattern to be copied, as the feature around which to organize an understanding of what is going on.

6.2

Parallel to the point that there is no fundamental line to be drawn between copying and discovery, there is also no clear line between novel and familiar situations. No two situations in the world will be in fact exactly alike. Hence, it is artificial to think of the world as presenting dichotomous cases, where an animal either uses old knowledge or has to learn something new. The tendency we have to see much more dichotomy in the world than there really is ("twoness") becomes a major theme in Chapter 14. In the most familiar situation, there must be at least some slight amount of novelty, hence the possibility of at least a bit of learning. Contrariwise, even in the most novel situation, the response a creature makes—whether it is a successful response or not—can only reflect some pattern of response it already has. Even what looks like random struggling is always highly constrained by cues in the situation. We see a retreat to loosely organized and changing combinations of elementary patterns, or to a stereotyped search pattern, but either still governed by cues in the situation. If a nonswimmer falls off a boat, he responds by doing something like swimming, not by blindly trying to scratch his ear, putting his hand in his pocket, and so on. A rat put in a maze, or a cat put into a house, responds with an exploring sort of activity, not with blind variation that might as well focus on rolling around, covering his face with his paws, and so on. Even in a totally confusing situation, the creature will exhibit something like a usual way of responding to a totally confusing situation.

Accordingly, even in a very familiar sort of situation, it is never exactly the case that nothing at all new is going on, and even in the most radically novel case there is always an important sense in which the animal is doing what it already knows how to do. We expect this to apply to patterns of ideas as well as to motor patterns. The common observation that novel ideas often seem to exploit analogies is misleading on this argument. Putting things that way implies that the use of analogy—in the broad sense of using a pattern already in the repertoire in a new domain—is an optional feature of discovery, as if there were any other way.

In contexts close to the familiar pole of a familiar-to-novel spectrum, it would be straining language to insist on calling what happens discovery by analogy—in particular, when all that is happening is that one is doing something that varies only trivially from what one has done many times in the past. My point is not to insist on calling that discovery by analogy. It is only to stress that there is nothing qualitatively different going on from what happens in the more markedly novel contexts. In terms of underlying processes, it is an essential point of the account here that there is just one, not two or more, psychological processes—say logical versus intuitive—underlying cognition [5.10], and this applies fully to any *P*-cognitive account of learning. Discovery must turn always on using patterns already in the repertoire in a new context that (perhaps only momentarily, by fruitful misperception) looks like a familiar one. A special case is (1) using a routine that leads to a thing to be copied, as a bee does in its stereotyped approach to a flower when it is memorizing its color, scent, and so on; or the less-stereotyped but still quite routinized way that you learn a new phone number by looking it up in a phone book. And another important special case is (2) the sort of copying we call training, where a creature is prompted to various bits and pieces of behavior in such a way that what is eventually produced is a more elaborate overall pattern (or set of patterns) than the creature could have managed if confronted with the whole thing at once. I will say more later about such routinized learning (1) and about training (2).

6.3

When a pattern is used in a new way, it gradually becomes better tuned to that new context. The variant of the old pattern that develops better fits the new context; and along with the variant pattern we will get a variant set of cues which will distinguish this situation from the sort of context which cues the original pattern. As experience with the pattern is gained, priming (of complementary) and inhibiting (of competitive) patterns and of subpatterns governing particular bits of the pattern will become increasingly refined; and the use of the pattern consequently becomes increasingly smooth, automatic, and undistractable: in a word, it becomes *habitual*. On the other hand, when an established pattern fails (when a misperception turns out to be fruitless, or when conditions have changed so that the old pattern no longer works), the balance-of-errors argument [2.3] favors a weakening of the effectiveness of the original cues unless the propensities involved are genetically engrained.

If the balance between inertia and flexibility (type 1 vs. type 2 errors) is inopportune, we could see either an erratic and self-defeating sequence that does not stay with any pattern long enough to make it workable, or (on the other side) a futile persistence in pursuing an unworkable pattern [2.3]. Consequently, we can expect that Darwinian selection would produce a variant of cognitive plasticity that reasonably suits the cognitive powers and normal life experience of a species. But an animal living in an environment radically different from that which shaped its genetic propensities will be vulnerable to mismatches between the problems it encounters and the balance it has inherited. There is no reason to suppose that the balance humans inherit from a million years of Darwinian tuning to a hunter-gatherer life will be optimal for such activities as social choice in a modern state or theory choice in science. Rather, the emergence of the state and of science could occur because the advantages that come with social cooperation on a large scale and with the development of science (hence access to its applications) are so large that even doing them badly will be better than not doing them at all. Of that, much more in the final chapters.

6.4

Very diverse work (from tests on normal humans to neurophysiological studies in slugs and other creatures with very simple brains) finds persistent hints about how the tuning of patterns and cues develops. Across the whole range of such studies, in one form or another, there is evidence of the processes commonly labeled (1) reinforcement, (2) extinction, (3) sensitization, and (4) habituation. The first two capture the increase (or decrease) in the probability that a certain pattern will be prompted depending on whether it is associated with a good result. The second pair amounts to a shorter-term (tactical vs. strategic) version of the first; and both pairs were characterized in the cognitive ladder discussion [3.2] as long- versus short-term versions of priming (1 and 3) and inhibiting (2 and 4).

On the balance-of-errors argument [2.3] learning would not ordinarily be fixed by one reinforcement unless the effect is so closely linked to direct survival that one-step learning has become genetically ingrained. Similarly, a balance-of-errors argument implies only gradual extinction (loss of a cue-to-pattern link because it failed to produce the usual result). There will be important biases, however, in the way that the tuning of pattern to cues proceeds. In almost any novel situation there will be some cues that are not reliable indicators of the context but nevertheless are immediately salient, either because they trigger innate responses (pain, for example), or because—at least when accompanied by other cues usually present in this context—they already play an important role in the life of the creature. On the other hand, there will also ordinarily be other cues the creature is capable of detecting, and that are reliably associated with this context but which initially are not salient all (they lack the features that make the previous class of cues salient). Then very extended experience would be needed (perhaps, as a practical matter, much more extended experience than would ever occur) for these cues to come to play a cognitive role commensurate with their objective reliability and sensitivity.

So the groping and polishing process that leads to smoothly automatic performance would turn on the gradual acquisition of an intricate network of priming and inhibiting relations between cues and patterns;

but we should be cautious about supposing that the details of this tuning (the exact nature of the cues used) are optimal rather than only workable. A parallel argument applies to the patterns themselves: we have no reason to suppose that the exact nature of the pattern that emerges will be optimal. Rather, patterns that have proved powerful, and so have become deeply entrenched through use (that have become habitual), may have perverse deficiencies which are then difficult to escape. An example on a grand scale will be provided by the Copernican material taken up in the closing chapters.

The simplest hypothesis that might account for the evolution of the *process* which yields the prompt translation of cues into fluently executed response remains the simple story already spelled out, turning on inhibiting and priming of cognitive patterns. Everything derives from the habituation and sensitization mechanisms that are now studied mainly in such convenient creatures as sea slugs, which have brains consisting of only a few hundred neurons. On the continuity argument, what plainly exists with respect to perceptions and physical habits provides the most plausible and parsimonious basis for interpreting what seems to be similar facilitating networks with respect to thinking and judgment. On that argument, we are committed to look for marked parallels between (for example) physical habits and habits of thought.

6.5

The way that chance interacts with *P*-cognition in learning sometimes takes a trivial and sometimes a striking form. A good deal depends on just what we understand as "chance." Even with no physiological evidence, we would suppose that there must be some level of detail at which chance groping becomes important. Looked at under a sufficiently powerful microscope, any process becomes unpredictable at the margin (for example, the smoothest line looks erratic). That is not always irrelevant to understanding things at a more aggregated level of analysis, but it often is. The designer of a computer sometimes needs to worry about quantum effects. A computer user, however, does not. Similarly, how far the polishing of a cognitive or behavioral novelty is ultimately based on reinforcement favoring some chance neural connections versus some more deterministic influence does not look like an important issue for understanding human judgment, though it obviously is an important issue for neurophysiology. At the level of observable behavior and judgment, whatever the balance between chance and nonchance effects, the outcome is known in advance: with practice the performance of a pattern (of behavior or of thought) becomes more and more smoothly automatic, and more and more integrated into more complex and extended patterns, with the whole becoming smoothly automatic, and the habits that develop (on the adverse side) becoming harder and harder to change.

After the performance is fluent, chance continues to have some role. Variation in performance (reinforced when it produces a good result, perhaps merely on an unlikely chance) continues to occur, governed by variations in the environment and presumably also by synaptic variation. But now what we see is as likely to involve degradation (picking up a bad habit) as improvement. That is why mature professional athletes often seek coaching. There is no comparable role for coaching of mature professionals among, say, musicians, where it is easier for the individual to critically observe his own performance.

It is hardly plausible that Newton would not have thought of gravity had that apple fallen a few minutes before he sat under the tree, or that Darwin would have failed to discover natural selection had he misplaced his copy of Malthus. But in both these cases, we have a discoverer who was very much looking for just what he found. I give a sketch of the Darwinian discovery in Chapter 10. But Columbus was not looking for what he found, and indeed died not knowing what he found [12.6]; the Copernican discovery (I will argue) very plausibly could have been delayed for a century if Copernicus had not stumbled on the key to it at just the time he did [12.5]. On the other extreme, while Roentgen's discovery of X-rays depended critically on pure chance, so much similar work was going on in physics laboratories of the time that it is very generally supposed that the discovery would have come shortly, even if Roentgen had missed his opportunity. There were probably other physicists who encountered happy accidents like the one that prompted Roentgen but failed to follow up.

So in these famous episodes, and beyond doubt also on more mundane matters, the role in discovery of chance varies greatly. Sometimes chance must have been essential, in the sense that, except for some fortuitous confluence of events, the discovery would not have occurred even approximately when and where it did. Often, though, chance is playing only the subordinate role it seems to play in the context of polishing a performance, or when the sort of chance involved is like that when a search pattern is executed that makes the eventual result foreseeable but not at some predictable step in the search.

6.6

A great deal of learning, even in humans, fits into a scheme in which learning can be seen as an extension of development [3.4]. For an embryo, out-of-the-ordinary chance effects must almost invariably be damaging. But later on the prognosis becomes better. The encounter with something for which the creature is not specifically prepared by its genetic endowment now comes in addition to, rather than in place of, what the creature needs to be viable; and Darwinian evolution would favor postnatal neural plasticity where it is likely to be helpful and favor rigidity where it isn't. In the processes of embryology and then postnatal development (of, say, a chicken), a certain amount of groping followed by reinforcement for some trial connections and decay for others (of tendons, muscle fibers, and, most important for our purpose, synapses) plays the tactical role I have mentioned before. This is true even though the whole thing will proceed (ordinarily) in a way and on a schedule such that an embryology text can tell us what is going to happen day by day, and even hour by hour.

After the chick is out of the egg, development continues, but now more often requiring acquisition of cues from the environment (rather than only incidentally subject to environmental effects). Once that is happening, we begin to call the development learning. But by and large, what an animal can learn is highly programmed (for a honeybee, colors of food sources, location relative to the sun and the nest, and a few other things). The animal is very good at learning certain things given the cues that trigger that learning and generally very bad at learning things that do not fit the innate propensities for learning. The extent to which that is so is very striking. It eventually proved to be a disaster for strict behaviorists (Seligman & Hager 1972). Sometimes the learning reflects a genetically programmed propensity, as when an animal gets sick after once eating a food and then never eats it again. And it would be arbitrary to draw a sharp line purporting to mark where development ends and learning begins. For mammals and birds, at least, learning often extends to some copying of the behavior prevalent in the social group, so that we see the beginnings of what we can call culture; hence the point that copying is (in evolutionary terms) an extension of discovery rather the reverse [6.1]. But it is only with human beings that within-species differences in what individuals learn clearly become something more than an occasional quirk of circumstance.

6.7

Apparent counterexamples excite a great deal of attention. One of the best-known cases concerns milk-bottle raiding by British tits. The birds peck open the tops of bottles delivered to doorsteps, stealing the cream. Yet this performance requires no assumption of unbirdlike insight to account for how that novel behavior might have started. The birds slash open the bottle cap with the motion they instinctively use to slash away bark to get at insects. And it is not hard to imagine serendipitous circumstances that might trigger the first use of this attack on a milk-bottle cap (as, an insect crawling across the cap). Similarly, serendipity can account for the endlessly cited case of food washing among Japanese macaques. (It is significant that this 30-year-old event remains so salient despite the blossoming of ethological studies in the decades since.) The creative act involved an island-bound troop fed by putting food on the beach. A young female (E. O. Wilson 1975: 171–72) began washing potatoes in sea water to get rid of the sand. The trick was gradually picked up by other members of the troop. Later the same young macaque made the even more surprising invention of separating rice from sand by throwing handfuls into the water, so that the mixed-in grains of sand sank and the rice grains could be skimmed off the surface.

But no abstract insight is required to account for these performances. For there surely must be occasions when a macaque would drop a potato, for example, because a grain of sand from the potato becomes painfully wedged between its teeth, or because it was startled by a snake. In this case the macaque was at the seaside. Picking up the potato again, it tasted salty and had less sand on it. Given the serendipitous discovery that it could be useful to put the potato into the sea (and we could easily construct half a dozen scenarios for reaching this point), no puzzle is posed by the gradual development of the rest of the potato-washing performance.

Further, given the potato washing, the macaque (recall that it was the same macaque) required only a repetition of the pattern that had worked before to discover that a handful of rice could also be improved by tossing it in the water. There is no need to suppose that the macaque foresaw that, if rice was thrown in the water, any sand in it would sink while the rice would float. That indeed would be genius for a macaque. But a decent respect for parsimony tells us to settle for the much-simpler explanation that the animal was repeating with one food a pattern that had already brought a good result in connection with another food.

Nevertheless, performances of this sort go beyond what can usefully be interpreted as an extension of development. We are beginning to see a more general-purpose capacity to learn, which becomes increasingly striking as the complexity of copyable patterns grows, the range of behavior that is plastic expands, and as the internal manipulability of patterns grows. I mentioned earlier that glimmerings of the judgmental step of the cognitive ladder [3.5] can be seen in the higher primates. But humans show the radically more powerful capability for making judgments about contingencies that are not immediately present (we can plan for tomorrow or review what we did yesterday). To do that requires the possibility of abstract thought, in the sense of thought not tied to the concrete situation the creature is facing right now.

6.8

Elaborating on the discussion of [3.5]:Think about what happens in the radically reduced environment of a board game, like chess. An expert chess player who cares to do so can develop competence in blindfold play, where the board is not seen at all. But even with ordinary play, variations the player considers before making his move involve an extension (thinking ahead several moves) from the position the player sees on the board. There is no cognitive discontinuity between the situation where a player is looking at the board and the situation where he is playing blindfolded. It is easier to manage the extension of the base position required in looking several moves ahead if the base position is available to anchor the mental sense of the position. But an internal representation of the board is required just as surely for competent play with a board as it is for play without seeing the board. Even when the board is fully in sight, what is explicitly in focus at any moment will usually be only part of the full position, and perhaps no more than is overtly imaged in the equivalent moment of blindfold play. Expert chess players, I have been assured, have a sense of tacit awareness of the whole position, though the overt mental imagery they experience covers—at any moment—only a part of the board.

As there is no discontinuity between blindfold and sighted chess, there is also no discontinuity between the mental imagery of diagrams commonly reported by mathematicians and actually looking at a diagram in a book or on a blackboard. For in mathematics and many other uses, looking at a diagram commonly involves "seeing" (with the mind's eye) various extensions, variations, or continuations of what is externally on view. So while there is certainly a great extension of capabilities in going from Koehler's ape performance [3.5] to being able to think about such a problem away from the concrete context, there is no cognitive discontinuity. No radical novelty at the level of neurophysiology is required.

But the extension from externally anchored to more remote abstractions is of enormous importance, as has often been emphasized by writers on the evolution of human intelligence. Objects of thought, in contrast to objects in the world, are not intrinsically constrained in how they are moved about in time or in space; nor are they constrained to be only what has been observed, as opposed to what has been observed altered in some way, and perhaps even turned into a created object never seen in the external world. So the range of things that can be searched, the fluidity of the search, and the temporal span of the search, are vast compared to what is available to an animal that deals only with the physical context that confronts it. An essentially unlimited range of combinations of objects can be set beside each other in the search for promising "looks like" perceptions. The process can continue for a long time (or intermittently over a very long time). Bundled-up interim results can be prompted to attention by a chance encounter (like Newton's apple, or Darwin's browsing through Malthus). Sometimes these encounters themselves are essentially in the head, as with Wallace thinking of Malthus while in a malaria delirium or Kekule's seeing his linked snakes in the dream from which he awoke knowing how the carbon atoms in benzene were arranged.

These strokes of chance occur only to individuals who have mulled their problem for a long time: in more biological language, it occurs only to people who have become sensitized to respond to any chance appearance of something that looks like it fits the problem they have mulled over for so long. Also, a piece of movement or a piece of argument can be learned by self-training because it looks like it fits a more ambitious pattern as yet only dimly in hand, as when a person decides he had better learn a little about some theory in an adjacent field. The material will then be available, perhaps at a critical moment. At each particular step of the process the individual is only moving from cues that have become familiar to patterns already in the repertoire. In later chapters we will have occasion to considerably extend this argument about what it is that prepares the mind of a discoverer.

What is most immediately relevant is that out of the process being sketched, many of the relations of cues and patterns *can be generated during the very search process itself.* The whole process acquires a radical open-endedness so that, in the case of humans, learning requires an analysis that talks about internal processes (about what goes on inside the head) and about the repertoire of patterns a person may be using inside his head.

On the continuity argument, we must expect the techniques and insights developed by behaviorists for studying animal learning to provide building blocks for work on higher learning. But beyond the ethological constraints mentioned earlier, once we reach the radical opening of possibilities of the peculiarly human steps of the cognitive ladder, features of learning emerge that have no useful counterparts in other animals.

6.9

Since learning depends on the patterns we already have (novel patterns must be built up out of those already in the repertoire), learning must also be constrained by those patterns. Once a pattern (whether of behavior or of thought) has become well-entrenched so that it is habitually prompted by some pattern of cues, it cannot be easy to turn it off. An act of will ("Turn off that habit") cannot put in place some standing order for that effect. By concentrating on my next golf swing, I may be able to keep my left elbow straight on that shot, but I can't just issue a general command: "From now on, keep that elbow straight." Essentially the same point applies to habits of thought, and with essentially the same set of issues (the need for drill to alter a habit, not just a decision to do so; the problem of relapse to the old habit, and so on).

So learning that requires eliminating an old habit is difficult, as every reader must know firsthand. But unless a creature has the capability for abstract thought required to know what a habit is, it would be impossible even to have the notion of wanting to change a habit. However, once the capability to manipulate abstract patterns evolves, and has been enriched (in particular) by language, we have the possibility of active learning (of search for new things beyond what is prompted by the immediate concrete situation), and we also have the possibility of searches that are helped by awareness of learning, capable of intuitions about intuitions, and hence of such explicit notions as doing better rather than worse, or getting something right rather than wrong.

Although there will be no exact line, there will be an essential distinction (for analysis of human judgment, though perhaps not for neurophysiology) between learning which can draw on cognitions about cognitions and learning which cannot. Call the latter *Level 1* learning, and the former *Level 2*, corresponding in a straightforward way with the levels of the cognitive pyramid introduced at the end of Chapter 5. All animal learning, on this account, and a great deal of human learning are at Level 1. The argument of this chapter has sketched how the emergence of increasingly powerful capabilities to use and manipulate internally generated patterns changes the character of human (vs. nonhuman) learning. It is what accounts for why a human being will take a lesson, practice a move, engage in mulling/rehearsing a problem at odd moments, and so on with all the things that human beings commonly do but which it is only comical to mention that other animals don't do.

So there is something essential to distinguish between learning that uses cognitions about cognitions (perceptions about perceptions, intuitions about intuitions) versus learning that doesn't (Level 2 vs. Level 1 learning). It is the distinction between cognition that can exploit representations only versus cognition that can exploit representations plus propositional attitudes. This chapter has sketched an account of Level 1 learning, turning at the end to how, without violating evolutionary continuity, that might lead to the Level 2 cognition considered in detail in Chapter 7.

Seven

Learning: Level 2

Newell & Simon have argued that expert knowledge—such as playing high-level chess or speaking a language fluently—implies something on the order of 50,000 stored patterns. Each of these Newell & Simon patterns must be built up from many smaller patterns. For example, a well-educated person will command a recognition vocabulary of something approaching 50,000 words. But each word is a pattern of phonemes, and each phoneme is a pattern of sensations depending on subtle discriminations of sounds, timing, and so on. The meaning of those words, on the other hand, is contingent on the larger pattern of words in which they are embedded; and it is the larger pattern that is comprehended (the utterance, not the collection of words). What a particular utterance means, in turn, will commonly depend on details of context (what is going on parallel to and what went on preceding the utterance), starting with tone of voice and accompanying gestures. Yet we talk almost as effortlessly as we see the images we see (just by opening our eyes) despite the ambiguous and erratic character of the information that reaches the retina. For both, we would be tempted to deny it is possible, except that somehow we do it.

What applies to language applies to other functions of the brain, such as those controlling physical activity. We go through life switching rapidly and effortlessly from one context to another, as a multilingual person can switch smoothly from language to language between (but not within) utterances, and often without noticing the switches. (If asked "Do you realize you changed languages?" the answer will often be "No.") We are paying active attention often to one context while handling others more or less automatically (for example, playing a piano or driving a car and talking at the same time).

The evolution of that capability must reflect strong and very prolonged Darwinian selection for more efficient, reliable, automatic (attention-free) ways to locate, choose, and sequence whatever patterns have been stored. Fluent performance must exploit the likelihood that the next move in a chess game will produce another chess position, not a bedroom scene, and that the chess positions that will arise at the next move, or the language situations that will arise at the next utterance, are some very small fraction of all that are in the repertoire. Given the last few spirals of *P*-cognition plus whatever now is externally available, some patterns must be easily prompted. And an extension of this priming argument says that competing patterns will become hard to prompt (inhibited). All this arose in the cognitive ladder discussion of Chapter 3.

So achieving fluent command of a rich repertoire means acquiring and exploiting intricate networks of priming and inhibiting relations among cues and patterns. But that implies that learning will be contingent not only on the obvious point that complicated things (like playing the violin) will be more difficult to learn than relatively simple things (like playing a chord organ), but on the extent to which the learning can build on—or on the other hand, requires disturbing—entrenched priming and inhibiting networks. As will be seen, this point plays a particularly salient role in the analysis of that special form of learning we are inclined to label "discovery." Difficulties that are commonly attributed to complexity will often turn out to be consequences of entrenchment.

When introduced in 1905, both Einstein's relativity argument and his photon argument exploited formalisms (for relativity, the Lorenz transformation; for the photon, Planck's radiation law) already familiar from the work of leading theoretical physicists. What made Einstein's argument difficult was that these formalisms had to be seen in a way that violated habits of mind deeply entrenched by work within the framework of classical physics. Those who had come to know the formalisms before Einstein's new interpretation had (naturally) done so in a way that made them compatible with that older framework. Much of this chapter will be a sketch of how that happens. As is perfectly characteristic of this sort of situation, Planck himself, though the first important convert to relativity, could not accept the photon for some years (Kuhn 1978); and neither of Einstein's principal precursors on relativity (Lorenz and Poincare) ever quite accepted the full consequences of relativity (Pais 1982:chap. 5).

How deeply entrenched a cognitive network becomes must vary with how salient a role the network plays in the life of the individual concerned. The more a pattern is used, and the more ways in which it is used, the more elaborate the neural connections required to mediate the behavior. Consequently, the harder disruption of that network will become. This yields the key point for the balance of this chapter. The most interesting puzzles about learning usually turn on understanding how someone could make a discovery many others had missed, or why some people continue to see as definitely wrong a new idea that—in hindsight—had already been shown to be very plausibly right. I want to argue that the critical problem in such cases will be to find what it is about the entrenched networks of cues and patterns that will account for what has occurred. In Chapters 11–14 I will go through an analysis of that sort for a series of questions connected with the Copernican revolution. But here we are concerned only with laying out the general considerations to be drawn on in those applications. The most fundamental point is that questions about learning will be misposed if the problem is taken merely as, "How did a person learn *B*?" For adult learners especially, the question should be, "How did he learn *B*, given the habits of mind (the entrenched networks) on hand when he faced the problem?" The critical steps will almost always concern how a person gets *from* knowing *A to* an altered cognitive state of knowing both *A* and *B*, in the sense of being able to use them in the way that people who know how to use them do use them.

There may be profound problems in managing such a transition, even when there is nothing intrinsically very complicated to be learned. For getting from *A* to *A* + *B* may disrupt the network of priming and inhibiting that facilitates fluent use of *A*.

7.2

Suppose *A* is Newtonian mechanics and *B* is Einsteinian mechanics. It is convenient to have both (the Newtonian being simpler and perfectly good for a vast range of situations, the Einsteinian being necessary for some contexts, including the context of speculating about what reality is really like). A contemporary physicist will know both. A particularly common case is that of a physicist who routinely uses Einsteinian physics in his research but routinely teaches Newtonian physics. But if you ask the physicist, you will find that he/she sees a Newtonian situation as just a Newtonian situation, not in some more complicated way (as a special case of Einsteinian mechanics) which might be construed as a burden for work on nonrelativistic problems. Our physicist can use the two inconsistent physics just as comfortably as you or I—and the physicist too—see the sun setting not the earth turning, though if I ask you to form a mental image of the earth in relation to the sun, moon, and planets you see things differently.

Switches of this sort (once fluency is attained) are almost as unlikely to come to conscious notice as in the case of the fluent speaker of languages. Hull (1973), commenting on how biologists seem to use both molecular and Mendelian genetics, remarks:

> They can operate successfully within the conceptual framework of each and even leap nimbly back and forth between the two disciplines, but they cannot specify how they accomplish this feat of conceptual gymnastics. Whatever connections there might be, they are subliminal. In a word, those geneticists who work both in Mendelian and molecular genetics are schizophrenic. The transitions which they make from one conceptual schema to the other are not so much inferences as gestalt shifts.

But such switches are commonplace and hardly are usefully labeled "schizophrenic." They occur not only in the sciences (in connection with such things as thermodynamics vs. statistical mechanics, micro- vs. macroeconomics), but reflect essentially the same fluency-enhancing inhibiting of competing patterns and priming of complementary patterns that occur with more routine activities like comprehending ordinary language.

In science, after a Kuhnian paradigm shift—but not during the transition for that individual—someone who has come to see things in the new way will be fluently capable of talking the language of the old view (as is often convenient for expository or polemical purposes), just as an astronomer who happens to be a yachtsman can work very comfortably with the pre-Copernican astronomical system still used in navigation at sea. But during the transition from knowing A to knowing $A + B$, hitherto unproblematical assumptions become unreliable—sometimes applicable, sometimes not so. What replaces the old conventions when they no longer work is still unclear. Standard reasoning-why continues to apply in some contexts but not in others; and the learner is not yet fluent in knowing which. Motivational as well as cognitive issues often enter, as when some aggressively defended piece of work becomes no longer tenable.

7.3

In another form, the transition issue arises in the following way. Right now, a creature may do better to smoothly execute a pattern of response it knows well versus clumsily execute a pattern that might eventually do better once it has done it enough to make it go smoothly. So to some extent (with some nonzero, noninfinite degree of cognitive inertia), the creature will do better to continue to rely on cues and sequences of cues that have worked well in the past. It would be very risky to always (or never) admit to the repertoire novel responses to cues which (until sufficiently practiced and refined) could cause him to hesitate or vacillate or stumble in closely balanced or ambiguous or challenging contexts. So we have a tradeoff between the effectiveness of the responses that might eventually be produced versus the fluency of the process that produces the responses right now. If there were all the time in the world, then the latter concern should not significantly constrain the former. But under the conditions that shaped the more ancient cognitive machinery out of which our own had to evolve, there was rarely much time at all. Indeed, it is only a creature that has reached what I have called Level 2 learning—only, among the creatures we know about, among humans—that such change can be managed. For it is necessary to envision the future situation and prefer it to the current one despite short-run costs.

A tennis book (Braden 1977:10) warns:

> Every good teaching pro has heard the complaint: "Jeez, I was better before I took lessons." Very often this is true. No matter what the sport, when you are trying to make corrections, there's always a force trying to bring you back to your old comfort levels . . . you want to do something the new way, but you want to maintain some of the old, and thus you get caught in the middle, vacillating between the two. This can be murder on your tennis game.

But it is also not easy for a person trying to learn a scientific or political line of argument that involves conflicts with existing habits of mind.

7.4

Since we have no introspective access to how a brain works, something that cuts against the grain of much-used patterns in the repertoire cannot be seen in those terms. Rather, reasoning-why to rationalize the sense that it would be uncomfortable or painful to try to learn such a thing will take forms like "The new thing is too complicated" (to be useful, or to be plausible), or "Looks wrong," or "Looks not important enough to bother with." Sometimes, of course, such reasoning-why is perfectly sound. But we see things that way even when doing so will come to look like mere self-deception. Slips and ambiguities of a sort that would be passed by without complaint in one context (or even endorsed as reasonable simplification) will be seen as clear signs of incompetence or fatal error in another.

Naturally, on this view, habituation to stubborn anomalies must be found not only in the case of inferior or otherwise unsatisfactory work but also in the response of even the most brilliant individuals to ideas that conflict with entrenched habits of mind. We are not surprised to find, therefore, that Leibniz and others of Newton's best-qualified contemporary judges (Bernoulli, Huygens) found it absurd to base a theory on action-at-a-distance (Westfall 1980). But the next generation had learned not to be bothered by that. That an ineffable force holds the earth in orbit across 93 million miles was no longer an absurdity but rather so unproblematical as to be hardly vulnerable to prompting a closer look. It took a cognitive shock in the other direction to allow that asking how gravity worked (the issue for Einstein's general theory) was a nonabsurd question. Similarly, particular examples of neglect of difficulties we see as insightful—such as Einstein's confidence in relativity in the face of its falsification by the best experimental physicist on the continent—cannot plausibly be produced by some different process than other particular examples we see as illusory (Pais 1982:154).

Polanyi's "tacit knowledge," therefore, is often in fact tacit blindness. We "know" what to ignore, including gaps in an argument that may later come to seem gross. Meanwhile, sensitization (priming) tunes a person to respond promptly to subtle cues that have proved to be reliable markers of circumstances in which a certain pattern of response is effective. The individual is not left open to every stray thought that might pop up. He is quickly focused and kept focused (in what is ordinarily a very useful way) on what he knows how to do. There is, using a phrase Kuhn introduced for the special case of science, an essential tension between what is best for doing well now versus what might be best for doing better later. And of course the interaction between the difficulty of learning a new thing due to its complexity versus difficulty contingent on entrenched networks works both ways. It is hard to learn new things that disrupt an entrenched network, but we are remarkably good at learning new things that fit such a network. A professional can absorb the ideas in a paper almost at a glance where a new student (with no entrenched network to either help or hurt him on this matter) must struggle for days. I mentioned an analogous effect in the context of chess [5.11].

7.5

On this account, when a person sees a contradiction of logic as intolerable, that discomfort could not be essentially different from the discomfort that until a few years ago many people felt with the notion that hockey players might wear helmets, or that people of a different race could stay at their hotel. Things normatively miles apart are often cognitively equivalent. That a reader of this book can be assumed to believe that taking logical inconsistency seriously is good and racial bigotry is bad does not say anything one way or the other about the empirical claim that all such "fits/doesn't fit" attitudes have a common cognitive basis and are context-dependent. Philosophers have a special professional commitment to closely reasoned justifications of views. They worry a lot (for example) about the lack of a really good piece of reasoning-why to support the belief that they are not brains-in-vats. But sensitivity to subtly flawed logic is not a kind of discomfort that is innate to the human brain. Not even philosophers exhibit much of it apart from the special professional contexts that elicit it.

Since there is no reason why only (normatively) good cues should enter the nonconscious mechanisms that guide our attention, affective responses, and intuitive seeing-that, clear intuitions may have (normatively) absurd foundations. The relation between cues and patterns is only one of correlation, not necessarily of causation [4.3] and sensitized cues will not necessarily be causally important cues or rationally defensible cues. Further, we have already had occasion [5.12] to emphasize that reasoning-why to defend an intuition may feed back to take on an importance it never had in forming the intuition.

The common explanation (by outsiders) for what obstructed the addition of helmets to the elaborate protective equipment already worn by hockey players was a commitment by players and coaches and fans to "macho" values. But if wearing a helmet was seen as unmanly in hockey, why was it a symbol of macho qualities for football players, racecar drivers, combat soldiers, or construction workers? If, however, cognition worked—as on the argument here it must work—by a logically arbitrary tuning of patterns of response to reliable associations with cues, then the intuition that helmets are inappropriate to (don't fit with) hockey players need not have any deep foundations. It could amount to something like the following: a reliable cue to discriminate hockey contexts from football contexts (another contact sport in which players wear pads) was that hockey players did not wear helmets. Logically, cognition could just as well rely on the more essential cue that football players do not wear skates. But given the propensity we have to focus on faces much more than feet, and given that no-helmet was until recently as reliable a cue as skates to discriminate hockey from football, it would not be a deep puzzle that such an association could take hold.

Of course the opponents of helmets would not see their own preferences as governed by such trivial associations. Nor would they be at a loss to produce what seemed to them perfectly good reasons for their intuitions (that helmets will hurt the game, cause as many injuries as they prevent, and so on). But their critics, looking at the same intuitions but not sharing them, see other reasons, principally focused on the macho theme. So what some of us may see as an honest report of why we believe what we do, our adversary sees as too implausible to be taken seriously and produces something like the macho argument as, consciously or not, underlying the preferences. (Note that the difference from folk psychology on this point is not that the cognitive argument is skeptical about conscious reasoning-why but that folk psychology isn't. Folk psychology has always had ample room for unconscious motivation: before we had Freud we had fate, witchcraft, spirits.)

Both sides in such matters report reasons that amount to a cuing from seeing-that intuitions to some familiar pattern of argument that seems to fit with those intuitions in a way that does not immediately prompt incompatible intuitions. For our own judgments, this usually occurs in a way that looks agreeable in terms of self-image, but it easily leads to an unflattering account of our adversary's intuitions as reflecting venality, stupidity, narrowness, or whatever else will serve to explain why they purport to see things another way. Since cognition works the way Darwinian pressures have shaped it to work, these comments are not contingent on what saints or sinners or logicians might prefer. As I have said earlier, we cannot suppose that an outcome shaped by vast periods of hunter/gatherer—and earlier, prehuman—evolutionary history will be flawlessly suited to life in modern societies. We have no reason to suppose that we live in the best of all possible cognitive worlds. On the evidence, it looks like we happen to live (unsurprisingly, after all) only in a pretty average example of all possible worlds.

7.6

In a sense, therefore, all reasoning is born as rationalizing (making up reasons, ex post). However, seeing-that/reasoning-why feedback allows the development of reflective versions of rationalizing, which may (not will) lead to judgments and reasoning that indeed do fit together in a way that makes normative sense. Suppose that a situation confronts an individual with contrary intuitions, or grossly conflicting cues, or social adversaries (to mention the most important contingencies) that trigger the "look again" *C*-scrutiny of the reasoning-why that purports to support a judgment [5.1]. On the argument of Chapter 3, inconsistency—sometimes seeing that *X* is true, and sometimes seeing that it can't be—would become a problem only when the contrary intuitions get in each others' way. We expect natural selection to shape cognition (produce brains) in a way that will ordinarily protect an individual from intuitions that trip over each other. But we have no reason to expect globally consistent patterns of inference or intuition. Rather, we can expect only a loose connection between inconsistency and what I will call cognitive rivalry.

A brain that required global consistency—one able to act effectively only if it held globally consistent beliefs—would be doomed in Darwinian competition with brains that could act on the pattern cued now without self-defeating hesitation or vacillation. We would expect intuitions to become cognitive rivals only if they are both adjacent (occurring within a common span of short-term memory) and interfere with the creature's ability to choose what to do next. Local inconsistency could produce cognitive rivalry. Global inconsistency alone could not. Nor even (other than briefly, with minimal consequences) would local inconsistency that did not cause difficulty for acting. We are not bothered by seeing a gestalt drawing as *X* one instant and as *Y* the next, since we have no occasion to act in response to the perception.

Cognitive rivalry is an intrinsically unstable cognitive state. What it means to call intuitions "rivals" (always for some individual, at some time, though I will not always make that explicit) is that something must change. In the language of the belief matrix [5.8], we have adjacent intuitions directly in conflict. With respect to some intuition we see I^+, and also (not simultaneously, but immediately) I^-. For short, I will say the person sees $I^+ \& I^-$ (direct rivalry). On the Darwinian argument about vacillation and hesitation [3.4], we could expect that the cognitive state I^+C^- (paradox, or indirect rivalry) must also be unstable but less so. For in contrast to direct rivalry, indirect rivalry does not easily produce immediate difficulty for acting, since ordinarily seeing an argument as right or wrong does not lead to an act: it is seeing the conclusion as right or wrong (I^+ or I^-) that does that. But the evolutionary arguments that are needed to account for how the conspicuous human propensity to produce and want to examine reasoning-why imply [5.8] that adjacent seeing of I^+C^- must also be unstable, though less so than direct rivalry. Common observation supported by the extensive literature on topics such as cognitive dissonance reflects these effects.

If we were only concerned with nonhuman cognition, the issue of local versus global consistency could hardly even arise. For then a creature could apply its cognitive machinery only to what is right here and right now. All that is essential is that, in any particular instance, cognition would work in a way that would ordinarily avoid self-defeating hesitation or vacillation. This does not require consistency even across two identical contexts separated by time, only within the very same context. However, once we reach human cognition, where the search for a pattern that comes to mind at the next spiral of *P*-cognition is not limited to the immediate context of action, the opportunities for getting into trouble are vastly expanded. Rather, to the extent that it somehow becomes possible for a creature to profit from abstract searches, without the costs of a commitment to global consistency, Level 2 cognition becomes a more feasible matter (less likely to get the creature into trouble), and enhancement of the capacity for abstract manipulation can be favored. So the balancing of errors [2.3] arises here in yet another form: a creature committed to global consistency could not afford Level 2 cognition; a creature that tamed every novel intuition could learn nothing from Level 2 cognition if endowed with it. Neither is a plausible Darwinian outcome.

7.7

Cognitive rivalry (direct and indirect) can be mediated by cognitive change which will (1) adjust or (2) mute or (3) inhibit conflicting intuitions. Seeing *X* and *Y* as rivals, I could escape if my intuitions were adjusted so that I came to see *X* and not-*Y*, or not-*X* and *Y*. But there are less drastic possibilities: I may see *X* and some variant of *Y* (say, *Y′*) which was no longer a cognitive rival. Or (drawing now on the earlier discussion) perhaps seeing *X* comes to inhibit seeing *Y*, so that these contrary cognitions could no longer easily be seen as rivals. Or rivalry could be muted by a piece of reasoning-why that allows a belief that although *X* and *Y* might look incompatible, they really aren't. Striking examples will turn up in the Copernican analysis later in the study.

A person may prefer milk to wine at breakfast, and the reverse at dinner. Quibbles aside, at Level 1 (cognition intrinsically anchored to immediate stimuli) these preferences can never be direct rivals, since I can never have breakfast and dinner at the same time. But they could in principle be Level 2 rivals, since then my preferences about what I like at breakfast and what I like at dinner can be made adjacent, in the sense they occur within a common span of short-term memory (or, using other language, they occur together in some slice of the "psychological present"). Prompted by conversation, objects in sight, or whatever, my preference for milk over wine at breakfast might pop up just as I am about to order wine in preference to milk at dinner. But here reasoning-why to accommodate (mute) the contrary preferences is very easily available. I have an independent intuition about my food preferences that allows for a taste for variety in foods contingent on the context. So in this case we have adjacent intuitions about what we prefer which nevertheless could not become more than momentary cognitive rivals. A piece of reasoning-why that mutes the rivalry is easily available. Rationalizing conflicting preferences about butter versus margarine would be somewhat harder and might even result in a change in preference (adjust, not mute).

Adjusting, muting, and inhibiting will often be combined. An intuition is adjusted enough to allow what otherwise might be transparently weak reasoning-why to blur the rivalry sufficiently so that what is left is no longer too salient to be inhibited. On the other hand, responses of this sort, given further indications of rivalry, will sometimes become unstable, with the person having a rather sudden sense of seeing the pieces fit together in a new way (conversion experience), though ex post it will be hard to recapture what was so new at that moment that could account for the switch.

7.8

As rivalry can be muted by reasoning-why, nonrivals can be made into indirect and even direct rivals by reasoning-why. How that will work will be influenced by the amount of information that can be handled within the span of attention and discrimination characteristic of human brains (Miller 1956). The range of that span is apparently something like the range of running speed or tolerance for little sleep. Some people can handle longer cognitive spans than others, as some can run faster or get by with less sleep than others. But the world's fastest runner or minimal sleeper is hardly likely to vary by as much as a factor of 2 from what an ordinary runner or sleeper can achieve; and it is doubtful that the range is much larger for basic mental tasks (Gordon et al. 1984). Even the brightest human easily loses the thread of a novel argument of any complexity. This must sometimes lead to logically absurd outcomes. If *X* and *Y* are seen as direct rivals when they are adjacent, the rivalry might be muted by an argument that is not even believed (which yields only C°, not C^{+}.) But it takes the mind off *Y* long enough for *X* and *Y* to be no longer adjacent.

The difficulty of grasping an extended argument is (naturally) contingent on the same factors that apply to learning in general: on the complexity of the argument, and on how close the argument is to familiar patterns of argument. Even a complicated argument, once familiar, can be seen in a few steps or a single step (chunking). Mathematics can serve as a model of the bundling-up effects. What is initially a complicated, hard-to-grasp argument leads to a theorem. The theorem is then used as a one-step bit of reasoning-why. In general (not just in the special case of mathematical argument), a person no more reviews step-by-step the supporting reasoning-why when he uses a one-step argument than he reviews a step-by-step proof before applying the Pythagorean theorem (even if he ever knew the proof).

The possibility of chunking means that the number of cognitively manageable steps in a piece of reasoning-why is not fixed. That is why the distinction between direct and indirect cognitive rivals cannot be a fixed distinction but must vary over time and across individuals. Indirect (via reasoning-why) rivals can come to be direct (one-step) rivals; or vice versa. Confronted with a direct rivalry ($I^+ \& I^-$), where I^+ is familiar and works, we expect to find some response that leaves I^+ essentially in place. So just to the extent that I^+ is deeply entrenched, we expect that ordinarily its rival will be tamed. But if I^+ is only weakly entrenched—either because it is itself novel, or because it recently had been working poorly—we may find a shift to I° from which more radical shifts now become far more likely.

7.9

Once a mathematical theorem is widely accepted, it takes on a social character, which makes it convincing to people who have never seen the proof and also to people who could not comprehend the proof if seen. We know the Pythagorean theorem is right in the same way we know that Columbus discovered America, or that people know that witches exist who live in societies where belief in witchcraft is endemic, or that for a long time everyone knew that the sun moved around the earth. By and large the easiest and even the most reliable reason for believing *X* is to be aware that everyone else believes *X*. We respond to that as we respond to subtleties of gesture, usually without explicit notice. But if we are pressed to say why we believe *X*, a one-step answer that often seems good enough is something like "everyone knows *X*."

When that is not good enough, many other one-step arguments are potentially available, often more plausible as logical reconstructions of the basis of a belief but less so as cognitively accurate reconstructions. There are "as if" arguments, "good enough" arguments, "take care of that tomorrow" arguments, "only appears contradictory" arguments, even "some things are beyond human understanding" arguments. As with the simpler cases discussed in [5.5], which turned on the meaning of particular words in context, or the appropriateness of particular inference patterns, these highly condensed arguments are not believed or disbelieved in some absolute sense. Rather, like the proverbs also mentioned in [5.4], they are seen as intuitive when, but only when, they look like they fit (a matter of pattern-recognition). Even when the initial argument was long and complicated and not easily accepted, a claim may eventually be seen as believable in terms of an "as if" (or other one-step) argument which comes to be seen as fitting the situation. Eventually *X* may be seen as believable because there is an "as if" argument that justifies it, even though we know that actually not-*X* holds.

Also, I may believe (as always, I mean: "find myself believing," not "choose to believe") a many-step argument after checking only some of the steps carefully, but with an awareness that John, who I have never known to be wrong on this sort of thing, assures me that it is OK. Even if I was competent to check every step and did so, I am hardly infallible. So the cognitive condensation of an argument from multistep to one-step could never guarantee that the multistep argument was flawless, or not so. Neither the possibility that the judgment will look sound in the long run, nor the reverse, can be assured by either a faithful step-by-step check or by a check shortened by reliance on authority or chunking.

The following important asymmetry then can be expected: once the reasoning-why becomes a one-step intuition, successfully challenging it will require a multistep argument, showing something wrong with the original argument. A crucial new fact of startling cognitive salience will sometimes shortcut this process. But in almost every practical case there is room for maneuver about the reliability of a (claimed) fact, its relevance, its interpretation.

In sum, arguments (reasoning-why) can mute rivals or make propositions not previously seen as such (even when adjacent) into rivals. But an argument that mutes rivals is cognitively easier (other things remotely equal) than an argument that makes rivals, since the first avoids disruption of entrenched networks, and the second makes disruptions. Eventually multistep arguments come to be seen as (or be replaced by) one-step arguments. What had been easily noticed direct rivals ($I^+ \& I^-$) become inhibited—when one is operating the other becomes difficult to prompt. In all these cases, we will see some such responses as insightful, others as illusory. However, though you and I will both have such responses, we will not always agree on what is insightful and what is illusory.

7.10

If a person felt he knew *X*, but his belief state [5.8] is now that of contrary knowledge ($I^- C^-$) the subjective sense is "I thought I knew *X*, but I was wrong." In the state of doubt ($I^\circ C^-$) the response is "Perhaps I was wrong." But if the state is $I^+ C^-$, the response is usually only "I need to fix that," for the actor still sees *X*, and allows only that he has not yet seen where the error lies in the counterargument, or where some adjustment will fix the paradox. But once at $I^+ C^-$, it is possible (especially when repeated efforts at "fixing" fail) to be confronted with a stubborn direct rivalry ($I^+ \& I^-$), and then to slip to $I^\circ C^-$, where the more radical changes in belief discussed a moment ago become plausible. As with seeing an ambiguous picture, a point is reached where modest tilting of the cues to favor one pattern rather than another can decisively tilt the perception. A particularly common sort of heuristic shift—from a focus on what is similar between two things to a focus on what is different—could shift the cognitive significance rather abruptly from one that supported the traditional view to one that supported its adversary (Tversky 1977).

At the end of Chapter 2 [2.8], you saw Lincoln in a blurred pattern of shaded blocks that had been teased out of a photo of Lincoln, but which also could have been teased out of a thousand other images. Subjectively, there was no sense at all of ambiguity, conjecture, or anything of that sort. You just saw Lincoln. The point of the demonstration was that it was predictable that the objectively underdetermined cues would trigger the perception of Lincoln. We are aware that the cues are feeble (we see Lincoln as out of focus). But the pattern that is cued contains so much more than is objectively on view that unless we are shown the picture in focus (as a rather featureless grid), we have no sense that we are seeing a pattern that objectively is not there.

So the Lincoln image, like the perceived difference in length in the Muller-Lyer arrow drawing [1.1], is a response to cues that are ordinarily associated with the pattern we perceive, but in this case are not. Nevertheless, it is easy to feel (in fact it is hard to overcome feeling) that Lincoln is somehow hidden behind the grid, like an actor behind a scrim. The arrows are routinely referred to as an illusion (the Muller-Lyer illusion, after its discoverers); but even in professional discussions, the Lincoln effect is routinely treated as a case of insight.

We can point to the tacit distinction that seems to account for that. It is easy to draw the arrows, but it would be hard to make a grid yielding the Lincoln image unless we actually started with an image of Lincoln. Therefore, it is understandable that we see the first as a quirk having no reliable connection with a drawing in perspective. But we see the grid as really abstracted from an image of Lincoln, which in fact it was. Nevertheless, there are contexts in which we might describe the way we see the arrows as showing insight, and contexts in which an image seen in a blurred grid would be called an illusion. An example of the first is given by the photograph in figure 7.1 (Gombrich 1982:19), where the rear lamppost actually is shorter than the first guardpost, although it does not look that way. (Satisfy yourself about that by laying a bit of paper alongside the two, marking the endpoints of the posts.) An example of the second would be an image seen under the conditions of the contrived Lincoln grid suggested in [2.9].

So the same perception or intuition might be seen as an illusion by one viewer and as an insight by another; and it might be seen by a given viewer as an insight in one context and an illusion in another. It takes special care to contrive contexts in which that is the case for close equivalents of the arrow diagram and the Lincoln grid. But even for those examples—

Figure 7.1. The rear lamppost (with its light) is shorter than the front fence post. From Ralph Evans, *An Introduction to Color* (New York: Wiley, 1948). With permission.

which are so well-known just because they are such striking examples of their type—it can be done, as illustrated in the previous paragraph.

Of course even if the arrows had been abstracted from a perspective drawing, one line on the page would still not be longer than the other. But, of course, there is also no actual face on the page when a person is seeing Lincoln in the blurred grid. Further, there is no clear difference between these cases and ordinary vision, where we are constantly seeing things not objectively accessible. There is a blind spot in each eye where the optic nerve enters the retina. But even if you cover one eye (so that it can't fill in the blank in the other field of view) you do not see the blind spot. What is then left of the insight/illusion contrast is a distinction between what philosophers since Bertrand Russell have called "propositional attitudes." The insight/illusion distinction takes on the character of an across-persons version of the belief matrix [5.8]. The entries in figure 7.2 below show one person's attitude toward another person's intuition. For example, if I (think I) know that your intuition is wrong, I see it as illusory. Since a creature capable of Level 2 cognition can see its own perceptions as an object of cognition, the across-persons attitudes can be self-applied. So we speak of ourselves as having an illusion, meaning that looking at the

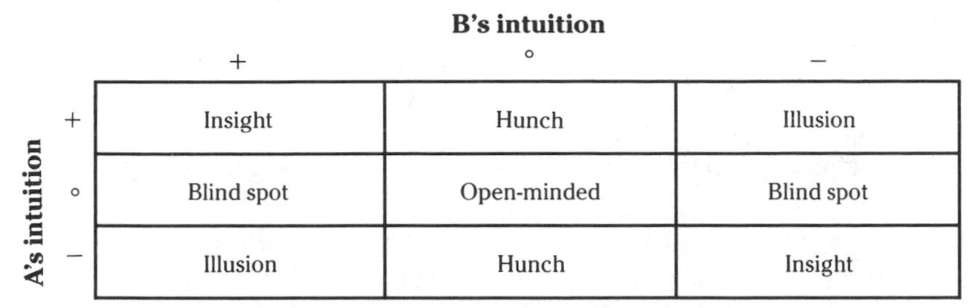

A's intuition \ B's intuition	+	o	–
+	Insight	Hunch	Illusion
o	Blind spot	Open-minded	Blind spot
–	Illusion	Hunch	Insight

Insight vs. illusion: B's response to A's intuition

Figure 7.2. A full "across-persons" version of the belief matrix would require an unreasonably precise account of 81 cells (9 × 9). The simplified version here considers only *B*'s *I*-response to (his perception of) *A*'s *I*-responses.

arrows (for example) we see as true something—the outward arrow as longer—which more information persuades us is false.

I have suggested labels for some of the interior states between the illusion and insight corners. But as in the original belief matrix in [5.8], the argument is not contingent on these labels of convenience.

7.11

Creatures that must act in the world to survive cannot be continually open to taking a closer look at their cognitions. Challenged insight/illusions (intuitions that circumstances or adversaries prompt us to check) must sooner or later be either dismissed (seen, when noticed at all, as illusions) or confirmed (seen as insights, and eventually just seen, as we just see most things in that unproblematical way). The possibility of a "look closer" response which will undermine even what had been a very clear intuition never totally disappears. But we must expect that an individual could have only a relatively few active conjectures (insight/illusions that easily prompt checking).

In particular, breaking a habit, including a habit of mind, ordinarily requires a Level 2 process, hence it is something that demands conscious effort [6.9]. Acquiring a habit, including a habit of mind, does not require conscious intent. Children, for example, do not need to decide or try to learn how to speak. They just learn, without noticing. But breaking a habit ordinarily requires intent—if not on the part of the learner, then there must be an intending trainer, as when a person tries to break a dog's habit.

Drawing on the earlier argument: learning that requires challenging entrenched networks will be resisted, never absolutely, but on a scale with how deeply entrenched the network has become by intensity of unproblematical use. We get the various propositional attitudes (belief states) I have characterized with the help of the belief matrix [5.8], and the various components [7.7] of response to direct and indirect rivalry (inhibiting, muting, adjusting). In the balance of the study I will try to show that all this can be put to work.

On this account of learning the process through which a person learns something new consists, at every step, of doing only what he already knows how to do. At every step the person is being prompted to a pattern that is already in his repertoire. For it is tautological that he cannot recognize a pattern that is not in his repertoire. Where, exactly, does learning then occur? The *P*-cognitive answer is that it occurs by way of changes in the relationship between cues and patterns that are stimulated by the very recognition of a pattern. Once I have seen a pattern in a given context, it becomes easier to see that pattern again in that context, and if the effect of seeing that pattern encourages me to look again and see it again, I will gradually become very good at seeing the pattern in the new context. Where, on the first occcasion, it may have been only very odd conditions that prompted me to see that pattern (just the right sort of lighting, just the right sort of context), I gradually learn to see the pattern under very much less-favorable conditions. At each prompting, I am still only doing what I already know how to do, but the neurological reinforcement by the act of recognition, and the augmentation of some cues prominent at this recognition, the decay of others absent, will gradually reshape the relations of cues and a particular pattern. That pattern itself will also evolve, since various subpatterns will become associated with the pattern in this new context, and some subpatterns previously associated with it will be lost in this context. So we get the gradual appearance of what amounts to a new pattern with its specialized set of cues, though (repeating yet again that fundamental point) at any step the person is only doing what he already knows how to do. As in the earlier discussion, I do not mean this to deny any role for purely blind variation in the synaptic structure [6.5]. But, as earlier [6.2], I want to stress the importance of the Lamarckian (responding to the environment) as opposed to the Darwinian (blind variation) aspect of this process of cognitive evolution. This line of argument will play only a subdued role in Chapter 8, where we will be concerned with merely static contexts (taking the cognitive repertoire as fixed). But it is fundamental for the balance of the study, where we will be concerned with the discovery and contagion of new knowledge.

Eight

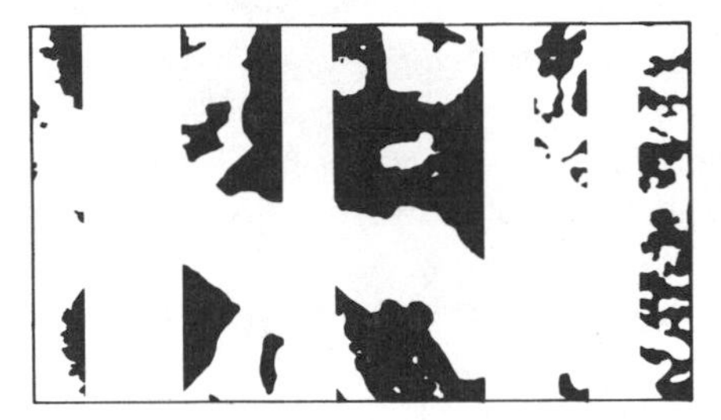

Cognitive Statics: Three Experiments

This chapter deals with what I will call cognitive statics, in which the repertoire of cues and patterns governing responses can be taken as fixed (no learning). At a superficial level, the argument might be read as supporting critics of the work on cognitive illusions [1.8]. I reach different accounts of various well-known experiments than Wason or Kahneman & Tversky have given. But the account here essentially follows their basic insights about the fundamental role of factors beyond the reach of standard logic in accounting for human judgment. Like critics such as Henle (1962) and Cohen et al. (1981), I will argue—and in fact try to provide striking, easily replicable demonstrations—that what has usually been taken to be incorrect reasoning leading to cognitive illusions in fact is better characterized as normatively plausible responses to a question different from what the experimenters intended. But in contrast to Henle and Cohen, the detailed conclusions I draw strengthen rather than invalidate the basic claim of the experimenters. For although subjects can be—in fact, I try to show, ordinarily are—giving reasonable responses to a different question, the different question can be wildly irrelevant to anything that plausibly could be construed as the meaning of the question asked. The locus of the illusion is shifted, but the force of the illusion is confirmed, not invalidated or explained away.

It is convenient to deal with experiments that have been thoroughly studied and are simple enough so that a reader can replicate them by taking a few minutes of class time to try them on students. I also wanted to avoid criticism that I had searched around for experiments that happened to suit the story I want to tell. Consequently, as a few readers who happened to attend seminars I gave early in this project (many months before the general *P*-cognitive argument had been worked out) may recognize, I settled on the three experiments that received special attention in a symposium on rationality in *Behavioral and Brain Sciences* (Cohen et al. 1981): Wason's selection test [1.3], plus two of Kahneman & Tversky's problems ("Linda" and "taxi").

The results, I think, are doubly striking: first, in being different from what has usually been said, and in a way that works across all three experiments; and second, in yielding an analysis that can be applied—which *is* applied in the balance of the study—to a rich and empirically important range of questions far beyond the scale of the toy problems immediately at issue. But I have come to realize there is a difficulty in using these much-studied experiments. For the very people who would be in the best position to judge, replicate, criticize, extend the results are likely to be either personally very committed to some view of the experiments—necessarily not the view I will be arguing for—or, if not, they are likely to be utterly bored at the prospect of working through yet another discussion of these very much worked-over experiments. Hence, this strained introduction to the material. The analysis I will give might prove to be in error. But there is no chance that, if even approximately correct, the analysis could fail to be important. We get a qualitatively different kind of account from what has been offered before.

All three experiments share the following basic character: for each, some elementary inference or intuition seems beyond the competence of a very large fraction of intelligent subjects: for example, undergraduates at good universities, or (only modestly less marked) faculty members at those universities. For Wason's selection problem the unavailable intuition is an elementary logical inference: "If *p* then *q*" implies "If not-*q*, then not-*p*" (modus tollens). For Kahneman & Tversky's "Linda" problem the unavailable intuition is that the probability that a pair of events, *A* and *B*, both occur cannot be greater than the probability of *A* occurring with or without *B* (conjunction). For the "taxi" problem the unavailable intuition is that the immediate evidence bearing on some particular case must be weighed against the background rates at which various cases of that sort arise (Bayesian inference).[1]

If it were true that these experiments showed what they have so often been claimed to show, we would have a very deep puzzle. For the least controversial notion in psychology is that some version of the law of effect is essential to an account of performance (Dennett 1983). On a plausible theory of human judgment, normal human beings should exhibit good intuitions about pervasive features of their experience. In particular, on the *P*-cognitive argument, the claim that patterns of inference are shaped by experience in the world is so fundamental that either people cannot lack intuitions (like modus tollens) that reflect commonly encountered features of experience, or the *P*-cognitive argument is wrong. Yet each of the cognitive illusions we are considering has commonly been interpreted in a way that violates that claim.

8.2

This gives us a straightforward agenda. On the *P*-cognitive argument, the subjects giving the anomalous judgments must actually have ready access to the intuitions they seem to be incapable of using. There are then two conditions which might make sense of the anomalous responses. (1) It should be the case that there are commonly experienced contexts in which each of these logically impeccable inferences is pragmatically inappropriate. (2) Additionally, there should be something about the questions such that they are seen "as if" in a context where the contested intuitions are in fact pragmatically inappropriate. On the argument here, and in agreement with the experimenters, the typical answers are perverse. There is a real puzzle about why these anomalies occur, not something trivial, such as a reasonable alternative reading which makes the usual responses a good response. The *P*-cognitive argument implies that for all such *stubborn* cognitive illusions, we can somehow expect that subjects are prompted by experience in the world to see the problem in a way different from what the language can be reasonably understood to warrant. There must somehow be ambiguity in the problem beyond any merely semantic ambiguity.

The point here is that ambiguity could be immediately in the language; but it could also be something more remote. For convenience, simplifying this spectrum into a dichotomy: the ambiguity could be due to semantic or to what I will call "scenario" effects. That the "selection," "taxi," and "Linda" problems have generated extensive discussion implies that it is an interaction of scenario and semantic (broad and narrow) ambiguities that accounts for the force of the illusion, since if there were some single factor explanation (especially a single factor that was narrowly in the particular language used) the puzzle of what is responsible for the anomalous responses would not have proven so difficult.

In general: since our experience in the world is one of dealing with things in context, not as detached set-piece puzzles, a naked puzzle can cause us trouble that goes far beyond what can be accurately accounted for in terms of its intrinsic difficulty. The more obvious aspect of this concerns difficulties that arise because we lack the context that ordinarily makes resolution of semantic ambiguities smoothly automatic. A less obvious aspect is that we will see the problem in terms of some wider context (in terms of some scenario or pattern of experience in the world), even though in this case there is no such wider pattern, just a naked puzzle. As we can tell the fifth note of some familiar melody only by going through the melody, we apply routines we know very well—but which fit some contexts and are inappropriate in others—in terms of some more complete sort of context in which they fit. In the impoverished environment of a set-piece puzzle, therefore, we may impute a wider context to the problem that is not only not there at all but perhaps is flatly inappropriate. It is that sort of possibility that we will be considering under the label of "scenario ambiguity."

8.3

Here are some particularly simple illustrations as preliminaries. The last word in each sentence of figure 8.1 is physically identical: in terms of very local cues (just applying to that word), the correct reading is ambiguous: the word could be "clay" or it could be "day." But in terms of broader cues (the whole sentence), the reading is unambiguous. It is an essential point that you do not see the word as ambiguous and then consciously decide that given the wider context it ought to be read "clay" in the upper sentence and "day" in the lower. You just see it as "clay" in one sentence and "day" in the other. If not for the contrived juxtaposition, it would not occur to you that there is any ambiguity.

An example a bit closer to the material we will be treating is this: suppose that I say (using another of Wason's inventions), "No head injury is too trivial to be neglected." An unalerted reader will understand the sentence to be saying just the opposite of what it does say. In fact, if you now go back and read the sentence carefully you will probably find that, although alerted, you misread it yourself. (Compare: "No head injury is too trivial to be treated seriously.") The misreading of the "neglect" sentence is almost universal. Yet no one will claim that this shows that ordinary speakers of English do not understand the word "neglect." Rather, the puzzle is why we somehow do not see the normal meaning of the word in this sentence, although we understand the word correctly and effortlessly in almost any other context.[2]

The resolution of that "neglect" puzzle must turn on how a person's experience in the world generates expectations about what kind of meaning makes sense of a sentence beginning "No head injury is too trivial to. . . ." There is no narrowly semantic ambiguity here at all. The language

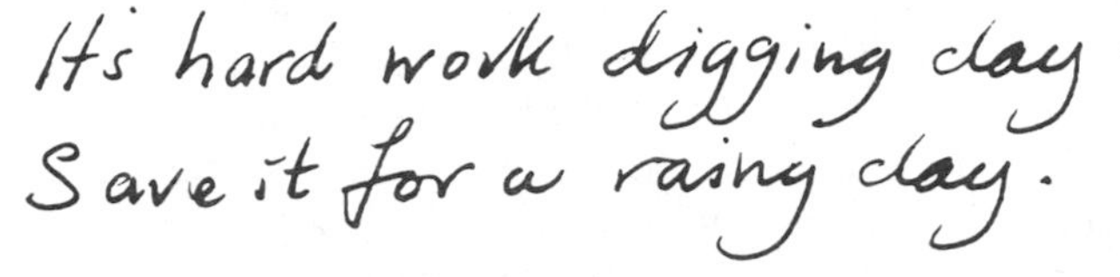

It's hard work digging clay.
Save it for a rainy day.

Figure 8.1. A simple illustration of visual context-dependence. From John P. Frisby, *Seeing: Illusion, Brain and Mind* (London: Roxby and Lindsey Press, 1979). With permission.

unambiguously says one thing, but we read it to say the opposite. So we are—unsurprisingly on the argument to this point—guided not only by immediate cues but also by a broader sense of the sort of context that is involved. If some context is very strongly primed [3.3] by feature(s) of a situation, we will tend to respond as if that wider context held (in computer language, that will become the default scenario), unless something else is very well-marked by more local cues. So unless an alternative is well-marked, we learn to override what will usually turn out to be local errors, such as saying "negligible" when you mean the opposite. How hard it will be to cue us away from the default scenario will (of course) depend on how deeply the habitual response is entrenched. But for simple misreadings—in contrast to physical habits and the richer conceptual habits of mind we will take up in later chapters—that should not be terribly difficult. If before giving the "negligible head injury" example I had led you to expect a line from a horror movie about a mad doctor, you would not have misread the sentence. But absent such explicit cuing, nearly everyone not only misreads the sentence, but then requires some work to become satisfied that the sentence really says what it does say.

Our procedure for analyzing Wason's selection task and then the Kahneman & Tversky illusions will accordingly go this way. For each of the anomalies, we will first consider possible ambiguities of language, then ambiguity of scenario. The interaction of the two will suggest what apparently is governing the anomaly and also what might be changed in the presentation of the question. We require a variant that corrects the anomaly in some way that leaves the logic of the question unchanged. So we hope to get (a) an account of what is causing the anomaly which is consistent with supposing that people acquire good tuning of intuitions to commonplace features of experience. We then want to devise (b) a test of that analysis by way of a variant of the problem which remedies the difficulty by attending to (a).

Wason's selection problem is, by a wide margin, the most extensively studied of all such problems. For surveys see Johnson-Laird & Wason (1977), Evans (1982), and the collection of papers by Wason and others in Evans (1983). Its basic form is starker than the playing card version in [1.3], using cards with only an *A* or *D* on one side and a 2 or 3 on the other side. The rule to be tested then is: "If the letter is an *A*, then the number must be a 2." But using more familiar materials (playing cards) as in [1.3] does not, of itself, improve performance.

Finding the answer to Wason's problem requires that a subject go from "If *p* then *q*" (here, "If the letter is *A*, the number is 2") to "So if not-*q*, then not-*p*" (the card that shows a 3 cannot have an *A* on the other side without breaking the rule). The very low rate of correct responses has usually been seen as showing that modus tollens—elementary though it is—is too difficult an inference for subjects to manage [1.3].

The only facilitating variants proven to be both reliable (replicable across researchers) and substantial (more than 50% correct responses) are those that use *rules* (not just materials, as in the playing-card version) which will look familiar to the subjects from practical experience. For example, if the problem is framed around the rule, "If someone is drinking beer, he must be over 18 years old," about 75% of subjects answer it correctly (Griggs 1983). Another example turns on a store rule that cashing checks above a certain amount requires the manager's approval on the back (D'Andrade 1979).

In either of these cases, the subject is unlikely to have experience carrying out the rule but is almost certain to have experience as the person under inspection (having ID checked or having a clerk examine a check to find whether it has the required approval). The subject will ordinarily have firsthand experience involving physical acts, though from the inspected, rather than the inspecting, side. We all have real-world experience about what needs to be done in such contexts. Wason (1983) summarizes his own view by quoting two remarks of Rumelhart (1981), who says (1) "Understanding the problem and solving it are nearly the same thing"; and (2) we see "exactly the result we would expect if our knowledge of reasoning is embedded in task-specific procedures rather than in general rules of inference." In other words, (1) once a person grasps the framing of the problem, he has a subjectively immediate intuition of an appropriate response; (2) this occurs in cases where the subject has real-world experience with the problem situation. A physical example would be: once you are put in mind of your kitchen, you know where the light switch is with no conscious sense of further processing.

Of these points, the *P*-cognitive account is close to the Wason/ Rumelhart view only with respect to statement 1. For unless the *P*-cognitive argument is wrong, then Wason/Rumelhart's 2 must be wrong. As you have a sense of where light switches are usually located, so that you can ordinarily find them with little effort even if a dark room is *not* your own kitchen but one you have never been in before, you have a sense of patterns of language that makes sense, which is not really "task specific" (not contingent on familiarity with the concrete case), though of course if you have that, responding well is that much easier.

With respect to statement 1, and using the logic$_t$/logic$_u$ distinction [5.3], a piece of language is seen as meaning$_u$ something, not as implying$_t$ something, given its literal meaning plus the context. When we come across the phrase "a likely story" in an ironic context, we do not read it in a literal sense and then make an inference that "likely" is intended as ironic in this context. We simply sometimes read "likely" as if it meant "unlikely." Similarly for patterns of inference. For the selection problem, a person typically looks at the problem, mulls it, and eventually sees what looks like his best response. The subject rarely sees that immediately; but when the intuition comes it is usually subjectively processless. The inference is part of the pattern; or, as Rumelhart and Wason suggest, "Understanding the problem and solving it are nearly the same thing."

In addition, the subject may also come to see the response as an explicit deduction if he takes a closer look, checking his answer. But the answer does not ordinarily originate in that way. Rather, the situation is similar to the case of a camouflaged image, where the person mulls the picture and finally sees the dog or face or whatever makes sense of the image. After seeing the image, the person may check the response, examining the picture in detail in a way that satisfies him that he was right to see what he did see.

But on the *P*-cognitive argument, there should be nothing *necessarily* task specific about how all that works. Rather, we expect that tuning of language to experience must extend to reasoning-why patterns of language (patterns of language that give inferences that experience usually confirms). As I have stressed before [5.8], if that were not so, then the very evident human propensity to take reasoning-why seriously would make no Darwinian sense. Human reasoning often will be grossly faulty by the standards of logic$_t$. Nevertheless, we can expect to find fluent command over a repertoire of patterns of inference governed by pragmatic use-conditions (logic$_u$). But the pragmatics are not necessarily what I have called Gricean [5.3]: they are not necessarily things that are ordinarily accessible to conscious notice, or things that would be acceptable, not odd or embarrassing, if noticed.

So we expect to find that human facility in handling patterns of inference is like the facility that is the essential feature of human language in general: we can use words and comprehend other people's use of words in new contexts, not merely in contexts like those we already know in any sense that ordinarily would be understood as "task specific." But to the extent the context is unfamiliar or impoverished or ambiguous, our understanding will be less reliable, since we can easily see the piece of language as looking like something in fact inappropriate here, or simply as looking incomprehensible (looking like nothing we recognize). We will be more vulnerable to error when we jump, and with less opportunity to notice contrary cues which prompt us to check.

8.4

The crucial semantic ambiguity in the selection problem turns on the multiple reading$_u$ possible for "if . . . then" statements. Running over the main possibilities:

(a) "If *p* then *q*; but *p*; therefore *q*."

Form (a) implies$_t$ and commonly also implies$_u$ its conclusion. But some variants of the selection problem to be noted later leave subjects almost as blind to (a) as the more familiar versions leave them blind to modus tollens. Responding to (a), we sometimes find that, after all, we don't believe *p*; or have doubts about believing "If *p*, then *q*" (Harman 1973). Or, even more often, we just don't see *q*. If challenged, so that we have to notice the omission, then we may produce one of the first two possibilities; or we may find that we believe *q* after all but somehow hadn't noticed it or bothered about it before.

A more commonly ambiguous form is the inverse:

(b) "If *p* then *q*; and not-*p*; therefore not-*q*."

Form (b) is false$_t$. But in ordinary language, "If *p* then *q*" very often implies$_u$ "and if not-*p*, then not-*q*." So (b) is always invalid$_t$ but very often valid$_u$. If I say, "If it rains, I'll go to the movie," any user of English will understand that as implying$_u$ that if it doesn't rain I won't go to the movie.

Similarly, with respect to the converse:

(c) "If *p* then *q*; and *q*; therefore *p*."

Here the inference is again false$_t$ but often (depending on context) valid$_u$. If I say to Jones, "If you go outside, you'll get soaked," and Jones gets soaked, you are likely to infer$_u$ that he went outside.

Finally, we have the contrapositive:

(d) "*p* implies *q*; and not-*q*; therefore not-*p*."

This inference here (modus tollens) is valid$_t$ but often invalid$_u$. In logic$_t$, "If it's a swan, it's white," is simply another way of saying, "If it's not white, it's not a swan." Reduce the two statements to their most elementary form (using only the connectives "&" and "not" to which all the logical constants in formal propositions can be reduced), and they yield indistinguishable patterns. Given the reduced form, it would be impossible to deduce$_t$ whether the unreduced form was "If it's a swan . . ." or "If it's not white . . ."[3] Suppose, then, you see a green worm. It's not white, therefore it can't be a swan. And behold, it isn't a swan. Therefore, you ought to see the worm as evidence for the claim that all swans are white. But you will not easily see the green worm from that perspective.

Now (with suitable provisos about unbiased sampling), seeing the green worm does support the swans-are-white proposition; but the extent of that support is like the extent to which squeezing a lemon into the Pacific raises the ocean level. Experience would hardly tune a person to seeing an inference that is so pragmatically irrelevant. But, as with the "negligible head injury," it is easy to cue a scenario in which modus tollens will be seen easily. For example, imagine you are explaining the "swan/white" relation to a small child looking at a pond containing some swans and some pelicans.

There are many other contexts in which form (d) is invalid$_u$. Suppose I say (again): "If it rains, I'll go to the movie"; and you go to the movie and notice I'm not there. It will not occur to you that (modus tollens) "therefore it's not raining." Similarly, we expect there are other contexts, not very different, in which (d) is effortlessly available. If I say, "If I finish my work, I'll go to the movie," then indeed you are likely to infer (modus tollens) that, since I'm not there, I probably didn't finish my work.

The slogan for an antidandruff shampoo (Sassoon) says: "If you use it, no one will know; they'll only know if you don't use it." The firm does not expect that buyers would have to consult a logician to realize that the slogan reflects a valid logical relation (modus tollens). The presumption is that the connection will be understood immediately (like the point of a good joke). On the other hand, the slogan does not sound redundant, as it would if the second clause were nothing but a tautology. But it isn't, since we can easily think of cases where the first clause holds but not the second. For example, suppose the reference wasn't to shampoo but to cocaine.

A final example, however, though it may sound similar, really has a completely different character: suppose John, a moderate drinker, is told by his wife, "When we go out with Sam, you drink too much." To judge whether his wife's claim is valid, he will be prompted to think of times he went out with Sam, and times when he drank too much. In the selection context, these are the analogues of checking the *p* and *q* cards (the usual response). As will be seen, this last ties closely to the main point of the analysis I will give of what is going on in the Wason problem. It turns not on whether modus tollens is pragmatically appropriate, given the language (semantic ambiguity), but on whether the usual Wason solution is appropriate in any sense, given the scenario.

A little thought will ordinarily suggest a pragmatic basis for your $reading_u$ of an "if . . . then" expression. For example, the "movie/rain" example sounds like one in which you would know whether or not it was raining (if it was, you presumably got rained on coming to the movie); so the $inference_u$ is likely to be, "So he must have changed his mind," if it is raining, and no inference if it isn't.

The situation is parallel to the way that we can recognize faces of people we have not seen for many years, or that we can recognize letters of the alphabet in fonts we have never seen. For all these examples, we are often able, ex post, to point to cues that help us recognize what pattern fits; but for none can we give a general account ex ante of how we do that.

However, if all that was at stake in Wason's selection task was some such context-dependent meaning of "if . . . then," it hardly would have become the intensely studied puzzle it has become. For example, using here a point that will play a role in what follows, if all that was happening was that the "if . . . then" language was being understood to $imply_u$ form (b) as well as (a), then the puzzle would hardly be of any great interest. But we can tell that something more is involved, since if that were what was happening then the correct response would be to turn all the cards. On the other hand, the various examples were intended to make it obvious that people cannot be taken to be routinely incapable of dealing with modus tollens inferences, or that they can avoid difficulty only in contexts where they have task-specific experience.

It then follows that there must be some broader (scenario, not merely semantic) ambiguity which contributes in an essential way to the difficulty. A semantic ambiguity cannot be very deeply hidden for it must turn on the language we have in front of us. But a scenario ambiguity can easily be hidden, since it can turn on a default scenario for which what is in front of us gives no clue, or even gives flatly contrary cues but not strong enough to offset the default scenario—as in the visual analogue of a scenario effect illustrated by the "sring" illusion (fig. 1.2).

Semantic difficulties, as illustrated here by the multiple $meanings_u$ of if/then language, become a more important factor when *interacting* with scenario effects than they could plausibly be by themselves. They add to the complexity of the puzzle and so increase the risk of error by simply giving the solver more loose ends to keep track of. But they also would have an important effect when a subject is prompted to check his intuition.

It is an important point that there is some possible reading of the language that might be *misunderstood* as confirming many different intuitions, so that it will more easily be the case that a subject, primed to see a certain response as making sense, will then find that a response looks right even though logically it is certainly wrong.

8.5

To see the nature of the scenario ambiguity in the selection problem, suppose you were told that each of a collection of cards has "swan" or "raven" on one side, and "white" or "black" on the other. The rule is: *"If it says 'swan' on one side of a card, it must say 'white' on the other side."* Of the four categories (swan, raven, black, white), you are asked to choose any that must be checked to find whether there has been a violation of the rule. And suppose, as in fact is reasonable considering the frequency of that pragmatic meaning, that you read the rule as implying$_u$ form (b) as well as form (a): ravens must be black as well as swans white.

Then, to be sure the rule had been obeyed, you could check the categories "swan" and "raven"; *or* "swan" and "white"; *or* "raven" and "black"; *or* "white" and "black." Any would be certain to uncover every possible violation (every case of "black, swan" or of "white, raven"), as is required given the reversible reading of the rule.

But these choices do *not* include the correct response to the Wason problem. With a rule in the form, "if *p* then *q*," the correct response to Wason's problem [8.3] would be *p* and not-*q*: here, "swan" and "black." But if you carefully review the wording of the problem in this section, you will find that that response would give you a redundant chance to observe a black swan, with no chance to observe a white raven. Further, the problem still does not yield the Wason solution if you read the rule more narrowly (not as reversible). For then the correct response would be: either *p* or not-*q*, but not both. Either would reveal any case of "black, swan."

You will have to go back and reread the problem statement carefully to see why. For it is a pretty safe assumption that you did not notice that the swan problem is not just another version of Wason but a different problem. The difference illustrates what I am calling a scenario effect. Similar scenario effects (interacting, as here, with some semantic ambiguity) will account also for what happens in the "taxi" and "Linda" problems later in this chapter.

The difference between the "swans" problem and the original Wason problem is between (I) a situation in which the inquirer is choosing *how* to search (for example, examine all cards with "swan" on one side, all cards with "white" on one side) versus (II) the situation in which the choice of how to search is already foreclosed and the choice remaining is limited to a selection within the constrained set of possibilities. Call Scenario I "open" and Scenario II "closed." In Scenario I the choice about how the possibilities are to be explored is open, as in the "John and Sam" example earlier [8.4]. But in Scenario II it has been foreclosed: you have to deal with them as they lie on the table, as in the Wason problem.

So however the rule in this section is interpreted, a correct response to this "swan" problem will parallel an *incorrect* response to the standard Wason problem; and the *correct* response to Wason's problem will be an incorrect response to the problem here. What is going on?

In a "swans" version of the actual Wason problem, the cards would be laid out on a table, and the only way you could check a card is by specifying it by the side you have been allowed to see. But if you attend to the wording, you will see that, as I stated the problem, you are allowed to choose categories, not particular cards. If you ask for "swan," you get to examine every card with "swan" on it, not (as in the Wason context) only those cards with "swan" on the side you have been allowed to see. So it is the *open* scenario that is appropriate to the problem as stated in this section, making the correct response to Wason's (*closed*) problem of [8.3] definitely wrong as the response to the swans form of the problem here.

Now suppose it were the case that the *open* scenario was far more typical of real-life experience than the *closed* scenario, or that ordinarily the person encounters the closed scenario only in a context that arises after he has moved past the open scenario. But in this set-piece puzzle, the open phase is skipped. In that circumstance, or perhaps even generally when the context is poorly marked or blurred or otherwise difficult, a person may be more easily prompted to see things in terms of the open than the closed scenario than the immediate cues warrant. That prospect yields an account of what is going on in Wason's problem that makes sense of the results without being forced into exaggerated claims about the intrinsic difficulty of modus tollens inferences. As a species that emerged from a very long period of Darwinian selection for good hunter/gatherer propensities, we have a lot of background in searching. Ordinarily, the search is "open" (we can choose how we will proceed) and only later becomes "closed" (we have to choose which of the possibilities the open search has uncovered we will concentrate upon).

8.6

Reviewing: we have two (main)[4] sources of difficulty. On the semantic side, it is commonplace to encounter pragmatic contexts in which modus tollens—form (d)—is not implied$_u$, and/or in which one or both of forms (b) and (c) are implied$_u$. To some extent, then, we can expect that the difficulty of the problem grows out of the ambiguity of language, especially language used in an impoverished context. But that difficulty is not as serious as the scenario risk that, in the impoverished context of a set-piece puzzle, a person will see the context as "open" (in the sense just spelled out), even though that is not at all warranted by semantic ambiguity.

But if this analysis is correct, then a quite vast literature on the Wason problem must be almost uniformly wrong. For nearly everyone who has participated in the discussion has accepted as uncontroversial the very point I want to reject: namely, that the heart of the problem is the difficulty of drawing modus tollens inferences. But exactly the point of what I have been saying is that it should be possible to correct the illusion while leaving the logical problem untouched (that is, while giving no hint of task-specific help in knowing what to do). So a reader has special reason to be skeptical of the account I have given. On the other hand, if that account were correct, then we ought to be able to produce several striking variants of the Wason problem:

1. As I argued at the outset of this chapter, but contrary to the usual results (for an exception, see Hoch & Tschirgi 1985:453–62), we ought to be able to give a version of the Wason problem that yields a high fraction of correct responses even though the rule remains pragmatically as well as logically indistinguishable from the arbitrary rule of the original problem. On this account, a particularly simple way to prompt a correct reading of both the semantics (read the rule as one way, not reversible) and the scenario (closed, not open) would be to frame the problem in terms of some task-specific familiar situation, as with the ID and check-cashing examples [8.3]. But that should not be the only way.

In addition, a number of secondary results should be available:

2. The usual *p* and *q* response does not, on the account here, seem to have any deeply rooted advantage over other pairs that also yield correct responses to a broad reading of the rule in the open scenario. Rather, on the analysis I have given, that pair looks like it might be picked so often merely by the sort of marginal salience effects that a stage conjurer manipulates when "forcing" a card. If so, then it ought to be possible to rephrase the problem to make some other pair of cards as salient or even more salient.

3. Alternatively, by rephrasing the rule to narrow the merely semantic ambiguity, it should be possible to make the *p* and *q* response rare instead of common, since on the account here such responses look like they ordinarily depend on the joint effect of (mistakenly) responding as if the scenario were open, compounded by seeing the language as implying$_u$ its converse.

4. Finally, we could hope to give explanations consistent with the analysis here of two apparently similar problems Wason found usually

easy, not hard, to solve: (4a) If the rule is altered to the form: "if *p*, then not *q*"; and (4b) if the basic problem is left unchanged, but the "easy" cards are eliminated from the array of choices. This last is especially remarkable. For if we drop the "*A*" and "*D*" choices that generate few errors on the basic problem, it turns out that subjects can then usually handle the hard possibilities (the "2" and "3") correctly (Wason 1983:61).

Proceeding backward through this list:

4b. That eliminating the easy decisions makes the hard decisions easy has a straightforward explanation in terms of scenario effects. For now the "3" is the correct response for *either* the open or the closed scenario. So there can no longer be any scenario difficulty. And with no scenario ambiguity, the semantic ambiguity also is easier to handle. So the cognitively hardest part of the problem disappears when the logically easiest part of the problem is suppressed.

4a. If the rule is "*p* implies not-*q*" rather than "*p* implies *q*," then the semantic ambiguity due to reading the rule as including its converse is minimal. On the usual account, this negative framing improves responses because subjects cannot handle modus tollens, hence fall back to childlike echoing of whatever cards are mentioned in the rule. For both "if *p* then *q*" and "if *p* then not-*q*," this yields the response "*p* and *q*," which is the usual wrong response to the basic problem but the correct response to the amended problem. So we get (on that usual account) a high proportion of correct responses, but in a way that (as with the familiar rule version) can be accounted for without crediting subjects with any competence to correctly handle a modus tollens inference. But on a *P*-cognitive account human competence to deal with simple logical puzzles does not fall *that* far short of normative standards. Rather, what accounts for the improvement is something like this:

With the "if *p* then not-*q*" framing, seeing the rule as reversible is no longer likely. If I say "if *p* then not-*q*," that ordinarily does not mean "if *p* then not-*q*, and if not-*p* then *q*." In fact, a pattern of language like that sounds so odd that you probably have to reread it to make sense of it. But "if *p* then *q*; and if not-*p* then not-*q*" rolls out as effortlessly as a familiar tune. On the other hand, salience effects are certainly real, as anyone familiar with the way magicians force cards will be aware. A subject's attention is drawn to the *q* card by a rule that says a *q* should not appear. So the "not-*q*" version both reduces complexity for subjects (by avoiding the semantic ambiguity) and makes salient just the card that requires close attention to get the problem right. By itself, this does not rebut the "childlike" response analysis. But if on other grounds—the analysis of (4b) and of the cases to follow—it looks like the scenario difficulty indeed is the heart of the problem, then we have no problem in accounting for the effectiveness of the not-*q* variant in the same terms.

3. If we rephrase the rule to make it less easily read to imply$_u$ its converse, the result in fact is to sharply increase the frequency of 1-card responses. The account here would make us expect that, and that is what occurs if instead of, "If the letter is *A*, the number must be 2," the rule is given as (for example): "Suppose the letter is *A*. Then the number must be 2." But now—lacking the salience given the "3" by the "not-*q*" version (4a), that alone is usually not enough to jar subjects out of responding to the problem as if the scenario were open. But from (4b), we can expect that not very much more help will be needed to prompt correct responses once the semantic ambiguity is preempted. I will mention how in a moment.

2. Suppose that we do not attend to the semantic ambiguity (3) but instead constrain the choice to selecting exactly two cards. Altering the instructions to "Circle two cards to turn over in order to check whether the rule has been violated" channels the responses even more emphatically into making the *p* and *q* choice. But now change that instruction slightly to read: "Figure out which two cards could violate the rule, and circle them."[5] This yields something akin to a cognitive Necker cube. Many subjects now gave the response not-*p* and not-*q*. In fact, more subjects give that response than the usual *p* and *q*. Apparently, this shift in wording (making "violations" more salient) prompts attention to cards *not* mentioned in the rule. So a logically irrelevant shift in phrasing turns out to be sufficient to shift the salient choices (with open scenario, reversible semantic interpretation) from the usual "*p* and *q*" response to the opposite pair. Note that given the (illogical) way subjects are seeing the problem, both those responses (plus two others mentioned earlier) are logically appropriate. So it is not surprising that superficial shifts in salience would shift the responses.

1. Finally, note 6 gives two variants that people seem to solve as easily as the well-known remedial versions (ID, check-cashing) though the rule remains completely arbitrary: (1a) Combine 3 above with the playing-card version of the problem of [1.3]. By itself, using playing cards with an arbitrary rule yields no noticeable improvement. And easing the semantic ambiguity but keeping the "if *A* then 2" rule shifts the nature of the response but does not produce a high rate of correct responses. However, the two together give good results. This presumably is because people are familiar with handling playing cards in the closed scenario, so that it is easier to "see" (tacitly) the scenario as closed with these familiar materials than with the special cards with numbers and letters. With the semantic ambiguity resolved (as discussed in 4a and 3), subjects now usually answer correctly, though the rule remains as completely arbitrary as before.[6]

An even more striking indication that the heart of the problem is misperception of the scenario—but one which I have not had the opportunity to test as fully—is that we can keep the problem exactly in its usual "If the letter is *A*, the number must be 2" form, and yet get good results by changing only the way in which subjects are asked to reveal their response. Instead of showing letters representing the cards and asking that subjects circle their responses, we give them a little packet of cards (some letter side up, others number side up) and ask that they sort them into two piles: those that need to be turned and those that don't. At one point in his survey Wason (1983:48) remarks that "my own opinion is that, without altering the material, there is nothing which could be done to induce the subjects to approach the task in a manner that befits its awesome difficulty."

But on the analysis here, it makes sense that putting people physically into the closed scenario—with no change whatever in the material content of the problem—would be enough to amend the awesome difficulty. And, given the second version in note 6, in fact subjects seem to lose their former inability to deal with the problem. It is hard (for me) to make any sense at all of the marked improvement that results from this shift in the physical mode of response, leaving both the logic and the abstract substance (*A*'s, *D*'s, 2's, 3's) completely unchanged in any way—other than in terms of the scenario effects sketched here.

What I will try to show in the balance of this chapter is that a similar analysis (turning on the interaction of semantic and scenario effects) yields comparable resolutions of the Kahneman & Tversky "taxi" and "Linda" problems. I have treated elsewhere (Margolis 1987), and will not repeat here, a closely related analysis of a different sort of anomaly, the Lichtenstein & Slovic "preference reversals."

Note well that in these toy problems, as in the history of science examples to come later, care has to be taken not to confuse explicit reasoning-why (which may or may not parallel cognitively effective reasons) with the intrinsically tacit seeing-that which immediately governs responses [5.10]. If a person were aware of responding to the problem in terms of the open scenario, he would correct it. Nor is he likely to be aware that he is responding as if the rule were biconditional. Rather, as in the remark on Polanyi's tacit knowledge [4.4], we need to extend Rumelhart's remark quoted earlier. Indeed, "understanding the problem and solving it is nearly the same thing." But so is misunderstanding the problem and "solving" it—ordinarily by an intuition logically appropriate to that misunderstanding, but sometimes completely absurd as the response to what the problem actually says.

8.7

One version of Kahneman & Tversky's taxi problem goes as follows:

A cab was involved in a hit-and-run accident at night. Two cab companies, the Green and the Blue, operate in the city. You are given the following data:

(a) 85% of the cabs in the city are Green and 15% are Blue.
(b) A witness identified the cab as Blue. The court tested the reliability of the witness under the same circumstances as existed on the night of the accident and concluded that the witness correctly identified each one of the two colors 80% of the time and failed 20% of the time.

What is the probability that the cab involved in the accident was blue? (Kahneman et al. 1982:156)

The predominant answer (by far) is .8, and the next most common is .2, which is only a kind of 2d-order mistake (the subject has made the usual mistake but then compounded it by grabbing, so to speak, the wrong end of that wrong stick). The answer a statistician will give is .41, which can be thought of as balancing (a) the 85–15 Green predominance in fleet size against (b) the 80–20 Blue predominance from the witness reliability. The usual response, therefore, relies completely on (b), ignoring (a).[7]

Commenting on the persistence of the effect, Bar-Hillel (1980) reports:

> A wide range of variations . . . was presented to a total of about 350 subjects, including (a) changing the order of data presentation . . . ; (b) using green rather than blue as the majority color . . . ; (c) having subjects assess the probability the witness erred . . . ; (d) having witness identify the errant taxi as belonging to the larger, rather than the smaller, of the two companies; (e) varying the base rate . . . ; (f) varying the witness credibility . . . ; and (g) stating the problem in a brief verbal description without explicit statistics. . . . Through all these variations, the median and modal responses were consistently based on the indicator alone, demonstrating the robustness of the base-rate fallacy.

And three years later a review by authors not directly involved in the base-rate research (Pollard & Evans 1983:124) concluded: "To summarize all the results . . . it is clear that many (possibly most) subjects generally ignore base rates completely . . . "

But as with the supposition that Wason test subjects cannot handle modus tollens except by a task-specific familiarity, these claims must be misleading if the *P*-cognitive argument is sound. Experience will in fact make base rates of questionable value in various practical contexts: especially if the environment may be changing, if the priors are not reliably known, and sometimes for other reasons (Hogarth 1981). But there should be no general propensity to neglect base rates. Indeed the force of scenario effects in the Wason problem, to the point of overriding logically unambiguous language, is best understood as a (tacit, and on this occasion perverse) Bayesian response to patterns of experience: perceptions are shaped not merely by local cues but by background experience about what sorts of contexts are most commonly encountered under the broader conditions cued by this encounter. We expect a *P*-cognitive analysis to reveal something about the taxi problem that blocks Bayesian intuitions that in fact are in the repertoire.

Further, for the taxi problem, a debate has developed in the professional literature that is as cognitively interesting as the original problem. On Wason's problem, there has been little controversy about the normatively correct response. But for the taxi problem many critics have insisted that in fact it is Kahneman & Tversky, not their subjects, who have failed to grasp the logic of the problem. Hence the persistence of disbelief among Kahneman & Tversky's critics within psychology and philosophy present a second cognitive issue, absent in Wason's context, about why a problem that seems trivial and wholly uncontroversial to statisticians (Bayesian or not, see n. 1) has turned out to be so stubbornly contentious among nonstatisticians.

8.8

On the model of the earlier analysis, we look for an explanation that turns on some scenario effect strong enough (given the semantic ambiguities in the problem) to dominate the way the problem is usually seen. However, the taxi problem is complicated (compared to Wason's problem). Solving it requires intuitions which—in contrast to the "modus tollens" pattern of inference—could not be acquired from ordinary experience in the world. For although qualitative Bayesian intuitions should be in the repertoire, what is required here is a quantitative probability, turning on a balancing of competing propensities. As such problems go, it is trivial. But like riding a bicycle, what will come to seem easy will be difficult until you have gotten the knack of the thing through practice.

Historically, we know that quantitative probability is not a notion that appears as an untutored intuition. A propensity for gambling seems to be a cultural universal, and the notion of betting odds can be found in primitive cultures. But quantitative probability is a remarkably recent idea. It seems to have occurred to no one prior to the seventeenth century (Hacking 1975). Cognitively, we can make sense of that, since what we gain from tuning to experience is a sense of how to judge acts: we must respond yes or no (take a particular chance or not). But a person ordinarily has no occasion to perform the act of estimating an explicit probability. Even in a situation that seems to a modern reader to intrinsically involve quantitative probability—such as deciding whether to offer or accept a bet at odds—cognitively all that is required, and therefore all that we can expect, is qualitative tuning to experience. A person has to make a yes-or-no decision which is like those that enable us to decide whether or not to pay (or accept, as a seller) a particular price for something. We can judge odds without explicit probability as we judge prices without using the explicit utilities that economists find it convenient to postulate.[8]

We get a clue to the unnaturalness for untrained intuition of quantitative judgment by noticing the difficulty inexperienced people commonly have even in far simpler contexts than the taxi problem, such as conversion between metric and English measure, where what is required is semantically unambiguous and also computationally trivial. Once practice has made the process familiar, a person cannot recall the initial difficulty, and certainly cannot recall why it seemed difficult, since what was critical was a habit of mind, not anything introspectively observable as the reason for the difficulty.

So on this account, the stubbornness which is such a remarkable feature of the taxi problem can be expected to result from scenario-plus-semantic difficulty (as in Wason's problem), compounded by the extra difficulty and complexity of its quantitative aspect. This complexity, we can observe, is sufficient to produce the "moving target" effects discussed in the context of long arguments [7.8]. We get, in the taxi problem, a context just severe enough to give us a glimpse of what so often makes real world controversies more stubborn than can reasonably be accounted for in terms of the logic of the situation, as that logic interacts with the (possibly conflicting) interests at stake. Even for professional readers provided with detailed analyses, seeing the problem correctly has proved remarkably difficult.

8.9

But a version of the taxi problem that elicits reasonably good rather than unreasonably bad judgments requires, most of all, only making the information in the last sentence of (b) in the statement of the problem [8.7] more salient and less ambiguous. In addition, parallel to more clearly cuing the intended meaning of the rule in the Wason problem, it helps here to change (a), so that the base-rate information does not look intrinsically flimsy compared to the careful statement about witness reliability. The following variant of (a) and (b), leaving the rest of the problem unchanged, largely eliminates the base-rate effect, despite the many claims that merely rewording the problem (making no substantive change) would not do that:

(a′) 85% of the cabs in the city are Green and 15% are blue; and the maintenance policies, management, and so on of the two companies are similar (b′) A bystander identified the cab as Blue. The police tested his reliability under the same circumstances as existed on the night of the accident. The tests showed that he correctly identified Blue cabs as blue 80% of the time, but misidentified Green cabs as blue 20% of the time.

Compared to (b), (b′) preempts one obvious semantic ambiguity and also—but less obvious—a scenario ambiguity.

1. Unless there were other complications, a reader could be expected to make the Gricean inference—that is, read statement (b) in the original problem as saying, without a conscious step of inference—that if a Green cab isn't correctly identified as green, it is incorrectly identified as blue (not as yellow, purple, or whatever). But (b′) makes that obviously intended meaning explicit. It removes the semantic ambiguity.

2. The revision also clarifies a possible scenario ambiguity, which arises as follows: unless you are very experienced in work with probability calculations, you will be vulnerable to confusing

(i) the probability of A, given B,

with

(ii) the probability of B, given A.

An inexperienced reader will have difficulty disciminating clearly between the two, partly because in ordinary experience it is not unusual for the two probabilities to be pragmatically the same. This occurs in the probabilistic analogue of any situation in which "if/then" language is reversible (so that form (c) of the if/then readings in [8.4] holds). But although in an impoverished or unfamiliar context it is easy to confuse (i) and (ii), in a familiar context you are not likely to be confused. Even if you found it awkward to distinguish (i) from (ii), you are not likely to have difficulty distinguishing:

(iii) the probability it is raining, given that it is cloudy

from

(iv) the probability it is cloudy, given that it is raining.

For the problem here, the original (b) obviously refers to:

(v) the probability that a cab is identified as blue, given that the actual color was blue and not to

(vi) the probability that the actual color was blue, given that the color identified was blue.

On this point, logically (b′) is strictly identical to (b): both describe situation (v); neither describes situation (vi). But (b′) makes it hard not to notice that the process produces false blue as well as true blue reports. In turn, that makes it easier to see that the question asked is not just a rephrasing of the information already given in the statement about witness reliability.

8.10

From the remarks of Kahneman & Tversky's professional critics, it is plain that they are not really confused about some *merely* semantic ambiguity in the problem. Nearly always, the critics go out of their way to make it clear that they are familiar with the calculations, which explicitly use the meaning (b′) emphasizes. Often they go through the calculation themselves (as in Cohen et al. 1981:328). Rather, what the critics want to argue about is not the calculation per se but what they somehow see as the irrelevance of the calculation.

To say that the critics did not realize the process produces false blue reports in 20% of the green cases would be like saying that you did not know the meaning of "negligible" if you misread Wason's sentence [8.3]. We need what I have been calling a scenario effect, sufficiently well-entrenched that it takes special care to present the problem in a way that keeps the entrenched default from dominating the explicit language. There is something the person knows, which at the point of giving a response he somehow is blind to.

In the head-injury problem, it happens to be easy to see what the scenario effect is, and indeed it is so transparently linked to the language that it becomes arbitrary whether we call it a scenario or a semantic effect. But in Wason's selection problem, it is not obvious at all. It is only because the *P*-cognitive line of analysis told us to look for such a thing that we uncovered a scenario ambiguity that might account for the anomalous responses.

The scenario ambiguity that seems to fit best with the features discussed in the last few pages would be a cause-to-effect ($c \rightarrow e$) versus an effect-back-to-cause ($c \leftarrow e$) ambiguity analogous to the *open* versus *closed* distinction in the Wason context. On such an account, the difficulty that nearly all of us who are not statisticians have in fluently handling the abstract "$p(a/b)$" versus "$p(b/a)$" distinction is linked to the habitual treatment of the $c \rightarrow e$ scenario as the default situation: absent strong contextual cues, of a sort that we know how to use from real experience, we tend to see probability linkages as going from cause to probable effects, not as inverse inferences from effects to probable causes. As with the *open* versus *closed* effects in the Wason context, the point is emphatically *not* that we can't handle the nondefault scenario: effect-back-to-cause situations are much too common to fail to tune intuitions to handle that pattern, given that the cues in the situation prompt it at the moment of response. Rather, the situation is analogous to the visual situation in which we see a familiar mispelled word (like *sring* in fig. 1.2) as what we expect. We can see the thing correctly, but unless the nonroutine feature is well-marked, it may be remarkably hard to see.

Given a verbal description, where the situation is not one so familiar that we can recognize it at a glance (see it as a single unit), the content of the description may be too far back (fig. 1, Introduction)—its spiral may be out of the sweep of the current spiral—to be effective in prompting the scenario that is appropriate when we finally reach the question to be answered. Although a parsing of the language shows that one scenario is appropriate, it easily happens that we still—a moment after the words have gone by—see a response that fits the pattern of the default scenario.

So even though Kahneman & Tversky's critics write as if they clearly understand that it is the $c \leftarrow e$ scenario that the language describes, as soon as the focus on this point is relaxed, the tendency to see the situation in terms of the $c \rightarrow e$ scenario returns. We then observe a reflex response that defies the logic of the problem. The *P*-cognitive argument gives us an account of how such a situation might arise: an account that (judging from comments on the taxi problem) statisticians, who have had their noses rubbed in this sort of difficulty by frustration in trying to teach the distinction, are likely to find more readily credible than philosophers or psychologists.

Note well that, as with the Wason problem, the scenario that seems to drive responses here is not in the least reasonable: the language in the original (b) can no more be reasonably read in a way that justifies the dominant .8 response than could the logically exactly equivalent language in the remedial (b′). The situation only begins to make cognitive sense in terms of entrenched habits of mind, which can be cued in an impoverished or otherwise difficult environment. Responses which are logically indefensible *look right* to the person suffering what is clearly a cognitive illusion. What is then peculiarly interesting about the taxi problem (compared to the Wason problem) is the extent to which very competent professionals, with plenty of time to study the problem and looking at detailed analyses of it, nevertheless can continue to see the wrong answer as right. They then exhibit our usual facility [5.11] at producing reasoning-why which they see as justifying the .8 response, exactly as the stubborn subjects Wason (1969) describes.[9] The remarkable stubbornness seems tied to the specialized and quantitative character of the inference required, which makes it hard to see—even though the calculation is very simple[10]—that in the conditions of the problem the 80% correct responses will be a weighted average of 96% correct responses when the report is Green against only 41% correct responses when the report is Blue.

It is not surprising, on this view, that once the problem is more emphatically stated even unsophisticated subjects, who have never heard of Bayesian adjustments, do respectably well on the problem. They can handle the $c \leftarrow e$ scenario in the (b′) alternative framing. For we do not expect that ordinary people will have failed to acquire (qualitative) Bayesian intuitions from experience in the world with $c \leftarrow e$ scenarios. If we think of everyday situations involving inferences backward from effects to probable background causes, it is easy to notice (as the *P*-cognitive argument would make us expect) that ordinary people in fact do not ignore base rates. If you are in a poker game, facing a possible bluff by Wilson, you will not need the advice of a statistician to make your judgment depend not only on how things look on this hand but also on what you know about Wilson's general propensity to bluff.[11]

8.11

Kahneman & Tversky's "Linda" problem brings out yet another aspect of scenario effects:

> Linda is 31 years old, outspoken, and very bright. She majored in philosophy. As a student, she was active in civil rights and in the environmental movement. Which is more probable:
>
> (a) Linda is a bank teller.
>
> (b) Linda is a bank teller and is active in the feminist movement.

About 90% of Kahneman & Tversky's subjects choose (b), though the mathematical probability that Linda is both a bank teller and an active feminist obviously cannot be greater than the probability she is a bank teller. The usual interpretation of the perverse responses has been that people use a generally reliable heuristic (estimating relative probability by representativeness) even when it leads to a result that appears to be downright stupid. What follows is, in effect, a reinterpretation of the Kahneman & Tversky analysis intended to emphasize the role of scenario ambiguity.

In general, we cannot expect good quantitative statistical intuitions [8.8], nor even good qualitative intuitions, for probability questions of a sort that do not arise in ordinary experience. But we would expect good intuitions to the extent that pragmatic conditions in the world would provide the required tuning to experience. We have unsound intuitions about the birthday problem [1.2], which turns on exactly how low probability events are distributed (how often two will fall in the same cell); but we should *not* have unsound intuitions for a qualitative probability intuition as commonplace as the conjunction effect. Rather, we can expect to find that somehow the question has been (mis)understood in such a way that the answer in fact commits no conjunction fallacy. The subject is wrong, but not in a way that contradicts tuning to the patterns of experience. Rather the error comes from the habitual, automatic responses that are due to such tuning. As with the Wason selection problem and the taxi problem, we look for sources of semantic ambiguity complicated by some scenario effect, finding:

1. As Kahneman & Tversky point out, the most obvious difficulty is that the situation can be seen as inviting a judgment about the probability of *A* with *B* versus the probability of *A* without *B*. In ordinary language, if I say, "Which is more probable, *A* or *A* + *B* (for example, "that Jim will order pie or pie a la mode?") you will routinely hear that (experience with language use will tune you to hearing it) as implying$_u$ that *A* and *A* + *B* are mutually exclusive. For the Linda problem, that would lead to a reading of *A* as "Linda is a bank teller, but not active . . ." Kahneman & Tversky considered that and found that a substantial improvement results merely from making it explicit that *A* means Linda is a teller whether or not she is active. However, this only raises the fraction of correct answers to about 25%, leaving a large majority of responses still in apparent violation of the conjunction rule.

Part of the problem is that semantic ambiguity makes it harder than might be casually supposed to offset the "mutually exclusive" reading. I am not commiting a conjunction fallacy if I thought it more likely that Linda will go skiing at Christmas (instead of going home) "if her parents say that is OK" than "whether or not her parents say that is OK."

2. Pragmatically, sufficiently small differences are negligible. Suppose the probability that *A* + *B* occurs is only trivially larger than *A* with or without *B*. In that case, judging *A* + *B* more probable than *A* becomes a fallacy$_t$—in contrast to the situation in (1)—but still not a pragmatically interesting fallacy$_u$.

Most important, however, is a difficulty that although focused on a particular word (so it can be interpreted as semantic ambiguity) is better understood as a scenario ambiguity:

3. "Probable" and its various cognates ("likely" and so on) have more than one sense, though a person has little occasion to notice the distinction I will stress.

What I will label probability$_g$ is probability in the gambling sense, which is clearly what Kahneman & Tversky intended but not necessarily what their subjects understood. What I will label probability$_b$, on the other hand, is synonymous with "believable" or "plausible." It carries the operational meaning in terms of *P*-cognition that an intuition is unlikely to prompt doubt; it is unlikely to prompt the look-closer gestalt discussed at length in connection with the belief matrix [5.8]. So probability$_b$ has deep psychological roots. It is not surprising that unless the context is well-marked as one for which experience prompts the probability$_g$ sense, it is probability$_b$ that is seen. In the *b*-sense, "probable" would include the cases where the item judged as probable$_b$ looks like a typical or representative outcome. If that is the meaning$_u$, then it is a tautology, rather than a fallacy, that if someone asked you which was more probable, *A* and *B*, which together make a likely story, or *A* alone, then it is *A* and *B*. For probability had been understood as meaning plausible, not understood as probability$_g$, followed by a misjudgment revealing an absence of reasonable statistical intuitions.

In an article about the psychology of uncertainty, Kahneman & Tversky (1982) mention that "a man who asserts that 'I think Billy John will win . . . the high jump in the next Olympics' will not be considered a liar if he prefers to bet against this proposition rather than on it . . ." Suppose now that we amend that by making it weaker, saying: "I think Billy John will probably win . . ." Obviously Kahneman & Tversky would still not consider that speaker a liar, though he seems even less likely to gamble for John rather than against him. The statistical theorist, M. G. Kendall (1956), comments (about medieval usage): "Early writers used *probabilitas* as relating to the degree of doubt with which a proposition is entertained. At the outset of our science the two things were distinct, and it is a pity that they have not remained so and that our language has tended to confuse them" (see n. 8). To preempt a misreading, probability$_g$ must be well-marked by cues that in the life experience of the individual are associated with the intended reading. (Merely saying: "This is probability in the gambling sense" is worth little.)[12] Usually, however, difficulties do not arise since probability$_g$ mostly coincides with probability$_b$, or the sentence is embedded in a larger context of language and activity in the world that clearly marks the gambling sense.

The probability$_b$/probability$_g$ distinction is an especially interesting one for the account here, since it invokes the psychological mechanisms of the knowledge/belief analysis that has played a role in the argument since Chapter 5. The sense of more probable as more believable turns on whether an outcome is less jarring, less likely to prompt a closer look (triggering the *C*-gestalt). If I say, "If Hitler survived the war, he probably went to Paraguay," you are not likely to regard that as a stupid thing to say. Yet the probability$_g$ that he went to Paraguay—one particular small country out of all the places in the world—could hardly be literally larger in your judgment than the probability$_g$ that he did *not* go to Paraguay. After all, it is not the case that most Nazis who went into hiding went to Paraguay. But an unusually large number (relative to the size of the country) did so, so that in the sense of "probable = believable" Paraguay and Hitler make a good story, unlikely to prompt active doubt (prompt a closer look). In Kahneman & Tversky's term: it looks representative. Here, on the lines of this analysis, is a remedial version of the Linda problem:

> Linda is 31 years old, bright and outspoken. As an undergraduate she majored in philosophy and was active in the environmental and civil rights movements. A personnel survey showed that of clerical workers in banks (including tellers) fewer than 1% have personality profiles that sound similar to Linda's.
>
> If you stood to win $10 if the statement you choose turns out to be true (whether or not the other statement is also true), which choice is more likely to win you the $10? Circle one:
>
> (a) Linda is a bank teller.
>
> (b) Linda is a bank teller active in the feminist movement.

Notice that there is a change in the revision corresponding to each of points 1–3. As with Kahneman & Tversky's experience, the "whether or not" change (1) by itself leads to a substantial improvement, but still leaves the large majority of responses in violation of the conjunction rule. The amendment intended to offset the negligibility (2) effect *avoids* saying anything that is substantively significant. The extra sentence added to the first paragraph, in its literal significance, merely reinforces what was already said. In particular, it is consistent with $p \approx 0$ that a person with Linda's personality profile is a bank teller. But its cognitive role is to make it more likely that the subject will be alert to at least the possibility that the probability$_g$ might *not* be zero. And in fact, empty though it is in logical significance, the sentence apparently fulfills that cuing function. Accompanied by the other corrections, the sentence is helpful, though by itself it has little effect.

These changes, however, as with secondary changes in the earlier problems, deal mainly with clearing the problem of sources of difficulty that are really extraneous to the main cognitive effect. For the more the problem is complicated by secondary difficulties, the easier it is for the subject to be distracted from things that cue the appropriate scenario, and hence for the default scenario to govern the response.

The cognitively crucial change here must deal with the probability$_g$/probability$_b$ ambiguity. Notice that this is not done by explicitly saying how the word is to be interpreted. As with the "reversibility" issue in the selection problem, explicit provisos of that sort are not an efficient way to guard against the misreading. Rather, what works best is ordinary language that pragmatically carries the intended meaning, not formal or legalistic language (unless the person is a philosopher or lawyer familiar with dealing with formal or legal language in this sort of context). That is, we need the sort of language that will be effective in prompting the intended meaning in the experience of the listener, as is consistent with the general argument about scenario effects I have been stressing throughout this chapter. Here, since the predominant reading of words like "probable," "likely"—even in many contexts "I bet"—is probability$_b$, the simplest remedial procedure is to work with that reading$_u$ rather than fight it. The main correction, therefore, works by attaching the word "likely" to winning the prize rather than to the A versus $A + B$ choice.

8.12

Now consider the argument of this chapter in more general terms. Miscuing anomalies cannot be very rare in ordinary life. But ordinarily things proceed smoothly, with very little awareness of such miscuing as does occur. Further, in ordinary life—but easily not so in contexts either narrower than ordinary life (as in set-piece puzzles) or broader than ordinary life (as in social choice or theory choice)—the appropriate scenario is both familiar and well-marked, hence easy to recognize. So in ordinary life, what I have been calling "scenario" ambiguity will be rare.

Merely semantic ambiguity must be common. But if the alternative meanings of a word are very distinct (such as right = correct, right = opposite of left), then the context alone will easily cue the appropriate sense. This is almost tautological, since if the usage is such that in this context it ordinarily did not make the intended meaning clear, then we wouldn't be in the habit of talking as if it did. If you are talking to a small child, or to someone with a weak command of your language, you adjust to that, without making a self-conscious effort. It will be reflected in your choice of words, in how rapidly you speak, in supporting gestures, and so on.

On the other hand, consider the case of subtler variation: for example, the word "political" in the sense of policy-related, versus "political" in the sense of social maneuvering to favor some policy, versus "political" in the sense of a judgment turning on broad policy considerations likely to be underweighted by routine bureaucratic decision making. Even when the intended sense is not clearly cued, whatever sense the listener perceives is likely to be reasonably consistent with the context. It is only when the disambiguated meaning makes a remark jarring or incomprehensible or contradictory that we have occasion to notice that what we understood was apparently not what the speaker intended. So ordinarily we are not made conscious of multiple meanings of words, multiple patterns of inference, and so on, either because the intended meaning is unambiguously cued (and its possible rivals inhibited), or because if an unintended but closely related meaning is cued, it can usually pass by without anything jarring enough to prompt a look-closer response. It is only occasionally that miscuing occurs that causes us any difficulty, prompts any "look closer" work. And in only some small fraction of these cases does the miscuing have sufficiently sharp consequences to be more than quickly corrected and forgotten.

Scenario effects are another matter. Even simple miscuing will be more troublesome when individual slips (or what will eventually be recognized as slips of judgment) are embedded in far more complex arguments; where those complex arguments are embedded in contexts out-of-scale with everyday experience; where motivational factors complicate matters; where over time important differences in the repertoires of various actors emerge; where partisans on both sides of an argument are pressed by the "win or lose" psychology that sets in, once commitments have hardened, to concede as little as possible to the other side. Such things are characteristic features of the dynamic (no longer static) controversies we will take up in the balance of this study. Then we can encounter cases where the issue is both out-of-scale with everyday life experience and contains important novelties, so that habitual responses can be highly inappropriate responses. The opportunity for unrecognized contextual effects akin to the scenario effects of this chapter can be something much more than an odd quirk that shows up in some contrived situation.

Nine

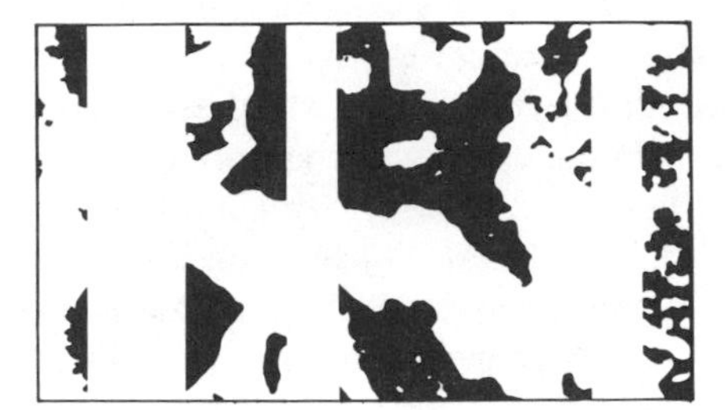

Cognitive Dynamics: Paradigm Shifts

Chapter 8 dealt with "static" cases, where what is involved can be restricted to variations in response given a fixed repertoire. We now want to extend the analysis to cases where the repertoire is changing. Here a person has to grasp a novel pattern, or see a pattern adapted to a novel kind of context where in previous experience it would have seemed irrelevant, so that the scenario and semantic effects of Chapter 8 are subsumed in a broader notion of how habits of mind constrain and shape cognitive responses in ways that are ordinarily outside conscious awareness and control. Kuhnian paradigm shifts in science provide particularly instructive examples, and the balance of the argument will be developed mainly through analysis of key episodes of the Copernican revolution.

For the static cases of Chapter 8 we could try to tease out how the interaction of cues and repertoires yields the results by looking for variations in typical responses to variations in the framing of a question. But when the issue becomes how the repertoire changes, fruitful experiments become much harder to conceive and carry out, and trying to learn from history becomes essential.[1] We cannot rerun history to see if Copernicus would still discover his theory if we changed his life this way or that. An analogue is available, however. We can try to see how differences in training, situation, and so on might account for why different individuals responded very differently to what seem logically the same cognitive opportunities.

Famous cases in science in which something seems to have gone wrong are not hard to find. On the contrary, as most readers will be aware, they are usual in the history of science. Conflicts within science are intense, even if we limit ourselves to cases in which ex post it is hard to see how partisans of what turned out to be the losing side could have been so sure they were right. Priestly, in almost his dying words, whispers his faith in phlogiston; Mach, having thought things over with the greatest care, confides that he can no more believe in relativity than he can in the reality of atoms; Lord Kelvin spends the last four decades of his long life confidently denouncing almost every important new idea that comes along (he was sure Darwin was wrong, sure that X-rays were a hoax, sure that the transmutation of elements was impossible).

We want to be able to account for such judgments in terms of a cognitive model that can also account for judgments that are alert to, rather than somehow blind to, the power of the new ideas. The puzzle is not to see why some people used a good cognitive process (hence got things right) while others didn't (hence got things wrong). Rather, the problem is to see how, within a community of people with comparable qualifications and all sharing basic cognitive processes common to the species, some see things in a new way that ultimately proves to be compellingly superior, and others could not see things that way.

9.2

Consider the novel sense of the word "limit" that must be grasped when a person studies calculus, or the novel sense of "demand" when a person studies economics. A bystander hearing these terms is not likely to feel puzzled. The technical concepts are close enough to ordinary language to make sense to a casual listener. It is only when the person faces a need to act—for example, to answer exam questions—that it becomes apparent that fluent technical use of these notions requires adding a new habit of mind to the repertoire, and one that needs to be discriminated from other habitual uses of these words already in the repertoire.

The difficulty of coming to fluently command the pattern must be contingent on how deeply entrenched are existing habits of mind that must be differentiated from the novelty. It will also depend on how closely related are the contexts in which the new pattern must now be used to the contexts in which until now the older pattern had been appropriate. In particular, it is when displacement of an entrenched habit is conspicuously required that the clumsiness analogized by the tennis coach [7.3] becomes conspicuously troublesome. Then a new habit of mind has to be acquired in a way that depends on fluent discrimination of cues marking contexts where the new sense is to be used, though until now some other (and in some contexts still essential) response was automatically invoked.

But whether the organism is a snail, or a rat, or a person, the cues that work best to enable fluent use of a pattern are cues that in the experience of the individual have been reliably linked by association to good results. Suppose, however, it is only after the novelty that displaces an existing habit becomes an established part of the repertoire that a person really begins to get good results from its use. If the novelty were very useful even when only crudely used, fluent cuing could be easily acquired. But if the novelty becomes very useful only after fluent cuing has made it easy and effective to use, then in order to profit from the enhanced competence of commanding *A* and *B* a person has to get through a period of diminished competence in which he suffers some clumsiness in using *A* without yet being smooth in using *B* [7.1]. Confronted with such difficulties, a person will easily find that the new sense looks not merely difficult and wrong but repulsive, with "all in pieces, all coherence gone" (Donne 1611).

Recapitulating: Miscuing problems can be understood as failures to cue an intended pattern available in the repertoire. But even miscuing where semantic ambiguity is compounded by scenario ambiguity can be remedied by reframing the question, as illustrated in Chapter 8.

There will be larger difficulties when the problem is not merely miscuing but requires adding a new pattern to the repertoire. This case necessarily includes not only learning the new thing but also learning how to fluently discriminate it from similar things already in the repertoire. Discrimination will be especially difficult when the discriminated older patterns are ones that continue to be important and that up until now have been central to some of the very contexts in which the new pattern is now to be used.

The most difficult cases will therefore be those in which the new thing operates in a domain that overlaps with deeply entrenched patterns in the repertoire, so that what was habitual and unproblematical and very useful must be finely discriminated and often inhibited. Interfering with such patterns will invoke a range of difficulties comparable to those that come with having to overcome well-entrenched, frequently used physical habits.

9.3

The *P*-cognitive argument commits us to the claim that at every step leading to a radically new idea the individual is only using patterns and relations of cues and patterns already part of the repertoire [6.2]. Even the most radical novel idea somehow must be built out of, or adapted from, or be assimilated to what is already in the repertoire.

Doing that must involve seeing a familiar pattern in a context where it ordinarily would not have been prompted. This can be accounted for in several ways: (a) there is something new in the situation, providing new cues or altering the salience of those usually present; (b) something has changed in the individual's repertoire from experience in another context; or (c) there is a fortuitous circumstance—some transient oddity of perspective, or some transient sequence that primes a particular response, which allows the individual to be prompted to a pattern that otherwise would not have been seen in this context.

But just to the extent that (c) is transient, it can hardly play more than a tactical role. By itself, a transient feature could only lead to a passing glimpse of the novel idea. Somehow, cues sufficient to prompt the pattern were encountered, but if the pattern conflicts with others habitually prompted in this context, it will simply be forgotten as we forget many odd (mis)perceptions every day unless the favorable cues are more than transient. It is only if the novel intuition repeatedly looks right (looks like it fits, looks like it makes sense of the situation) that instead of being forgotten as a momentary illusion the novel insight may be fixed in memory.

If the situation is one in which the new pattern conflicts with no deeply entrenched pattern, then the story ends quickly: the individual has made a small discovery which fits comfortably with—or at least does not uncomfortably jar with—the existing repertoire. But for the polar opposite case of radical discovery, there must be a story to be told about how the individual could come to see the novelty as making sense: as insight rather than illusion [7.10].

The most challenging component of an account of radical discovery, consequently, is to provide cognitive continuity, explaining how some pattern already in the repertoire could come to be seen as looking right for (look like it fits) some radically novel context. We can imagine the discoverer as a rather conservative climber, tempted on by a succession of new vantage points, without realizing how high the ultimate peak is, and so eventually seeing the peak only after he has already climbed so high that there now seems no turning back. The discoverer finds himself close enough to the top to make the final ascent, which at this point involves nothing cognitively more difficult than many of the efforts along the way.

Overall—sketching out now a general schema for the appearance (*discovery*) and spread (*contagion*) of a radically new idea—for a new idea that requires displacement of some well-entrenched habit of mind, we should be able to divide the story in a convenient way into what I will call (1) an *uphill* phase, (2) a *consolidation* phase, and (3) a *downhill* phase:

The *uphill* phase ends when the discoverer sees the new idea as conflicting with what he had previously taken for granted, yet somehow believable (as insight)—in contrast, quite likely, to earlier occasions on which he or others caught a passing glimpse of the idea seen in a way that looked implausible or narrow or like some minor variant of a familiar notion. *Consolidation* starts from that point and reaches its own climax when the discoverer realizes he feels he knows (in the sense of the belief matrix [5.8]) that the new idea is right. From that point a *downhill* phase begins which consists of polishing and supplementing the reasoning-why that might persuade others to accept the new idea. The division is a natural one in terms of the belief matrix, and a corresponding division can be expected to provide a natural schema for giving an account of contagion. In the discovery context we move to belief, then on to knowledge, then on to justification; in the contagion context we move from "It's absurd" to "It's wrong," and then from "It's wrong" to "It's obvious."

9.4

The concrete illustrations in the balance of the study are intended to give a usefully sharp sense of how this uphill/consolidation/downhill schema (and its contagion counterpart) works by putting it to use in particular cases. The present chapter sketches out the general argument, starting with this problem: an important role is obviously going to be played by the notion of some "deepest" entrenchment that has to be overcome. So we need some operational notion of how to recognize depth of entrenchment in a way that amounts to something more than an ad hoc claim that fits whatever story we want to tell. The way that is handled is that we can look at the record to see what people are so much taking for granted that they do not seem to even realize how far their judgments depend on it, in contrast to points freely discussed in ways that reveal that even opponents of the idea can see how the intuition could be wrong. Whether or how often the question is explicitly mentioned is not the key, but how far the possibility that it could be wrong is discussed. I have given an analysis of this in the context of economic theory (Margolis 1982:Chap. 6). In what follows here, we will have occasion to consider the question in the Darwinian and Copernican contexts. The important point here is that, in fact, there is an independent (of the particular argument being made) way to assess depth of entrenchment.

9.5

The sharper the ultimate conflict with entrenched habits, the more noticeable and explicit insight/illusion tension [7.10] is likely to be, hence the more important the role of the consolidation phase. There the new idea sometimes looks right; but sometimes it doesn't look like it could really work or make sense. A person can for a time see, alternatively not simultaneously, both sides of the dilemma. But what is seen is not always what the person was looking for or wanted. As in looking at a novel gestalt figure, it will be difficult to "command" the way things are seen. Considered in this way, the "uphill" phase, to the extent that the discovery is radical, is almost certain to be built out of a series of milder discoveries, from which the perception of rivalry with a deeply entrenched intuition that climaxes phase 1 only gradually is reached (Gruber 1981b). In particular, we do not expect to find—it would be embarrassing for the *P*-cognitive view to find—a case in which rivalry between a relatively local pattern and a central pattern was resolved by abandonment of the central pattern. On the contrary, what we expect is that potential or passing rivalry with central intuitions is almost always merely tamed; it is the more local pattern that is amended or habituated or muted.

Even once rivalry with the most deeply entrenched intuition is reached, successful consolidation will require some combination of reshaping of the new idea and reshaping of the entrenched intuitions that makes sense of the new situation. Consolidation could not be routinely successful. A new idea that seems believable will often have to be abandoned under the stresses that arise in trying to consolidate it; and sometimes it will be partially tamed, in a way that eventually comes to look like a failure of nerve. The discoverer eventually comes to have an altered set of intuitions that merely blurs the rivalry [7.6–7.10].

It is easy to underestimate the significance of this consolidation phase of a discovery. In hindsight, the crucial discovery has already been made, and the habits of mind that once had made that discovery hard have long since been reshaped (if not forgotten), so that when we look back at the history it is hard to recapture some reasonable sense of the cognitive rivalry faced by the discoverer and his near-contemporaries. Further, because of the often tacit aspect of habits of mind illustrated by the scenario effects in Chapter 8, the crucial cognitive rivalry does not necessarily play a salient role in the arguments about the discovery. I will try to show how the most fundamental cognitive rivalry between the Copernican and Ptolemaic views was *not* the point that was the focus of explicit debate (the sun moves vs. the earth moves). But in the Darwinian case, the deepest rivalry had to be explicitly faced. Through the concrete cases, I will try to show something about what distinguishes one sort of case from the other.

But in all cases of radical discovery, whether the deepest rivalry is explicitly salient or not, it could hardly be the case that only one isolated habit of mind will change when a very deeply entrenched habit is effectively challenged. Rather, consolidation will require a sequence of changes in intuition. Once what had previously been intuitive is so no longer, other intuitions will be engaged and will enrich and also complicate the process (Kuhn 1970: for example, 64). Some of this will happen during the "uphill" phase, but never all. Rather, much of the cognitive work that reshapes the intuitions of the discoverer takes place after he already believes he is right but does not yet feel sure. It is the fruitful resolution of the puzzles the discoverer comes to face after he believes he is right that builds the subjective sense that he *knows* he is right.

Finally, in the downhill phase, what is salient is the discoverer's interest in making the new ideas believable to others. This phase is downhill in the straightforward sense that now further puzzles involve intuitions less well-entrenched than the radical idea has now become. Work is focused on how to explain the idea to others in a way that has some chance of convincing them. In the language of earlier chapters, the discoverer's "seeing-that" has changed, but he now faces the problem of enhancing the "reasoning-why" to support that way of seeing. Yet the downhill phase cannot be simply identified with the context of justification in the common discovery/justification dichotomy. For reasoning-why is being constructed throughout the whole process, as the discoverer is repeatedly prompted to check and to rehearse [5.9]. What will become the justification is being built simultaneously with the making of the discovery and is an essential part of the process of making the discovery. The discovery itself is being refined during the downhill phase, which will always produce further development and clarification of the discovery (new seeing-that), not just the working out of arguments to justify a discovery that now can be treated as a static part of the discoverer's repertoire.

Summing up: Discovery will not (of course) come out of the blue, but out of some cognitive sequence that produces the required rivalry between a novel idea which is itself becoming entrenched by the use the discoverer is making of it and an already well-entrenched habit of mind. For a radical discovery the process will necessarily start from some puzzle that does not immediately challenge deeply entrenched habits. In Kuhn's language, revolutionary science will grow out of an episode that starts as normal science. At the climax of the uphill phase the discoverer has seen the new idea as something he *believes* but which conflicts with something else he finds it hard to disbelieve. At the climax of the consolidation phase he feels he *knows* the new ideas are right. Then work on developing the new view will continue through a downhill phase, usually with substantial consequences for the discovery itself, but now with the discoverer (subjectively) confident his idea is right and focused mainly on how to make the idea plausible to others. Pasteur said that discovery comes to the prepared mind, and Whitehead remarked that many people will have noticed some crucial thing before someone finally discovers it. What the uphill story is about is how a mind comes to be prepared, not just to glimpse but to discover a radical idea in the strong sense emphasized in the "uphill" discussion.

In a mathematical analogy [4.10], the uphill phase yields a conjecture; consolidation (if successful) yields what the mathematician feels comfortable calling a theorem; and it is ordinarily "downhill," though not necessarily easy, from there to a publishable proof. Kuhn (1977:165 ff.) has stressed that "discovery is a process, and must take time." The schema here amounts to a gloss on Kuhn's view, trying to add detail and structure to his account.

9.6

The consolidation phase may lead to outright abandonment of a novel insight, but in a way that is a success. Many promising new ideas turn out to be untenable. Abandoning interim solutions that do not work out (cannot be consolidated) is, in fact, a common part of the process that leads to challenging some even more deeply entrenched intuition. It is only ex post that the most deeply entrenched rival can be identified, as it is only when you get to the top of a mountain that you know there are no more ridges to climb. A balance between stubbornness and flexibility is required to thread the line between timidity and a kind of boldness that ultimately is only crankiness.

But for an idea that works, complete failure in the stage of consolidation could hardly occur. For the discoverer is highly motivated to work on his own idea. Then, seeing richer ways in which the idea works, and more aspects of the problem which are hard to make sense of in terms of older views (both of which are there to be noticed if the idea is essentially right), reinforces that motivation. At the same time, that further work entails just what is required to make habits of mind of the intuitions which make the novelty easy to use. The novelty itself now becomes entrenched by the mulling and rehearsing that is intrinsic to the work of consolidation; and exactly what we mean by the idea being essentially right is that, as a person comes to have fluent command over it, the ways in which it is fruitful grow. A person might fail to impressively consolidate an insight that someone else will later show to be essentially right in the sense just mentioned. But complete failure must be very rare.

Complete success is another matter: the novelty itself may be partly tamed, as was conspicuously the case with Poincare's relativity theory (Pais 1982), or the early Planck quantum theory (Kuhn 1978). And older intuitions may be adjusted in ways that do not look defensible in hindsight. Perfect discoveries—especially perfect radical discoveries—such that looking back we see no significant errors of either the first kind or the second kind—must also be rare.

9.7

Measured in terms of time (in contrast to risk of failure), it will be the downhill phase that is most likely to be particularly long. For it has no intrinsic point of completion. The argument can always be polished further, and new evidence sought, though now always "downhill": the difficulties no longer are seen as putting in doubt the essential idea, for the individual is by now (in Kuhnian language again) well-entrenched in the new paradigm. The work, for the discoverer, has become normal science puzzle-solving, where difficulties are seen as failure to work things out, not as casting doubt on the essential ideas. Since the downhill phase is intrinsically open-ended, with no definable conclusion other than the act of publication, it often terminates only because circumstances effectively remove the option of further downhill work—for example, awareness that rival discoverers are on the same track.

So we see the long intervals that often lie between a fundamental insight and a published version of the argument (20 years or more for Copernicus, Newton, Darwin). The case of Galileo is instructive here. Over a period of a few months he moved from the first building of a telescope to the publication of revolutionary claims. But here he did not rely on an argument but on what a person could see with his own eyes. More than he anticipated, opponents nevertheless denied his claims (asserted the telescope was unreliable, did not show what Galileo said it showed, and so on). But within a very few years serious opposition collapsed as more and more people, using more and more different instruments, saw what Galileo said could be seen. But his mechanics relied on arguments, and he nurtured those arguments and work on demonstrations that might support them for decades before he published the *Two New Sciences* at the very end of his long life.

9.8

Radical discovery certainly requires a discoverer with gifts such as intelligence, curiosity, energy, independence of mind. A capacity for self-criticism is obviously essential, since just to the extent that a discovery is radical the discoverer is likely to lack the usual supply of critical *help*. He requires boldness to proceed in a way that does not make sense to others, and self-criticism if that is to lead to a fruitful rather than cranky result. But only a small subset of exceptionally well-endowed people make an exceptional discovery.

Clearly it takes some self-confidence to commit major efforts to ideas that—by the defining characteristic of radical discovery—look basically wrongheaded to almost everyone else. We can observe that major discoverers have usually gained some reason for self-confidence by less radical but still striking discoveries (for example, Newton's work on colors, Einstein's on brownian motion, Darwin on the origin of coral reefs). Copernicus, before reaching the heliocentric insight, worked out a new model for the orbit of the moon which corrected a glaring incongruity in Ptolemy. But even if confidence-building experiences can always be found, that would not differentiate discoverers from others who have had confidence-building experiences but who did not go on to some more fundamental display of originality.

Another conditioning factor, often discussed in the literature of discovery, concerns the advantage of being a partial outsider. In the account here it plays an important role. Other things equal, a habit-displacing discovery is more likely to be made by someone (a) who is not so deeply entrenched to that habit, or (b) who is not so deeply reliant on its fluent use as other members of the community competent to make the discovery. A qualified version of this point applies also to the critical early stages of contagion.

For (a): unless a would-be discoverer is in fact somewhat entrenched to the habits of mind he will eventually challenge—unless he was reasonably fluent in the patterns of intuition needed for work in the field—he could hardly be an effective discoverer. Without being a partial *in*sider (sharing enough of the fluency of the complete insider to be competent to do significant work in the field) he could hardly come to appreciate the problem he will eventually solve; or command the technique to solve it; or distinguish deep from superficial anomalies confronting the standard view; or adequately appreciate what that standard view accomplishes, hence what the new view must compete with. The total outsider might have the qualities of mind apparently required for discovery: intelligence, energy, independence, and the rest. But that alone is likely to produce only futility if accompanied by self-criticism; and it produces crankiness when it is not—a Velikovsky, not a Darwin.

But for the partial (not total) outsider, anomalies which have been habituated by people fully established in a field will not be habituated by a person whose special focus of interest is connected with just that anomaly [7.2]. Other habits of mind deeply entrenched for nearly everyone in the field, but not so (or less so) for the partial outsider, can become familiar in a way that leaves open the opportunity for radical innovation.

The cuing required for this discrimination is an automatic consequence of becoming fluent with the pattern *after* seeing it is a pattern that must be discriminated from (not assimilated to) another useful pattern in the repertoire. The discoverer and his more "inside" counterpart each sees what he sees, as you (for example) see the meaning of the words you are now reading, with no consciousness about—and with no access to consciousness about—just what cues are guiding you. The cues that are coming to be significant during learning are no more conscious than they will be once the pattern has become fluent. But learning a pattern that you see as working in some contexts, but not in others, will automatically occur in a way that discriminates between the two. Cues associated with both, which would be well-marked indicators if one of the patterns were never present, will be ambiguous when both are. So the learner will become tuned to different (and more highly differentiated) cues than a person who has been habituated to (has been trained by experience to ignore) the very contexts in which a widely shared intuition fails. An inverse version of all these points applies to (b) above, as will be illustrated for the Copernican discovery [12.3].

9.9

But as there must be a continuum of cognitive entrenchment (not a crisp dichotomy), there will also be no sharp dichotomy which puts each person either inside or outside a community of persons sharing a certain set of ways of seeing its phenomena. Within or on the borders of a broad community (for example, economists) there will be subcommunities (Chicago economists, Keynesian economists; mainstream economists, Marxist economists; senior professors, graduate students) who partly share and partly don't share various habits of mind, and who vary in depth of entrenchment to habits they do share.

A social aspect of shifts in intuition then enters in the following way. Even for physical or perceptual habits, and even for a self-taught individual, patterns (how to ski, how to improvise at the piano) will be influenced by what he observes others doing. But the social aspect to habits of language and intuition must be more powerful. For such habits are largely acquired and used in the context of communication within a community. Even the private mulling of an individual will always be influenced by, and very often consciously directed to, whether other people would find some piece of reasoning-why comprehensible and plausible. Habits of mind central for an individual will tend to be habits shared across the group of individuals who make up a community: in a fundamental sense, what defines a community are such shared intuitions. The most challenging cases of discovery are consequently those for which a shift in intuition occurs which requires displacing a deeply entrenched habit of mind shared across an entire community. These community-wide cases will be the most challenging ones for accounting both for discovery (how some individual managed to break free of those intuitions) and also for how such an initially merely idiosyncratic shift in intuition spreads across a community (contagion).

A common special case of the partial outsider effect occurs when a novel technique or argument from one field happens to be the key to solving an important problem in a neighboring field. That problem may be one that people in the field have effectively learned to ignore. In 1950, the most obvious problem in biology was to discover the nature of the gene. But just prior to the discovery of the structure of DNA, mainstream biologists did not see that as a salient research problem. A persistently unsolvable problem becomes de facto unimportant (people in the field no longer see it as a relevant problem). So biologists had learned to treat genes as primitive, as physicists between Newton and Einstein learned to treat gravity as primitive, not as something that itself would be the focus of analysis.[2]

Disrupting a presumption against treating some problem as worth trying to solve is not often a matter of challenging an explicit claim of the field; rather the challenge is to a habit as logically indefensible as the opposition to hockey helmets [7.5], or the scenario effects in Chapter 8. It is something about how brains work, not something about the external world, that accounts for the tacit commitment to ignore certain questions, and also for the difficulty of shedding that commitment after pragmatic justification has eroded.

With DNA, Watson was too young to be a central figure even in the fringe group within biology most likely to have been alert to information on the physical character of the gene (the phage group); Crick was a physicist; both were influenced by the speculations of another physicist (Schroedinger's *What is Life*). They solved the problem by exploiting information and techniques foreign to mainstream biology and genetics (X-ray crystallography and balls-and-sticks model building). Despite the almost extravagant elegance and promise of the structure, two years slipped by before the mainstream community took their result very seriously (Crick, personal communication). But the existence of a substantial community beyond mainstream biology competent to have an opinion about the significance of what Watson and Crick had done, and more competent than mainstream biologists to judge the technical merits of their solution, made it hard to ignore their work for very long. So the existence of a broader community with competence to deal with an issue makes discovery more likely (by providing both a source of potentially relevant ideas and a supply of potential partial outsiders to apply them); and it also makes contagion faster (by providing a supply of competent judges open to the new idea even while the insiders aren't).

Absent that overlapping of expertise across specialist communities, contagion must be more difficult. Cournot's pioneering work in mathematical economics was ignored for 30 years (Schumpeter 1954:505). His mathematical colleagues were not knowledgeable enough in economics to recognize its importance (merely as mathematics, it was obviously not very important), while the use of mathematical models in economics made no sense to the mid-nineteenth century economists confronted by it. Mathematical modeling seemed as obviously inappropriate to them as it seems obviously indispensable to today's economists. And as the Crick/Watson versus Cournot contrast illustrates (continental drift in 1965 vs. Mendelian genetics in 1865 would be another pair), the difficulty of the habit-breaking modifications required for both discovery and contagion has to be understood relative to the communities involved as well as relative to the individuals involved. For the contagion stage as well as discovery, it makes a very large difference whether a habit of mind that must be broken is shared across the entire community potentially competent to deal with the issue, or holds only within a subcommunity.

9.10

Here is a taxonomy of this intrinsically social aspect of the discovery and spread of belief.

On the smallest scale, discovery amounts only to a matter of idiolect—a strictly individual matter of a person working his way out of some blind alley, like the person stuck with some mathematical slip he cannot easily see but which almost anyone else can. Miscuing has occurred, and the correct pattern is temporarily inhibited. The individual knows from the result that something has gone wrong, but he cannot see what it is. He is in a state of paradox, in both the colloquial sense and also in the technical sense of the belief matrix [5.8]. He has a small crisis of his own, but there is no crisis for a community.

On an intermediate scale, the problem is paradoxical for some subgroup within a community. Someone has proposed a new idea that conflicts with habits shared across the group, as occurs very often, since any particular research team or specialty grouping within a larger field will develop habits of mind specialized to that group. They will learn to think about certain issues in particular ways, and usually to see a rough boundary for the class of phenomena relevant to their work. People within the group sometimes act as if *X* was true, and sometimes as if *X* was false, and seem not to notice the inconsistency. Phenomena inconsistent with such shared commitments come to be habituated or otherwise tamed [7.7].

Within the subgroup, it is hard to see those habits, or tacit commitments, as a problem (except for the individuals who have already broken from them), or even to notice that there are commitments from which to escape. Rather, the novel idea that breaks with the commitment looks obviously wrong. If someone outside the group points to a situation where *X* is importantly false in a context where the group is in the habit of accepting *X* as if true, the outsider's claim will look wrong or trivial, so that the criticism is seen as merely showing how the outsider fails to understand things in the right way. But for the subgroup case, the committed community is (by definition) part of a larger community for whom the novelty will not look intuitively wrong.

On the other hand, even within the subgroup, entrenchment will not be as deep as for the community-wide case, since members of the subgroup will (as part of the wider community) have many occasions to see things in some other way, so that it cannot be profoundly difficult to prompt rivalries—prompt awareness of situations in which something that is being taken for granted for work within the subgroup looks wrong. So for small subgroups—in contrast to the case where the novelty is radical for the entire relevant community—we can fairly easily get what would be recognized as a revolutionary idea if a cognitive change of an equivalent sort had swept across an entire community.

9.11

Now consider some points about the contagion stage. For the generations who come after a revolutionary episode is over, the new view is merely the orthodox view, and there is no entrenched network, tuned exclusively to the older view, to be overcome. Rather, it takes Kuhn's hermaneutical effort to recapture something of the original difficulty.

But what of learners of the new view in the first generation? In particular, how are the early converts to a new view recruited? Naturally, the striking cases are those that involve challenges to intuitions shared across the whole community (in contrast to the subcommunity case), and in what follows I will always have that case in mind.

To make the contrast between self-guided discovery and externally aided contagion as stark as possible, consider the difference between what a dog can learn on its own and what that dog can learn by training. The trainer can work with whatever the dog already does, gradually building and putting together bits and pieces of the final performance. In the human case as in the canine, and for the discoverer as well as the learner, reaching a new pattern of responses requires a series of steps such that—in the sense I have been stressing [6.8]—at each step the creature is only doing what it already knows how to do. But a creature being guided by a trainer has a vastly easier task than a creature groping its way on its own.

The drill and going-over-examples characteristic of teaching provide the required disruption and reorganization of entrenched networks of cues, arranged in a way that does not demand more of the learner at any one step than he is able to manage. Further, as with the mathematical proof [5.9], or with a mountain path, the first discoverer, having no guide, easily gets caught on what had seemed a promising line of approach, abandoned only after repeated failures to work out the problems. But once the ascent has been made, it is possible not only to show a newcomer how to go without getting into blind alleys but also to simplify (often greatly) the pioneering path.

This contrast between the difficulty of having to discover a new thing versus the relative ease of being taught is so sharp that we still cannot really account for the difficulties that mark Kuhnian paradigm shifts by supposing that the habits of mind are so deeply entrenched as to be really irreversible. The more plausible view is that it requires a determined effort to overcome deeply entrenched habits of mind (as with deeply entrenched motor habits), not that it is impossible to do so. The problems that arise are then not due to some intrinsically insurmountable difficulty of learning the new ideas but to a combination of the difficulty of learning the new thing, as compounded by the difficulty of coming to see the new ideas as worth the effort that learning them seems to require. There is a triple problem: the old guard is very deeply entrenched; and its most influential members are the leading members of the mainstream community, who have (on the one hand) many demands on their time, and (on the other hand) constantly rely on their fluency in the mainstream way of seeing things. Further, they would be training themselves to be able to see how wrong they have been.

So it is not surprising that we get Planck's dictum: for radical discoveries the new view triumphs less by converting the old guard than by outliving it. However, that can be no more than half the story. For the new view would be left in command only if, as the old guard disappears, those who remain are converts to the new view. So we still face the question of contagion, and in particular the question of how the early converts are attracted. But now a good deal of what needs to be said is implicit in what has just been said about the old guard, turned around.

9.12

The very argument that says that a radical discoverer is likely to be a partial outsider implies that as a teacher he is not provided with students to train; rather, the learners have to somehow select themselves. Obviously, no one will drill himself into seeing something in a way that looks worthless, or perverse.

We look for people who are themselves partial outsiders drawn to the field by the new ideas; or who if inside the field have at least somehow been spared from becoming active opponents of the new ideas (as Lyell's close personal and professional ties to Darwin restrained him from quickly committing himself to the anti-Darwinian side). Or they will be young, so that they learn of the new ideas before they are deeply entrenched in the old. Near the end of the *Origin,* Darwin remarks that "I by no means expect to convince experienced naturalists whose minds are stocked with a multitude of facts all viewed, during a long course of years, from a viewpoint directly opposed to mine . . . but I look with confidence . . . to young and rising naturalists, who will be able to view both sides of the question with impartiality."

To encourage that, the discoverer will try to communicate a sense of how things would be for someone who made the effort to learn the new view, using the kind of language that is appropriate for the effort at motivation as well as persuasion that is required here. So we get the eloquent perorations with which Darwin concluded his book or Copernicus (*de Rev:* I.10) concluded his heliocentric argument, and much else that is as much poetry as logic. A good deal of comment bearing on this misconstrues the role of rhetoric in the work of men like Copernicus and Galileo, as if the rhetoric were tricking people into believing what they could not reasonably be persuaded to believe. The situation is better understood in terms of a complementary relation between the general sense of things that might persuade a person to want to bother learning, and the carefully worked-out arguments that can stand up to close critical scrutiny.

However, given a critical mass of support, things become much easier. The likelihood that a person will feel comfortable with—will fail to look closer at—an initial aversive response to a new idea ("looks wrong," "looks too complicated," and so on) will be influenced by the human sensitivity to what other people are doing and saying. Once the new view has won enough support to be respectable, if still controversial, anyone open at all to that view knows that he will have what is becoming a useful piece of understanding (for being on top of things, perhaps even for effective work) if he takes the trouble to learn the new ideas. Before then, a person has no way to be confident he is not merely wasting his time.

Further, after but not before the novel ideas take hold sufficiently to be a matter of common discussion, a simple one-step piece of reasoning-why becomes available with respect to what at first might seem implausible aspects of the new view. The person knows he is studying something that has already survived careful study and even calculated attacks. So he has reason to believe that if there is a fatal error in the argument, it is not likely to be something that can be seen by someone who has not yet even mastered the basic argument. What for the very first audience of the new view would have seemed paradoxical becomes only a puzzle for those who study the matter after it has acquired a measure of social respectability. What is seen as cognitively paradoxical stops further progress. (It must be attended to before further work on that topic can proceed.) But what is seen as only a puzzle can be set aside, to be taken care of later, or taken care of by someone else.[3] So the first recruits face repeated choices about whether to bother putting more effort into grasping the new arguments, repeated occasions on which the novel argument might be found to be more trouble than it is worth. But later learners can far more comfortably merely finesse, rather than face, difficulties with "I'll come back to that later" or "I'll get someone to explain that to me."

9.13

How is radical innovation, in fact, possible at all? Prior to the Copernican case, there is no example of radical discovery and contagion, such that deeply entrenched habits of mind within a community are radically disrupted by the force of arguments. No one in the years when the Copernican view was still unconventional could become a Copernican without following and being tremendously impressed by an argument (spelled out in [11.4]) about why the Copernican view made sense, and similarly for a long line of scientific discoveries since Copernicus. So although it would be naive and also contrary to the argument of this study to suppose that good arguments are alone sufficient to persuade, it also requires a certain entrenched blindness to deny that remarkably good arguments have played a crucial role in all the widely discussed cases of revolutionary scientific shifts. It is only since the Renaissance that we can point to cases where a community is actually persuaded to come to see things in a radically new way by reasoning—in contrast, say, to the shattering effect of conquest, or to the evolution of habit and custom over a period of many generations.

So something new to the world has appeared in recent centuries, which promotes (or at least permits) a pervasive flow of innovation contingent on radical persuasion. The obvious candidate for that something new is the emergence of printing. More generally, we have seen a broad advance in technologies which radically facilitates communications and travel, hence the flow of information. The consequence has been a widespread availability of information not monopolized or controlled by individuals committed to a particular view. In turn—contingent now on political circumstances, so that the next step occurred in Europe but not in China—this allowed for the emergence of competing centers where a new view might appear, and to whom a new view might be promoted. Although a detailed account would take us far beyond the scope of this study, I want to sketch in a little of this line of argument.

For both discovery and contagion, a radical change in the opportunity for innovation must come with the availability of very extensive information and reasoning-why embodied not only outside of individual brains [5.13] but also outside the control of specially privileged individuals. This availability is obviously a property of the societies involved, not of any individual. It turns on the scale and character of Popper's World 3. The argument on this matter extends the one I have made about the relation between Level 1 and Level 2 learning [6.9].

Recapitulating that earlier argument: On a general Darwinian view and on the particular *P*-cognitive view of this study, we expect no absolute discontinuity between what animals and humans are able to do with an internal representation of the external world. But a radical increase in competence comes when this ability extends from thinking ahead to the next move to thinking several moves ahead. We are then changing from a situation in which the internal representation depends on the concrete external anchoring (a chess board, for example), to a situation that enables "seeing" not only the actor's own moves but also a sequence or set of possible changes in the environment contingent on what the actor does with his next moves. Using the chess analogy, this enlarged capability requires the capacity to somehow represent a position different from the position concretely anchored by a visible board and pieces. But as that capacity extends, much else becomes possible. Freed from tight contingencies of time and place, the potential for searching for patterns that might fit (the capacity for search which appears when that capacity for abstract—for not concretely anchored—representations becomes well-developed) is vast compared to what can be done when abstract (unanchored) objects of cognition are not available.

But then a further radical release from previously binding contingencies occurs in the transition from necessary reliance on direct emulation (copying what a teacher is doing, or saying), and from limitation to what can be held in memory (even exploiting the division of labor within a community), to the availability of written materials. As in the concrete-to-abstract transition, this involves no qualitative break with the earlier situation but nevertheless vastly broadens the scope of search for a pattern that fits. Hence, the tremendous cognitive importance of the appearance of literacy [5.13].

But when literacy is extended to the cheap availability of copies, a further radical change in opportunities occurs. Here the step has two essential aspects. Not only is the pragmatically available range of searchable things greatly expanded for a potential discoverer, but the discoverer is no longer constrained by what will win the approval (or at least avoid the overt disapproval) of a teacher. The nature of the student-teacher relationship is such that, even more than either teacher or student consciously realizes, students become trained to see things as their teacher sees them. They are under some compulsion—not all of it overt—to learn to see things in a way that strikes their senior as insightful, not illusory.

Such apprenticeship effects are important today. But as seen today they must be very modest compared to a time when access to information could only be obtained under tutelage, and hence would be wholly contingent on maintaining the tutor's confidence. With the mass production of copies of information, the opportunities must increase enormously for an individual to be able to independently seek out and exploit information, techniques, arguments that might bear on his problem. A possible discoverer can now much more easily harbor and develop a radical view until it reaches a stage where it can be presented with some chance of persuading others. Similarly, it is only when communication becomes cheap that a radical discoverer can reasonably easily locate potentially sympathetic listeners (by reading what they have put in print) and communicate his arguments to them (again taking advantage of print and its cognate technologies). Opportunities of all these kinds become vast compared to the situation in which a person who was not himself deeply entrenched to habits that inhibit some radical discovery would almost have to be the student or assistant of someone who was.

So it is not a coincidence that the first great example of radical innovation (the Copernican discovery) followed closely on the appearance of printed editions of major astronomical texts at the end of the fifteenth century. The availability (due to printing) of hundreds of copies of Copernicus' book then played an important role in the contagion of the Copernican idea (Gingerich 1973). As will be seen in the concluding chapters, there are aspects of the story more striking and direct than the appearance of printing; but the appearance of printing is an important and probably crucial piece of the story. After all, 1400 years is a long time to wait.

An additional point, though, comes as a social analogue of individual confidence-building. Eventually—assigning a reasonable date is a cognitively as well as historically interesting question we will have occasion to consider—the Copernican discovery could be seen as a *demonstration* that radical discovery was in fact possible: that what "everyone knows" could not only turn out to be wrong but could be challenged by the initiative of a single bold innovator. The flood of radical innovation that marks the seventeenth century was not independent of what the Copernican example taught.

Ten

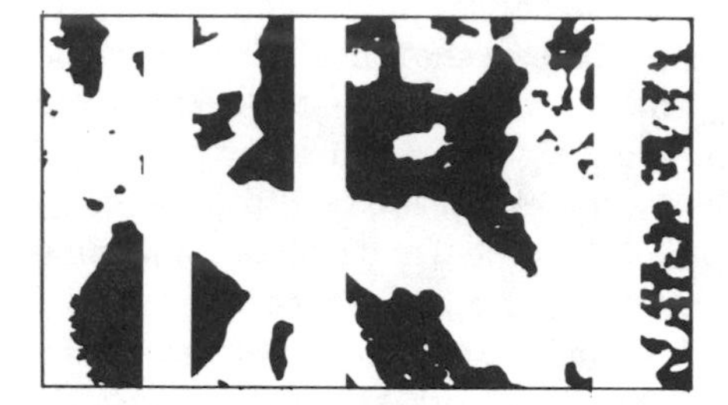

The Darwinian Discovery

This chapter will be brief, sketching out enough on the Darwinian discovery, together with a little on the discovery of continental drift, to illustrate and add some detail to the uphill/consolidation/downhill schema of Chapter 9.

The Darwinian argument made two key claims: (a) that new species evolve from existing species, and (b) that natural selection can account for how that happens.[1] Obviously (b) is meaningless without (a); and on the record, it is relatively easy to come to believe (a), and hard to go further and believe (b). The test described in [9.4] shows that evolution—in the pre-Darwinian sense in which the phrase "transmutation of species" was used—ran counter to no deeply entrenched habit of mind but natural selection as the primary mechanism behind evolution did. Evolution as a general idea had a certain intrinsic plausibility. For 2000 years prior to Darwin, the notion that species can be transformed appears repeatedly (Osborn 1894). After the great age of the earth began to be understood in the 1700s, evolutionary ideas proliferated. After Darwin, evolution became a thoroughly respectable belief within a few years. But natural selection as the process that drives evolution won no such easy acceptance; rather, Darwinism in this far stronger sense had to go through a prolonged period of eclipse (Bowler 1984).

By the time Darwin's book appeared in 1859, the idea of evolution (in particular, evolution understood as a gradual transmutation of species) had a double status. Among the general reading public, it was a familiar and intriguing possibility. But among experts, and in particular among experts in England, it had the status of a popular fallacy. Evolution was expounded in Britain only by people outside establishment science. It was the eccentric business of isolated amateurs like Darwin's grandfather Erasmus, or the Edinborough publisher Robert Chambers, each arguing in a way that struck the experts as full of misinformation and loose thinking. The prevailing professional view is reflected in the way that Lyell, whose gradualist account of geological change was a primary influence on Darwin, took great pains to warn readers away from supposing that this view of geology could sensibly be extended to the living world. So antipathy to talk of evolution was a habit of mind somehow entrenched primarily among experts.

But with respect to natural selection (b) we have a very different situation. Nothing more than one or two passing glimpses of natural selection as a creative (rather than only conservative) process have been found prior to Darwin. Even after Darwin, natural selection as the primary mechanism underlying evolution did not become solidly established within biology until the middle decades of this century. For the wider educated public it may be short of "solidly established" even today.[2] So where evolution was a popular idea, and it was the experts—even more narrowly, British experts—who especially needed convincing, natural selection was hard for the experts to believe, and even harder for the wider public. In terms of the discovery schema, therefore, we would expect to find some more deeply entrenched and more widely shared pattern of intuition (deeper compared to that which inhibited belief in evolution) that had to be overcome to be able to see natural selection as the force that drives evolution.

10.2

What inhibited expert belief in evolution before Darwin is not hard to find in terms of the "habits of mind" argument we are pursuing; and what inhibited belief in natural selection is easily seen even without any such prompting.

What makes the idea of evolution (in contrast to the idea of natural selection) intuitively easy is that anyone can see that living things come in groups that look very similar, with groups of groups that also can be seen as similar, and so on through some approximation of the Linnaean scheme. More detailed familiarity (with small details of internal anatomy) elaborates and refines on this commonsense intuition. Further, we are all familiar with the seedling gradually transformed into a tree, a baby into an adult, and so on. Hence, a sense that things that look to be related emerged from a common source by some sort of unfolding or development process reflects a pattern of experience that we all know very well. It has, in fact, always taken some indirect argument—something other than a direct appeal to common sense—to explain why the idea of evolution as the origin of species should be rejected. The sorts of arguments used have varied enormously over the long and wide range of anti-evolutionist argument from Aristotle to today's creationists. But it is an essential point that evolution is not an intrinsically counterintuitive idea; rather, it is an idea that people have to be trained to see as wrong. It did not go beyond passing thoughts so long as (in Europe) the earth was seen as being only a few thousand years old. But with the emergence of the fossil record in the eighteenth century, more explicit talk of evolution became common (Mayr 1982:chaps. 7 and 8).

Natural selection as the agent of evolution, however, is profoundly counterintuitive. For it implies something which on its face seems preposterous. Living things by and large are wonderfully well-formed to do everything they must do to carry on their lives. But on the Darwinian argument, good design of astounding ingenuity (like the structure of the eye) appears without the aid of a design. The brain appears from a process that has no ideas, no goal, no intent, no intelligence.

So what most needs to be explained in an argument about evolution is how firm belief that it is an absurd idea could ever have become entrenched among experts in the face of obvious family resemblances across species. What needs to be explained about natural selection is how anyone could ever have broken free of the intuition that good design implies a good designer.

10.3

Among Darwin's colleagues the antipathy to evolution seems to have had two reinforcing sources. The more fundamental of these turned on the role the notion of "species" played in professional work. But that was aggravated by something easier to grasp, which was the sheer discomfort at having to deal with what looked to the professionals (and what looks to biologists today) as amateurish and unworkable accounts of evolution. Contrary to what would follow if evolution were an intrinsically counterintuitive idea, the professionals found dealing with evolutionary arguments frustrating just because the idea so easily looked plausible to a wider public. From classroom to dinner parties, a professional found himself explaining, with what must have been (from the popularity of such views) often frustrating difficulty, why Chambers or Spencer or Lamarck was unsound. On the *P*-cognitive argument, that experience would train professionals to a general aversion to evolutionary arguments, and it would also polish and entrench the patterns of argument that attacked evolution. Within that expert community, then, there was a marked distaste for evolutionary talk that grew out of their very expertise, as a hundred years later the distaste for talk of drifting continents that was predominant among earth scientists was linked to their very expertise. The general public was easily tempted to draw what turned out to be the correct conclusion from the suspiciously good fit between the east coast of South America and the west coast of Africa.

But a deeper basis of expert aversion to evolution was the essential role that "species" played in day-to-day professional work. How to draw the line between good species and mere varieties was the most fundamental problem in biology. If species appeared by gradual transformation of existing species, then the "essentialist" premise of the whole discussion became paradoxical. Both before and after Darwin, the notion of species is crucial for biology. But what the notion means has been radically transformed (Mayr 1982:45–47), in just the way that Kuhn (1977:xi–xiii) has stressed in emphasizing the transformation of the notion of "motion" with the abandonment of Aristotelian physics, or as we can see in the transformation of the notion of "utility" in economic theory around the turn of this century (Steigler 1966:65 ff.). To an outsider, the distinctions in any of these cases tend to sound quibbling, and prompt puzzlement akin to what students commonly face when they first encounter the fussy way in which mathematicians deal with the commonsense notion of "limit" or economists with the commonsense notion of "demand" [9.2]. Nor would puzzlement have been preempted if the novel meaning were labeled by using a different word, since the same cognitive propensity that led the expert community in all these cases to work out their sense of the new notion as a development from the old would hold whether a new term was used or not. In the case of "utility" in economics, an effort in fact was made to use a different term ("ophelimity"), with a ludicrous result (Margolis 1982:14). So if you are not already familiar with the Darwinian history, you must expect some difficulty in grasping how fundamentally the notion of species was transformed after Darwin.

In the new Darwinian sense of species, it is a breeding population that evolves, and the distinction between a true species and a well-marked variety becomes intrinsically arbitrary. Spending a lot of intellectual effort on trying to say exactly (rather than roughly) where the line should be is like trying to draw a precise boundary between good movies and bad movies or, for that matter, between good science and bad science. In Kuhnian language, a salient problem for pre-Darwinian normal science was to decide, in this or that particular case, whether a well-marked variety was in fact a good species; and a salient larger goal was that out of experience in distinguishing particular cases it would be possible to find a general principle that marked that species/variety boundary. This problem became a nonproblem on the Darwinian view. After a while, one of the merits of Darwin's argument would come to be seen as ridding biology of this search for a nonexistent sharp line of demarcation.

But if you work on a problem, you acquire a certain fondness for it. You think it is significant and justifies devoting the effort you are giving it. You will not easily come to like an argument that implies you have been wasting your time, since the cognitive conditions under which you would be spending a lot of effort on something could only be when doing so seems (not necessarily for reasons you can consciously articulate) worthwhile. It will then look wrong to dismiss that problem as empty. This will hold all the more readily when the arguments for that view are in fact sloppy, come from amateurs, and yet perversely seem to find ready belief among your friends.

Reviewing: When Darwin wrote, the species/variety problem was central to the work of naturalists. It was a problem that they argued about and puzzled over a great deal. A tacit commitment of that kind of work is that there exists some well-defined boundary between species; otherwise there was no point to the discussion. In turn, that made talk of "mutability" of species look wrong to the professional community, though for reasons that at bottom are cognitively not much different from (not much more logical than) the aversion to hockey helmets taken up in [7.5]. Hence, the association between sloppy or amateurish arguments and evolution was powerfully reinforced and underpinned by the salient normal science role of the species/variety distinction. In all, it makes sense that it was the experts, not the educated public in general, who had deeply entrenched inhibitions about evolutionary belief.

In contrast, no such expert/layman cognitive distinction would arise on the presumption that good design implies a process that can in some way look ahead and see what will work. So here the expert's predilections would be powerfully based on pervasive everyday experience, which makes everyone understand why in the event of a crime we ask who had a motive, or in the event that a string is tied around our finger we are prompted to try to remember why it is there.

10.4

It is not hard to account for how Darwin escaped the aversion of his colleagues to taking evolution seriously. Lyell's evolutionary geology provided a model for evolutionary thinking about life. Though Lyell himself went out of his way to distance his gradualist geology from any extension to biology, Darwin's five years on the Beagle led him to a solid conviction that in fact gradual transformation of species had occurred. Darwin did not become an explicit evolutionist until after he returned to England. But on his own later account, confirmed by his journal notes while on the Beagle, he was struck over and over by close relations between species in neighboring climatic zones and in neighboring fossil strata compared to much larger differences between contemporary animals facing similar conditions of life, but in separated parts of the world. The patterns of life on the planet made sense in evolutionary terms, but it was full of unmotivated coincidences if evolution hadn't occurred. A pattern that makes sense of things will look right; and a pattern that works will be used a great deal, which yields the automaticity that is the operational definition of entrenched belief [9.6].

Given a firm belief in evolution, the problem of how the evolution works becomes highly salient. Further, especially for an Englishman of Darwin's time, what sort of process to look for also is highly salient. We have in Darwin someone who is an evolutionist and not only in the speculative sense that earlier men had been evolutionists, but an evolutionist growing out of detailed field experience. Further, he was a British evolutionist, who could see himself as the inheritor of the tradition that yielded Newton's physics, Adam Smith's economics of the invisible hand, Lyell's evolutionary geology. He would be very deeply imbued with the Newtonian view that the world works through wonderfully simple natural laws, not by detailed or extraordinary intervention of some force guiding the process from outside the laws. And Darwin's scientific bible (as he says) was Lyell's geology, in which the marvels of nature appear from the long-continued effects of processes going on all around us. And as an English gentleman living off his investments, married to a daughter of a famous family of entrepreneurs (the Wedgewoods), he would be very much aware of Adam Smith's economics of the invisible hand, in which a marvelously efficient grand system emerges from a process with only local incentives to individuals, none of whom need be aware of, much less intending, an effect on the grand design which is emerging without a designer.

Finally, the most salient confidence-building experience [9.8] for Darwin itself helped set a pattern for the future. Before he had ever seen a coral reef, he contrived a theory to account for how they developed, which turned on the competing effects of subsidence of a volcanic island and the response of the coral animals who produced the reef (Gruber & Gruber 1962).

The rest of the story need not be told in any detail, since although in terms of the *P*-cognitive argument some points of interpretation would depart from previous accounts, the story remains much as it has been told before (Gruber 1981a; Mayr 1982). The Malthusian population argument was familiar to every educated Englishman of Darwin's time. The climax of the "uphill" phase [9.4] comes when Darwin sees that a variant of that Malthusian argument could yield an account of the evolution of new species by natural selection. Eventually Wallace, collecting specimens in the jungles of Malaya (and like Darwin 20 years earlier, already a convinced evolutionist puzzling over how evolution might work), would have the same explicitly Malthusian flash of insight Darwin recalled.

Gruber (1981b) has been able to show that the flash of insight Darwin recalled when "for amusement" he happened to reread Malthus did not come at all out of the blue and did not produce immediate conviction. There followed what I have called a consolidation phase. But there is a cognitively significant sense in which Gruber (and Mayr 1982:478–79) understate the importance of the Malthus incident when they point out that everything Darwin knew after that can be found in his notes before it. For before it can be found only bit by bit, not as a coherent pattern, seen as such by Darwin. It is certainly implausible—as Gruber and Mayr make clear—that Darwin would have failed to discover natural selection if he had mislaid his copy of Malthus. But that reaching the climax of the uphill phase was so sharply fixed in memory was cognitively significant. It marked a point of no return. For Darwin, channeled so firmly by the examples of Newton, Smith, and Lyell, and his own coral-reef argument, and with the Malthusian argument a familiar one (even if a copy of Malthus was not handy), reaching that point sometime around when he did reach it has a convincing inevitability. For what he noticed at that point looked so obviously like just what he had been intently looking for over the preceding months. For other great discoveries—for Copernicus on the account I will work out—the discovery comes as a surprise. It is not something that was being looked for but something serendipitously encountered. Darwin and then Wallace, like Crick and Watson a century later, found what they could immediately recognize as looking remarkably like just what they were looking for. What prepares the mind of a discoverer to recognize the power of a novel idea, however, is not always so apparent.

10.5

Continuing to follow the discovery schema [9.5]: Darwin moved on, following the Malthus incident, to a lengthy consolidation phase. He did not move effortlessly from seeing the new idea as plausible to the conviction that it must be right. Seeing that natural selection could account for how species could evolve does not show that they did so evolve. But in the language I have been using [7.6], Darwin in 1838 faced the direct cognitive rivalry that completes the uphill phase of the discovery schema. He had an elegant theory, very much in the spirit of the work of Newton, Adam Smith, and Lyell, that *might* account for evolution, but in a way that lacked any principle or force that would encourage variation that solved problems the creature faced. The process does not *aim* toward solving even short-term problems. Yet Darwin wanted to account for attaining what look like long-term goals, so that something as complicated as a wing, or eye, or brain could finally emerge. The Smithian argument in economics provides a counterexample to the presumption that good design implies a designer, but not one that would plausibly give confidence by itself that something as intricate as the structure of an eye could appear without the aid of design. On the other hand, Darwin would also have the universal intuition that a pattern that very neatly fits—a blending of what Newton found in physics and Lyell in geology—is not likely to be all wrong.

On the record, this tension was not easily resolved. Gruber finds that, for a few months at least after the Malthus flash of insight, Darwin wavers, and even sometimes seems to set aside natural selection to consider some alternative theory. It took four years (from 1838 to 1842) before Darwin was convinced enough to write out a 20-page sketch of the argument, and two more years went by before he wrote the first version intended for publication (1844). He was away from home at the time, and perhaps was a little startled at how convincing the argument had come to seem. He *immediately* wrote to his wife to tell her he thought his species theory would prove to be of lasting importance to science, and to tell her how she was to see to it that it was properly edited and published in the event of his sudden death.

So there was in this Darwinian case an uphill phase climaxing with the Malthus incident in which Darwin saw the radical idea as looking right. A consolidation phase followed (climaxing with the 1844 draft), in which intensely working over that idea, and finding that despite all the problems on the whole it seemed to work better and better, produced conviction. A long downhill phase then completes the schema. For another 15 years Darwin marshals detailed arguments, works out as far as he can further answers to objections to the theory, tries to win allies through informal communications, and finally publishes only when Wallace's independent discovery forced his hand.

10.6

A century later—now illustrating the subcommunity case [9.10]—plate tectonics solved another "mystery of mysteries" debated since Darwin's time, of how to account for striking relations in biology and geology across broad stretches of ocean. As with the origin of species, there was no established theory to be replaced. Here, even more than for the Darwinian case, the deeply entrenched habits that had to be overcome were not in support of an explicit prior theory but were shaped mainly by opposition to similar ideas the community had learned to reject. But the cognitive basis for expert aversion to drifting continents was not underpinned by anything like the species-versus-variety focus of work that is so prominent in the Darwinian case. Consistent with that weaker cognitive basis, there was never the firm expert consensus averse to the idea that held for evolution in Britain prior to Darwin. Instead, the intensity of the opposition to the drifting-continents view has to be understood in good part in terms of the polarizing effects that will be taken up in Chapter 14. Wegener's version of the theory proposed a mechanism that technically sophisticated readers rejected as certainly wrong. But competing proposals postulating land bridges at whatever locations could profit from one were no better. When evidence for a surprising new mechanism began to appear in oceanographic data gathered in the post–World War II years, its earliest exponents found a small but ready audience, which grew modestly until the mid-1960s, when supporters found the compelling magnetic anomalies in the seabed the theory had told them to look for. This was followed by an extraordinarily rapid collapse of serious opposition. The existence of a community, much broader than only geologists, which found the issue fascinating and which contained many people undeniably competent to follow the technical arguments, played a conspicuous role in the rapid spread of belief. If the issue had been debated only within geology, resistance would have been strong, and the debate much more prolonged.

The broader community point is also relevant to the fairly rapid acceptance of evolution within the expert community after Darwin published, despite the inhibitions I have described. For although Darwin intended his book as "one long argument" for natural selection, in the course of developing that argument he also provided a very readable, richly detailed volume of natural history supporting the necessary premise that evolution had somehow occurred, one way or another. Darwin's evidence for evolution was comprehensible to readers with no special expertise; and in contrast to the earlier arguments of Lamarck or Chambers, it was not vulnerable to technical counterarguments that reassured an expert that at least his brighter dinner companions or students could see his point.

Though both Darwin and Wallace had vivid recollections of the Malthusian moment when they first clearly saw how natural selection might account for evolution, neither had any such vivid recollection about a flash of insight yielding belief in evolution. That contrast makes sense if the general notion of evolution was cognitively always available [10.2]. No climactic moment of cognitive rivalry would occur since we are dealing with a familiar idea that gradually comes to look increasingly plausible. With the accumulation of field experience and thinking about that field experience, an idea that was never radically unthinkable grew into a habitual way of seeing things. We expect the discovery schema to be applicable in principle to any discovery, not only to radical discovery. But even something as important as evolution need not leave a cognitively striking story behind. Unless the idea is cognitively radical, in the sense of challenging some deeply entrenched habit of mind, the uphill phase can go by unmarked by the critical turning points that make for a good story. Darwin eventually came to see the Galapagos experience as critical, but only in a sort of rational reconstruction. In real time, it slipped by with no conscious sense of a turning point.

But when an idea is radical (as shown by the widespread "it's absurd" reaction it prompts in its first audience), it should be possible to provide a striking uphill-consolidation-downhill account. Similarly, it should be possible to give a corresponding account of the contagion of such an idea, and particularly of the critical early stages of the contagion [9.12], before the idea becomes fashionable enough that puzzles come to be about the holdouts rather than the converts.

A striking paper by Hull et al. (1978) provides an opportunity to test part of the argument of this chapter. Hull found no significant difference between older versus younger scientists in propensity to follow Darwin, in conflict with what would be expected on Planck's principle [9.11], and contrary also to Darwin's own expectation. But on the argument of [10.3] we would expect that for acceptance of evolution (the situation is more complicated with respect to natural selection) the "Planck" effect would apply only to older versus younger *biologists,* since it was only a biologist (not a scientist in general) who would be entrenched to the species habit of mind.

Eleven

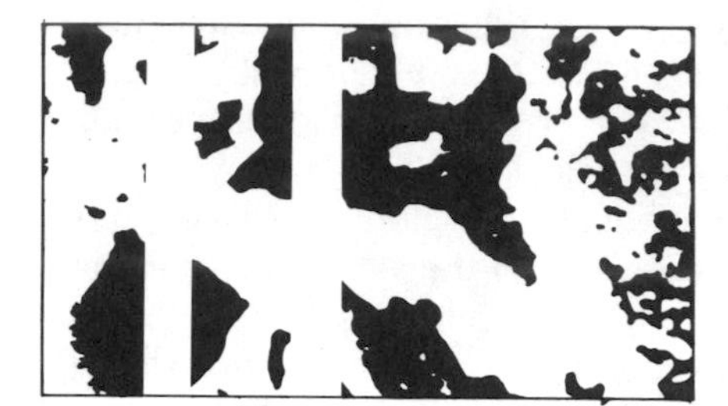

The Copernican Issues

An early sketch of Copernicus' theory ties his idea to dissatisfaction with a particular detail of Ptolemy's theory: the use of equants. But as will be seen, the two things (Ptolemy's use of equants and Copernicus' heliocentric theory) have no logical connection, so there is a puzzle here. This unexplained link imposes a prior constraint on whatever account we work out for the Copernican discovery. We want an account in which the uphill phase of the discovery somehow starts from Copernicus' concern with the equant—so that although the equant is *logically* irrelevant, we can see how it was not *cognitively* irrelevant.

What follows in the next several pages is a minimal sketch of the technical background to Copernicus' work, developed in just enough detail to let us reach the equant issue. A reader with a taste for the technical material ought to go on to the work on which this account is based (Price 1959; Neugebauer 1983:491–507; Swerdlow & Neugebauer 1984).[1] For many more readers, even the bare sketch of the next several pages will give more technical detail than can comfortably be handled. But although the story can certainly be followed fairly well even by a reader who skips these pages, a reader who takes the trouble to work through them will not regret it. A number of points that turn on the technical details are not only interesting as history but (as, specifically, I will try to show for the equant) cognitively interesting as well. The balance of this chapter will then consider just what Copernicus discovered, and how that differed from what Aristarchus discovered 1800 years earlier.

11.2

In Ptolemaic astronomy, planets have double orbits (fig. 11.1). The earth is at C'. The major orbit (the deferent) is offset from (eccentric to) C'

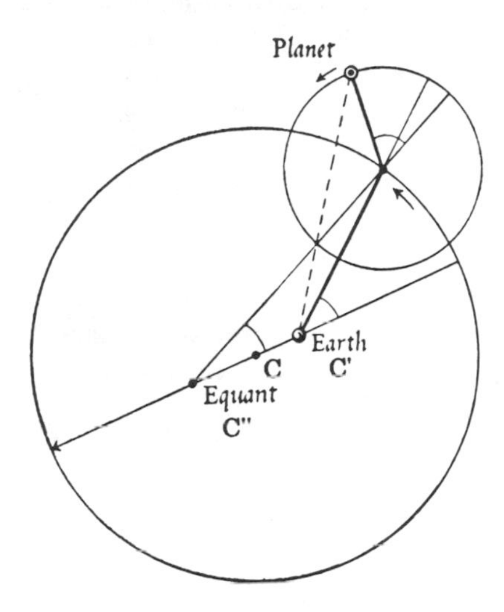

Figure 11.1. The Ptolemaic scheme, showing the major epicycle riding on (actually, see fig. 11.3, riding within) the deferent. But one of these orbits (the epicycle for the outer planets, the deferent for the inner planets) turned out to be merely an image of the earth's own orbit around the sun and so disappears in a heliocentric model. However, the missing orbit reenters a model designed to calculate observations by an observer on the moving earth. Hence the observational equivalence for naked-eye astronomy of the Ptolemaic and Copernican schemes despite the radical difference in cosmology. Ptolemy captured the nonuniformity of mean movement (abstracting from periods of retrogression) by the joint effect of making the orbit eccentric—hence the Earth is not at the geometrical center of the orbit—and by governing the mean motion from the equant—offset opposite to the Earth, as shown in the figure. Although Copernicus avoided use of the equant, Ptolemy's device in fact works just as well for a heliocentric system. On the scale of this drawing, planetary orbits would be indistinguishable from the circular figure shown. For the heliocentric case, the sun would take the place of Ptolemy's fixed earth at one focus of a nearly circular ellipse, and the epicycle would disappear. The required variation in orbital speed required by Kepler's second law (equal areas in equal times) would be virtually indistinguishable from that provided by a Ptolemaic equant at the vacant focus. From Derek J. Price, *Equatorie of the Planetis* (New York: Cambridge University Press, 1955). With permission.

so that its center is at C. The deferent then carried the secondary orbit (the major epicycle), which in turn carried the planet, all as shown in figure 11.1. The *equant* was then a third point, C'', symmetrically opposite the earth. Its role was to govern the rotation of the deferent. The equant was the point from which the angular movement of the deferent would appear *equal*. The deferent itself, consequently, would appear to sometimes slow down, sometimes speed up as viewed from an observer anywhere else, and in particular for an earthbound astronomer observing from C'.

In the Copernican models, there are deferents and eccentrics and epicycles but no equants—the effects Ptolemy gets by using the equant Copernicus gets by using small secondary epicycles. So Copernicus is more faithful than Ptolemy to the principle of compounding all heavenly motions out of uniform motions, as endorsed by all astronomers and cosmologists from Plato on.

This Copernican concern with the equant has often been read as revealing an instinct for the trivial. For from a modern perspective Ptolemy's equant is a brilliant idea. The setup gives a remarkably good approximation of the mean motions of the planets through the zodiac. Copernicus' alternative does no better. And the neatness of the arrangement (C'' and C' symmetrically in line with C) hints at some deep insight into the nature of things. You can get an intuitive glimpse of how it could be that this device worked so well by noticing how much Ptolemy's scheme looks like a nearly circular ellipse with C' as one focus of the ellipse and C'' the other focus. Almost Kepler's first step in putting the Copernican system into the shape that Newton would exploit was to drop the Copernican epicyclets and return to the Ptolemaic equants. In the light of that history, Copernicus' effort to improve Ptolemy by doing away with the equant looks merely fussy. A reader new to the Copernican literature will be struck by the unflattering character of some of the best-known works (Koestler 1959; I. B. Cohen 1985; Neugebauer 1983:491–506; Price 1959; Swerdlow 1973). Koestler titles his analysis of Copernicus "The Timid Canon." I. Bernard Cohen bristles at the idea there was any revolution that deserves to be called "Copernican."

But the first point that needs to be made about the equant and about Copernicus' interest in dispensing with it is that in the context of his times, Copernicus' initiative with the equant was not at all timid or pedantic. On the contrary, it provides a fine example of Kuhnian normal science. That the complex but always recurring motions of the planets were somehow compounded out of simple uniform circular motions was taken for granted by every important figure from the time of Plato through to Newton, with the main and almost sole exception of Kepler. Kepler abandoned it but only after intense labors trying to save it. His results were mostly ignored for decades until after Newton derived them from a more general principle. The failure to reduce observations to the compounding of uniform circular motion was seen (in just the way Kuhn has characterized normal science) as a failure for astronomers, not as evidence that the presupposition was wrong. Copernicus showed the boldness to take on and the talent to deal with what astronomers of the period regarded as the most fundamental defect of the Ptolemaic system.

But the equant problem could have led Copernicus to heliocentrism only by some roundabout route. For the problem that prompted Ptolemy's use of equants has no connection with the problems that heliocentrism solves. Use of an equant is a way to account for what Ptolemy called "the anomaly with respect to the zodiac," which is the variation in motion a planet exhibits contingent on where it is against the background of fixed stars. What caused this variation remained unknown until resolved by Newton at the end of the seventeenth century. A second and far more conspicuous anomaly—what Ptolemy called "the anomaly with respect to the sun"—takes the form of occasional backward looping (retrogression) of the planets against the background of the fixed stars. Retrogression can occur anywhere on a planet's path, linked (as Ptolemy's label implies) with the position of the sun relative to the earth, and independent of the planet's position in the zodiac. In Ptolemy's scheme, the equant combines with the eccentric to deal with the first anomaly; the second (retrogression) is dealt with by the major epicycles.

For Copernicus, there is no occasion for a device to account for retrogressions, since on the heliocentric scheme the looping motions are only a parallax illusion due to the motion of the earth. So the major epicycles disappear. The anomaly with respect to the zodiac, however, remains to be accounted for. You can see its effect by looking at a calendar and confirming that the seasons (number of days from winter solstice to spring solstice, and so on) are not uniform. Rather, the sun moves through the zodiac in a somewhat unequal way, so that winter (in the northern hemisphere) is several days shorter than summer. Similar anomalies occur for each of the planets, and any theory of the heavens had to somehow account for them.

Once all this is noticed, there is no puzzle about why Copernicus cared about the equant, and especially no puzzle about how he happened to start from work on the equant. The equant or something functionally equivalent to it is required to account for the anomaly with respect to the zodiac. And because of the well-entrenched belief that heavenly motions must be compounded out of uniform motions, which the equant violated, it was a perfectly normal concern for a sixteenth-century astronomer to consider how the equant might somehow be dispensed with. Ptolemy himself had left the problem completely up in the air. He reiterated the axiom of uniform motion; and then he used equants, without comment, as if he did not notice that he was violating his own stated principles. For upward of 1000 years, recurrent indications appear that astronomers would like to dispense with this inconsistency. But for us, the puzzle is how work on the equant could lead Copernicus "uphill" toward a revolutionary discovery that has nothing to do with the equant.[2]

That the normal component of Copernicus' work turned out to be more readily acceptable to sixteenth-century astronomers than the revolutionary component makes Kuhnian sense. Nevertheless, the extent of the disparity between astronomers' immediate response to his excising the equant and the decades-delayed response to his heliocentric ideas shows us how profound a cognitive barrier had to be overcome to see the world in the Copernican way. For 35 years after Copernicus published—and for seven decades after he circulated the *Commentariolus*—nothing was done that was actually built on belief that the earth moved.[3] Copernicus was recognized as a major figure in astronomy ("a new Ptolemy").[4] But Westman (1980) found just 10 explicit Copernicans up to 1600, most of whom—Bruno, Kepler, Galileo, and several more—came very late in the century, two generations after Copernicus' book. I will argue in Chapter 13 that the cognitively crucial shift had taken hold by 1590, even though there were still very few Copernicans. But even 1590 is almost half a century after the argument appeared. Even after Copernicus had spelled out the remarkable argument I will review next, and even though his book seems to have been studied by every important astronomer for the balance of the century, it took four decades before any substantive impact became apparent.

11.3

We next want to turn to the question of *what Copernicus discovered.* Here it is essential to stress a simple point that is at the heart of the cognitive argument I will make but which is often understated and sometimes ignored. The point is that what Copernicus discovered was not a fact (the earth goes round the sun, not the reverse) but an argument about what would follow if the earth went around the sun. And there was nothing in that argument that depended on any fact not known since Ptolemy. The idea that the earth might go round the sun was proposed 1800 years before Copernicus (see the discussion of Aristarchus later in this chapter). But the Copernican argument about what that implies was totally new with Copernicus. To the extent that Copernicus persuaded anyone that the orbiting earth was a fact (and he did persuade a few people, starting with himself and, decades later, Kepler and others), it was because the argument about what would follow from the heliocentric supposition was so striking that it became easier to believe even that the earth was in orbit than to believe that the Copernican argument revealed only an astonishing set of coincidences.

Figure 11.2 is a diagram Kepler published nearly 400 years ago (1596). Near the bottom of the diagram, unlabeled, is the earth, and closest to it, the moon. Moving outward, we then have the rest of the Ptolemaic ordering: Mercury, Venus, sun, Mars, Jupiter, Saturn, each labeled with its Latin name. But Kepler's diagram of the Ptolemaic system is not drawn quite the way a Ptolemaic astronomer would have drawn it. For expository convenience, however, I want to first go over the argument made by Kepler in 1596 and a half-century earlier by Copernicus. Then we must go back to consider what Kepler left out.

In figure 11.2, you can see that the sun and moon each have a simple orbit, but each of the planets has the double orbit already illustrated with figure 11.1. Each planet rides on a secondary sphere (the major epicycle), which in turn orbits the earth. In figure 11.2, the arc bearing the name of the planet gives the path of the center of the planet's epicycle. The deferents were conceived to be spherical shells which contained the epicycles and carried them around the earth, as explicitly shown in figure 11.3.

This scheme is sufficiently far from our own habits of mind that only a reader who takes the trouble to work through a more detailed account than I have presented here is likely to feel secure with the ideas, though as with other things, once you have gotten the pattern of the thing it will be impossible to recapture whatever it was that made the scheme seem difficult when first seen. Any reasonably well-educated person in the sixteenth century might know as much as Kepler's diagram shows, but rarely any more than it shows. An astronomer, however, would know the system intimately, including many intricate details (minor epicycles, eccentrics, and so on) with which an ordinary well-educated person would not be familiar. An important point to note, even if you have not yet grasped the picture, is that the alternative representations in figures 11.1 and 11.3 are not physically inconsistent.[5] Rather, figure 11.1 is a schematic version of figure 11.3. The explicit spherical shells carrying the epicycles of figure 11.3 are replaced in figure 11.1 (and in Kepler's fig. 11.2) by a line showing the path of the center of the epicycle as it is carried around the sun. But the deferent is actually a rotating spherical shell thick enough to contain the epicycle, even when it is represented in a diagram as a circle with the epicycle shown as a smaller circle cutting across it.

11.4

In Kepler's diagram, between Mercury and the Earth is the moon, which has no major epicycle; and the sun (again, with no major epicycle)

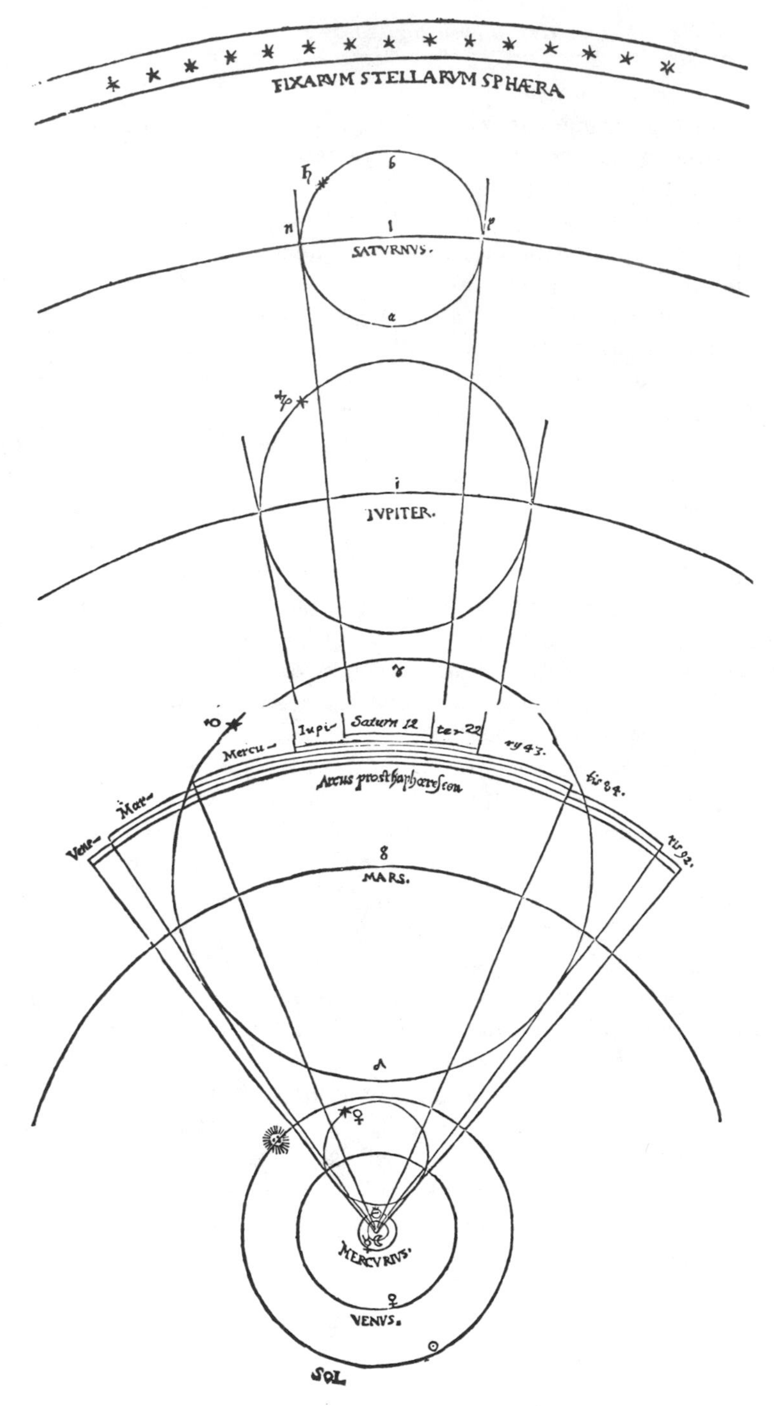

Figure 11.2. Kepler's (1596) diagram illustrating the logical structure of the Ptolemaic world. Compare with the Copernican analog, fig. 11.5. From Derek J. Price, *Equatorie of the Planetis* (New York: Cambridge University Press, 1955). With permission.

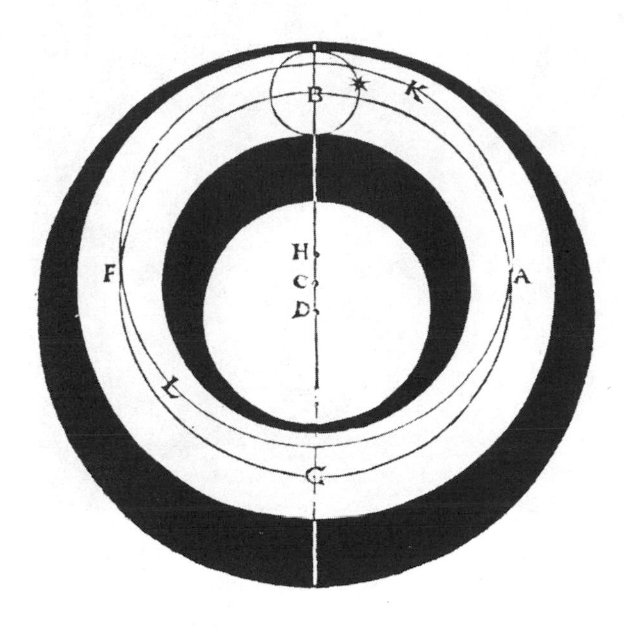

Figure 11.3. A fifteenth-century diagram explicitly showing the nested-sphere conception for Venus. Reprinted from *Vistas in Astronomy* 17, *Copernicus: Yesterday and Today,* Arthur Beer and Peter Beer, general eds. With permission.

comes between Venus and Mars. The moon and sun always proceed in a direct course around their orbits. But the planets sometimes loop backward (fig. 11.4). The epicycles account for that, since when the planet is on the part of the epicycle nearer the earth, the rotation of the epicycle gives the planet a motion that subtracts from the primary motion (it is moving backward with respect to the direction of the deferent). By adjusting the size and rotation of the epicycle, retrogressions can be arranged which just match the observed behavior of each planet.

In this Ptolemaic setup, all the planets are synchronized with the motion of the sun but not all in the same way. Mercury and Venus are synchronized in their primary orbits. Their epicycles move around the earth with the sun, so that these two planets always are seen close to the sun. Mars, Jupiter, and Saturn, on the other hand, have their secondary motion (not the primary) linked to the sun. For these planets, it is the rotation of their epicycles (not the deferents) that is synchronized with the sun. If you draw an arrow from the center of the epicycle to the planet, and another arrow from the earth to the sun, the arrow for the planet is required to

Figure 11.4. A planetarium display shows the loops that a planet traces among the fixed stars if observed over a sequence of evenings when the planet is rising as the Sun is setting. Permission to reproduce photo by Lessing-Magnum from Planetarium in Deutschen Museum.

move at whatever speed will keep it always pointing in the same direction as the sun's arrow. This guarantees that retrogressions will always come when the sun is on the opposite side of the earth from the planet (the backward looping always occurs when the planet is rising as the sun is setting).

Looking at how much information can be compactly summarized on a single, not terribly complicated, diagram gives some sense of the ingenuity and impressiveness of the Ptolemaic system. As seen by a Copernican, however, the Ptolemaic system describes much, but explains little. In particular:

A. Why do the planets have looping motions (retrogressions), but the sun and moon do not?

B. Why do the deferents of the moon, Mars, Jupiter, and Saturn orbit independently, while the deferents of the sun, Venus, and Mercury are tied together?

C. Why do the epicycles (not the deferents) of Mars, Jupiter, and Saturn move with the deferent of the sun? Why are these three planets linked to the sun in this way, Venus and Mercury in another way, and the moon neither way?

D. There is no observation that can test the Ptolemaic ordering. For the Ptolemaic theory gives definite results only about angular positions. The theory would work exactly as well if we shuffle the planets in any arbitrary order. So far as the mathematics of the Ptolemaic system are concerned, Mars could be put inside Mercury, or Mercury outside Saturn. Is there no way to know definitely the order of the planets?

E. Given a choice about the size of the deferent for a planet, an epicycle can be uniquely chosen to produce retrogressions that fit the observations. But the Ptolemaic scheme gives no hint at why the retrogressions take the particular values that are observed.

Altogether, on the Ptolemaic arrangement, the epicycles decrease in relative size as you move away from the sun in either direction; but there is no reason why that must be so. As I will be sketching out later, Ptolemaic astronomers could give an account of the symmetrical ordering of the planets, with the sun dividing the two classes of planets and with the epicycles getting relatively smaller as you move away from the sun in either direction. But a critic could claim that the ordering had been chosen to make the arrangement symmetrical, and then reasons found to give some plausibility to what is really an arbitrary arrangement, full of unexplained coincidences and puzzles.

With that background, we can now say what Copernicus discovered: namely, if you put the sun, not the earth, at the center, then all the questions and puzzles of points A–E are resolved. It then follows that:

1. Viewed from the earth, any planet in an orbit closer to the sun than the earth could never be separated from the sun by a large angle. That Mercury and Venus behave that way, and the other planets do not, simply tells us something about the location of those planets. Contrary to Ptolemy, Venus, not Mercury, must be closer to the earth.

2. If, however, the world is Copernican, then the sun, Mercury, and Venus no longer have a common mean period of one year. The sun is not in orbit at all, and when the periods of Mercury and Venus around the sun are calculated (88 days and 8 months, respectively), the intuition which had always been used to order Mars, Jupiter, and Saturn only (the greater the period, the greater the distance) now turns out to be applicable to all the planets. The earth, with its orbit of a year, comes where it should, between Venus with its period of eight months and Mars with its period of two years.

3. In the geocentric scheme, we can only measure the angle between the sun and a planet, taking the earth as the vertex. But if the world is heliocentric, a second angle is available, so that the distance of each planet relative to the sun/earth distance is definite. Of itself this argues nothing for the heliocentric view. It simply reflects the fact that more information is available from a moving planet than from a fixed position at the

center. By itself this implies nothing about whether the earth really is moving. But the additional inferences that must hold if the Copernican postulate is true are striking. The earth/planet/sun angle can only be 90 degrees when Venus or Mercury is at maximum elongation from the sun; so a single observation is sufficient to determine the distances. A more complicated procedure is needed for the outer planets. Still, just two observations will determine the distances for each. So while in Ptolemaic astronomy, Jupiter need not really be between Mars and Saturn (rather than, say, outside Saturn or inside Venus), in the Copernican scheme the ordering of the planets and the relative size of each orbit are completely determined by a few observations. For the outer planets, the ordering from calculation coincides with the traditional ordering by the intuitive relation of period and distance. But now this intuition extends (point 3) to all the planets.

4. The most striking peculiarity of planetary movements now simply disappears. Retrograde motions are merely an appearance. No planet actually reverses direction, so there is no puzzle about why some orbits exhibit retrogression and others don't. The earth cannot overtake the sun, which it orbits, and it cannot be overtaken by the moon, which orbits the earth. But from the perspective on any one of the planets (say the earth) all the others will appear to have periods of retrogression—as a tower on land, seen against a background of mountains from a ship traveling east, will appear to be moving to the west.

5. The planets exhibit (apparent) retrogression if, and only if, they rise as the sun sets (or the reverse) because what that means is that the earth and the planet are in line on the same side of the sun. But that defines the case where the earth is overtaking an outer planet, or is being overtaken by an inner planet. The appearance of retrograde motion by that planet must take place then, and can only take place then.

6. However, it is not only the qualitative relations that are explained if the world is heliocentric. For given the known periods and the calculable distances of the heliocentric scheme, we can determine not only when retrogressions must occur but also how extensive they must be and how long they must take. For each planet, these calculations turn out to neatly yield the results required to match what is observed. So a complete account of retrogressions—quantitative as well as qualitative—is rigorously implied by the Copernican axiom; in every observable detail, what is seen coincides with what the calculations require.

Copernicus began *de Revolutionibus* with an explanation of why he supposed it reasonable to look for an alternative to the Ptolemaic system; and by way of his guidance to Rheticus (1541), he had provided a more extended version of his reasoning. As in Kepler's argument in 1596 and in Galileo's in 1632, what is emphasized above all is the systematic coherence of the heliocentric view. In the letter to the pope that serves as the introduction to *de Revolutionibus,* Copernicus climaxes his argument by asserting that the Ptolemaic system cannot "discern or deduce the principal thing—namely, the shape of the world and the unchangeable symmetry of its parts. With [Ptolemy] it is as though an artist were to gather the hands, feet, head, and other members for his images from diverse models, each part excellently drawn but not related to a single body; and since they no way match each other, the result would be monster rather than man." In contrast, Copernicus offers a scheme that proceeds from a few observations to compute "the shape of the world, and the unchangeable symmetry of its parts." Everything follows "as if linked by a golden chain" (Rheticus 1541). In contrast, *de Revolutionibus* contains scarcely a word explicitly about equants (Swerdlow and Neugebauer 1984).

11.5

Some sense of the change in gestalt (the sense of how things fit together) that is involved in the Copernican move might be gotten from thinking through how the world is put together in a Ptolemaic way (as in fig. 11.2) versus in the Copernican way (fig. 11.5). Beyond the basic assumption that the earth is at the center of the system, an explicit Ptolemaic account needs assumptions to determine the order of the planets; to allow for the existence of epicycles (violating Aristotle's commitment to homocentric movement); to account for the coordination of the epicycles of the outer, but not the inner, planets with the motion of the sun; and to account for the coordination of the deferents of the inner, but not the outer, planets with the movements of the sun. Then observations (beyond those required on the Copernican scheme) were needed to fit the model to the world.

Given sufficient observations and assumptions, you can then start at the presumed center (the earth) and build up the three-dimensional structure of the world as the successive planets are added, using the principle of snugly nested spheres. The inner edge of the sphere of Mercury coincides with the calculated maximum distance of the moon; the outer edge of the sphere of Mercury is then determined by the ratio of epicycle radius to deferent radius required to fit the observed retrogressions. The outer distance of the sphere of Mercury then determines the inner distance of the sphere of Venus; and so on out to the fixed stars, presumed to lie just beyond the outer edge of the sphere of Saturn. All this is put together rather like constructing a spherical layer cake (see Van Helden 1985 for a detailed discussion).

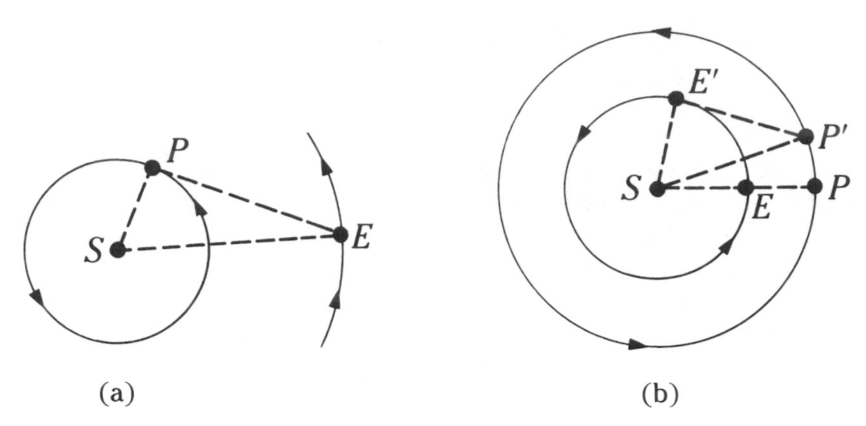

Figure 11.5. These sketches (from Kuhn 1957) for an inner (*a*) and outer (*b*) planet illustrate the "golden chain" logic of the Copernican scheme: a few observations (relative to the observations and supplementary assumptions required to give three-dimensional structure to the nested-sphere Ptolemaic scheme) determine the ordering, relative distances, and retrogressions of the planets. Reprinted from Thomas S. Kuhn, *Copernican Revolution* (Cambridge: Harvard University Press, 1957). With permission.

But the Copernican scheme works and *feels* (going over in your mind how things fit together) like a radically different sort of thing. A few measurements (relative to what had been required to work out a complete Ptolemaic scheme) are sufficient to determine the rest of the system by strict deduction, "as if linked by a golden chain." The order of the planets, their retrogressions, their distances relative to the earth/sun distance, are all fixed by the internal coherence imposed on the system by the single assumption that the sun is at the center.

This Copernican argument was a memorable performance. It could scarcely have been forgotten even if it were false that the earth orbits the sun.

However, if the argument is so striking, and so simple, and based essentially on features of heavenly movements that must have been known to numerous astronomers for centuries before Ptolemy, then why had no one else ever noticed? The obvious parts of the answer must be that the Copernican idea contradicts common sense, contradicts Aristotelian physics, and contradicts the Bible. But we have good reason to deny that these things by themselves can account for why the Copernican discovery was so long in coming and so slow to be taken seriously by technically expert astronomers (by the sort of people who could actually do anything with the Copernican idea). The reason is that the daily rotation of the earth on its own axis—without supposing any annual orbit of the sun—also contradicts Aristotelian physics, the literal interpretation of the Bible, and common sense. But cognitively the rotation of the earth was not a terribly difficult idea.

The idea that the earth rotates daily—as the alternative to the standard idea that the rest of the universe rotates daily in the opposite direction—had been proposed many times over the centuries (see, for example, Duhem 1985:505–6). It always had a certain intuitive appeal to offset the theological and physical arguments against it. Once a reason had appeared for taking it seriously (in the way I will come to in Chap. 13), it proved to be easy to swallow for many people who found the whole Copernican scheme undigestible. In terms of the earlier argument about how to recognize depth of entrenchment of a habit of mind [9.4], therefore, the commitment to Aristotelian physics and the commonsense intuition that the earth is not moving cannot account for the thousand-year lag before the Copernican idea emerged. For daily rotation of the earth was an idea that we find repeatedly invented over that period and which the record shows looked fairly plausible even to people who rejected it (like Tycho).[6]

The harder part of the Copernican idea was that the earth is a planet orbiting the sun. On the record, it must have been very much more difficult to see the earth as a satellite of the sun than to conceive the earth as turning on its own axis. Although a very few references to the notion can be found, there is no evidence that anyone between Aristarchus (c. 250 B.C.) and Copernicus presented even a vague sketch of technical development in astronomy based on the idea that the motion of the sun might be an illusion due to an annual orbit of the earth (Neugebauer 1982:693–98). And the half century delay before the Copernican argument took hold, following the many centuries of delay in its discovery, shows that there was some very powerful cognitive difficulty that kept people from seeing an orbiting earth as making sense. I want to proceed with unpacking the difficulty in two steps: first dealing with the interval from Aristarchus' proposal of the heliocentric idea 400 years before Ptolemy, and then for the period from Ptolemy (c. A.D. 150) to around 1590, when we first see clear evidence that the Copernican argument has been fully grasped. For the cognitive story is quite different for (roughly) the period up to Ptolemy and the period after Ptolemy.[7]

Logically, the Copernican argument was available long before Ptolemy; although cognitively it was not, but for a fundamentally different reason than holds for the many centuries of further delay after Ptolemy. A person might see the heliocentric idea, as Aristarchus actually did, but he had no cognitively feasible way to reach the golden chain of relations that could make that heliocentric idea look plausible. After Ptolemy (I will argue), the golden chain was there almost asking to be discovered by any competent astronomer: yet for a very long time no one saw it. Somehow the Ptolemaic system served as both the seed of the idea and as a very tough hull that prevented the seed from sprouting. I want to tease out what made that hull so hard to crack and how (finally, after more than a thousand years) Copernicus became the man who cracked it.

11.6

Aristarchus' heliocentric argument has come down to us in the form of a few lines in Archimedes' "Sand-Reckoner," which reports just two points: first, the appearance that the sun goes around the earth could be an illusion generated by the earth going around the sun; second, if the earth does go around the sun, then the distance to the fixed stars must be enormous, since otherwise a person would be able to observe a shift in the apparent location of the stars as the earth moved around its yearly orbit. The distance to the stars must be so enormous that the distance from the earth to the sun is negligible in comparison. Archimedes' context was one in which he wanted to demonstrate how to deal with fantastically large numbers. But if Aristarchus said anything more about the heliocentric idea than has just been mentioned, no hint of it has survived, and that is hardly plausible if he had offered anything so striking as the Copernican golden chain argument of [11.4].[8] So we have a double question: How did Aristarchus manage to see the heliocentric possibility? And given that he had seen it, why did he miss its striking implications?

The first point has a fairly standard answer. For although we only know one important thing about Aristarchus other than his heliocentric proposal, that happens to be enough to suggest how the discovery might have been made. The one other thing we know about Aristarchus was that he worked out an ingenious piece of reasoning to produce an estimate of the size of the sun.

Aristarchus' method was in fact very crude, since to give an accurate result it required an observation far more precise than could be obtained even for some decades after the invention of the telescope. Nevertheless, naked eye observations alone allowed Aristarchus to set a lower bound to the size of the sun that was very large compared to earlier estimates. At their most generous, earlier estimates made the sun as large as the earth. But Aristarchus put the sun's diameter at seven times the diameter of the earth, making the sun's volume 250 times the volume of the earth. The correct numbers are in fact 20 times larger for the diameter, hence $20^3 = 8000$ times larger for the volume. Yet Aristarchus' estimates were large enough to make him the first person to realize that the sun is immense compared to the earth. At that point, a salient pattern of intuition for Aristarchus becomes easy to point to. Common intuition tells us that little things often swarm around big things (birds around a ship, ants around a piece of food, and so on); but big things are not known to swarm around a little thing. Aristarchus therefore faced, as no man before him had faced, the startling idea that the earth is very small compared not just to the size of the universe but compared to a single object in the sky. If anyone could have the intuition that the earth might orbit that object rather than the reverse, Aristarchus was the person.

But the idea had no lasting impact. Lacking the golden chain of relations that Copernicus was able to produce, the core idea survived only as something akin to Zeno's paradoxes: an argument that was harder to answer than you might suppose considering how ridiculous its claim was. Aristarchus' discovery of the immense size of the sun could not shock people into thinking the earth might really be in orbit because a particularly effective taming [7.7] was available for this idea. For if you think of the heavenly bodies as having natural movement around the center of the universe, and the earth happens to be at the center, then the big/small puzzle disappears. The big sun orbits the small earth only incidentally, because the sun is in orbit around the center of the universe, and the earth happens to lie at the center. From then on, *if* a person came to consider the heliocentric view, he could see the disparity in size between the earth and sun as something that makes better sense in terms of the heliocentric world. But unless new techniques let someone calculate a more accurate size for the sun (so that a new cognitive shock could be produced), recognizing that the sun was as big as Aristarchus estimated could not produce the cognitive rivalry we require. That was something a person already knew and had reasoning-why to deal with.

11.7

But given that Aristarchus saw the heliocentric insight, why did he apparently fail to see anything of the Copernican harmonies? A simple answer is available, though as far as I have found it has not been previously discussed. Aristarchus lived before Ptolemy made it cognitively feasible to see the consequences of the heliocentric idea. When Aristarchus was alive, the standard view of the heavens was that enshrined in Aristotle, which accounted for movements in the heavens as the interaction of rotating spheres centered on the center of the universe (hence, also, centered on the earth). Ptolemy's universe was crucially different.

There were 55 nested spheres in Aristotle's universe. All the planets moved in a region between the sun and the fixed stars. But beyond the sun there was no three-dimensional quality to Aristotle's world. Each planet was contained in a nest of spheres whose combined motions were intended to approximate the observed movement of the planet. But the nest of spheres accounting for the movement of a particular planet need be no thicker than the minimum required to have room for the planet itself, which would be trivial on the scale of earth/sun distances. Therefore, beyond the sun the Aristotelian world had no intuitive sense of depth. There were occasional bits of speculation about a three-dimensional structure in Plato and elsewhere, but they were only that: a line or two of speculation. Consequently, there was no occasion to work with these distances, to check them, to train students in how to derive them, to become familiar with diagrams in which the space between planets played an essential role. Hence, in the way I have often stressed, there was no way that an astronomer would acquire the habit of mind of seeing the starry vault with marked qualities of depth.

But in a universe where the planets could be what they naively looked to be, a person could not be prompted to the Copernican insights by intuitions well-entrenched by everyday patterns of experience with perspective and parallax effects. Rather, the tacit habit of mind would see the planets as moving *among* the stars. Such a person could not just "see" (in the mind's eye) that if the earth moved around the sun then the looping retrogressions of the planets could be created by the resulting parallax illusion. To naive intuition, the vault of heavens is a surface. There is no sense of depth, and it is only experience and drill and training (not an isolated line or two of speculation) that could overcome naive intuition and give a person the three-dimensional sense of the heavens that is required to put the Copernican argument within cognitive reach. In the essentially flat vault of Aristotelian cosmology, retrogressions would only appear—even in Aristarchus' heliocentric world—if the planets really do loops backward, just as the parallax effect from a ship would not appear if the tower and mountains were only a painted backdrop.

But Ptolemy's world, unlike Aristotle's, intrinsically has depth (fig. 11.2); there must be room for the epicycles. That depth is the key to both aspects of the puzzle we are exploring. It is the intrinsically three-dimensional character of the Ptolemaic heavens that makes the Copernican argument accessible in principle to any Ptolemaic astronomer who saw the heliocentric possibility; but as I will try to show next, it is also that three-dimensional character that kept the heliocentric insight from astronomers for more than a thousand years. In the metaphor I have used: Ptolemy provided both the seed of the great idea and the tough hull that made that idea very hard to get to.

The most deeply entrenched habit of mind Darwin had to overcome was tied to everyday experience, not expert experience [10.2]. But the reverse held for the Copernican case. The intuition that good design implies a designer reflected very widely shared experience in the world; but the naturalists' aversion to evolutionary arguments in biology was tied to the specialized experience of that expert community [10.2], and it was also easier to overcome. In the Copernican case, both sorts of entrenched intuition can be noticed, but this time it is the expert intuition (not the everyday one) that was most deeply entrenched. The argument can be framed as follows:

What made it difficult to discover the heliocentric argument, and then difficult for the first generation of post-Copernican astronomers to see its force, was that a Ptolemaic astronomer had a very solidly entrenched sense that his science yielded deep insight into the nature of the world. The Copernican golden chain argument, once discovered, did not suddenly make sense of what had previously seemed only a collection of oddities. A Copernican, starting with Copernicus himself, would *come* to see the issue that way, and so far in this chapter I have presented it that way. For once you can see things in the Copernican way (once, in Kuhnian language, you can see the world in terms of the new paradigm or gestalt), the Ptolemaic way of seeing things comes to look so contrived that it becomes hard to imagine that it could ever have been taken seriously.

But if you only knew the Ptolemaic gestalt, it would not look contrived at all. Rather, it would look very powerful. To start, although we now think of Ptolemaic astronomy as full of empirical difficulties, for a Renaissance astronomer, it was a magnificently successful enterprise. Although much has been made of the need for reform of the calendar, what was involved was a discrepancy of 11 days which had accumulated over 15 centuries. So the error was about 18 hours a century.

Similarly, Tycho would later comment on his disgust with Ptolemaic astronomy when as a student he found that a rare conjunction between Saturn and Jupiter occurred several weeks after it should have occurred, according to the Alphonsine tables. But in the way Kuhn (1977) has described, a scientist's judgment about whether a theory exhibits "reasonable agreement" with observation reveals something about his expectations, not something about an absolute standard of reasonable agreement. In the sixteenth century, that tables prepared 300 years earlier could forecast a conjunction with an error of only a few weeks spoke much more for the power of the theory than for its deficiency.

So it is essential to notice that there was an enormous and very reasonable presumption in favor of Ptolemaic astronomy. Improving it was an admirable objective; throwing it out and rebuilding astronomy on some radically different basis was not even a plausible objective.

Given that powerful presumption, and the intricate training and practice it took for a person to become a technically competent astronomer, we would expect to find that astronomers had learned to see the Ptolemaic universe in a way that made sense to them, and learned to tame its various difficulties. And indeed a remarkably neat sense of that had developed. So in the bootstrapping way that is characteristic not only of science but of belief systems generally, any occasional doubts about the empirical adequacy of the theory tended to be stilled by its theoretical elegance, and any doubts about its theoretical elegance by its evident empirical power. If you were deeply involved in astronomy, seeing things in the Ptolemaic way would become automatic, effortless, and comfortable. For the Ptolemaic astronomers (surely including Copernicus before he became a Copernican), the Ptolemaic system made sense of the heavens, as for chemists before Lavoisier phlogiston made sense of combustion, and for physicians before Harvey, Galen's liver theory made sense of blood.

For people deeply entrenched in any of these viewpoints, an argument framed in terms of another, incompatible, viewpoint would look not merely wrong but as not even possibly right. Only with difficulty—with the kind of highly motivated, concentrated effort and drill it takes to overcome an entrenched physical habit—would it become possible to see the Copernican view as making sense. Otherwise, the Copernican argument would look incoherent; and the Ptolemaic astronomer would not see that the perception of incoherence was not something internal to the Copernican argument but an incompatibility between the Copernican argument and how astronomers thought they knew the world was organized. The Copernican argument would look like a proof that $2 + 2 = 5$, where the cognitive response is one of knowing the argument is absurd even if you cannot put your finger on where it is wrong.

11.8

An explicit version of the Ptolemaic way of seeing things would go something like this, starting with the ordering of the planets: indisputably, the moon is closest to the earth, since it eclipses all other heavenly bodies; only the moon is close enough to show observable parallax over the course of an evening. On the Ptolemaic view, it was then not merely a coincidence that the moon also exhibits the most complex movements. Today we would say that its orbit is perturbed by the three-body gravitational interactions with the sun and earth. But in Aristotelian terms, the complications were due to the moon's proximity to the corruptible sublunar world. And of all the bodies beyond the moon, Mercury has by far the most erratic movements. It is the only visible planet, we can say today, whose orbit is elliptical enough to vary in an easily noticed way from a circular orbit. On the Ptolemaic argument, just as the complicated movements of the moon show the effects of its closeness to the earth, the somewhat less complicated movements of Mercury show that it too is close enough to the earth to be vulnerable to sublunar influence.

Next must come Venus, whose orbit is linked to the sun in the same way as Mercury's but without the complications. Then comes the sun, dividing the two classes of planets and located where it can govern both classes. Finally, Mars, Jupiter, and Saturn follow in the commonsense order suggested by their increasingly long periods of revolution (2, 12, and 30 years, respectively). The ordering shown in figure 11.2, consequently, would look intrinsically plausible to a Ptolemaic astronomer. But beyond the reasons already given, there was another, very striking, confirmation of the ordering.

A Ptolemaic astronomer, if we had one alive today, would perceive at a glance that Kepler's diagram (fig. 11.2) is wrongly drawn, and the slip is one that a Ptolemaic astronomer could no more make than you could draw a giraffe but neglect to give it a long neck. Kepler shows the heavenly spheres floating apart from each other. This is not clear for the inner planets, where the diagram is so crowded the point is ambiguous. But for the outer orbits, it is clear that the figure is drawn as if there are gaps between the spheres of the successive planets. The cosmology of the Ptolemaic world, however, requires that the inner boundary of the sphere of a planet coincide with the outer boundary of the sphere of the next closest planet [11.5]. The universe is built out of the resulting structure of snugly nested spheres (Kuhn 1957:81–82; Van Helden 1985:16–40).

Aristarchus' calculation of the earth/sun distance, as rederived by Ptolemy using a second kind of observation, confirms this by a remarkable fit between (a) the space available between the moon and sun and (b) the space required to insert Mercury and Venus to yield the Ptolemaic symmetry. In the Copernican scheme, each planet stays at approximately the same distance from the sun. The variations in earth /planet distances (roughly 7:1 for both Venus and Mars, less for the other planets) occur because sometimes the earth and another planet are on the same side of the sun and sometimes on opposite sides of the sun. The distance variations come out to be exactly the same on the Ptolemaic scheme (as they must, given the observational equivalence; see n. 5); but now it is because the major epicycles carry the planets closer to or farther from the earth.

So in the Ptolemaic scheme, there should be a sequence of planets whose combined ratios of greatest-to-least distance from the earth fit the available space between the nearest approach of the sun and the furthest distance of the moon. Further, the period/distance relation says that this sequence should start with Mercury and Venus and would most neatly end with just that pair, since the planet that must come next (Mars) is qualitatively linked to Jupiter and Saturn and different from Mercury and Venus (point 3 in [11.4]). According to the surviving copies of Aristarchus' calculation, the sun/moon ratio is between 18:1 and 20:1. The actual ratio (we now know) is about 400:1. Ptolemy's calculation used a different argument, based on another sort of observation, but reached a result close to Aristarchus': about 17:1. This, remarkably, is almost exactly what is needed to allow Mercury and Venus to fit neatly between the sun and moon. In the form all this reached Europe by way of Moslem astronomers, the rough edges had been smoothed out, and the fit was exact. Hence, the very planets that most plausibly (on the earlier argument) would be fitted between the moon and sun—namely, Venus and Mercury—can be calculated from observations to yield a combined ratio which just matches the space available between the sun and moon. Neugebauer remarks (1982:111–12) that "a most implausible accident . . . became the solid foundation of the planetary shell structure [the nested-spheres] which dominates mediaeval astronomy."

The symmetry that then appears in the Ptolemaic arrangement—with the sun between the two kinds of planet and with the epicycles diminishing as you move away from the sun in either direction—becomes an elegant confirmation of the ordering imposed by the nested-sphere logic. We then also get a very detailed and concrete picture of the world, familiar to every literate person through the works of Dante and many others. For although a Ptolemaic astronomer cannot calculate detailed dimensions directly from observations (as in the Copernican scheme), once you have seen the nested-sphere logic you can calculate the distances just as well as in the Copernican scheme.

The distances so calculated are about 50% larger than those calculated on the Copernican argument. The Copernican calculations were essentially right (we now know); and the Ptolemaic distances were wrong. But within the limits of naked-eye astronomy, which yielded only angles not (directly) distances, not the slightest indication of that state of affairs could be discovered. Since there was no way to *empirically* discriminate between the Ptolemaic and Copernican theories, what this discrepancy showed a Ptolemaic astronomer was that something must be wrong with the Copernican scheme. Not only did it violate what was (thought to be) known about physics, but if the theory were correct, then the distances should yield a snug fit. The Copernican scheme, though, failed totally to do that. The Copernican solar system was only about two-thirds as large as the Ptolemaic (Van Helden 1985). Yet, since the major epicycles disappeared, deferents thick enough to carry the remaining minor epicycles left huge gaps in the heavens.

11.9

All of this looks completely different from a Copernican perspective. From that perspective, there is something downright phony about the remarkable Ptolemaic fit between the space required for Mercury and Venus and the space available between the moon and sun. It turns on a measurement far more precise than naked-eye astronomy could possibly provide, and in a way such that the estimated distance changes very rapidly with even a slight change in the measurement (Van Helden 1985:18–19). The measurement required must determine with great precision the size of the angle opposite the right angle at the base of a very, very long slender right triangle. This triangle has the sun at the apex and the earth and moon at the base. The slightest error throws the Ptolemaic calculation far off. Ptolemy's own version of this calculation is actually done in a way that implies that no annular eclipse of the sun could ever occur, and later astronomers knew that was false (Neugebauer 1982:104).

So it is an important point that the remarkable nested-sphere coherence turned on cognitive coherence that was not in fact logically coherent. Seeing things in the Ptolemaic way implied tacitly knowing what can be ignored—we will see some other examples later—as well as what must be noticed. So did seeing things in the Copernican way, but much less flagrantly. So a comparative not absolute judgment is needed to balance the Copernican and Ptolemaic claims. But the comparison strongly favors Copernicus. Compared to what was possible in the Ptolemaic system, a Copernican could give a much more explicit articulation of his system (such as provided by Copernicus himself, and later by Kepler and Galileo) without exposing logical weakness. Relatively (not absolutely) the Copernican view was rigorously coherent, and the Ptolemaic view was not. Rather, once a person was (cognitively) able to see the two as rivals [7.6], the Copernican argument looked solid and the Ptolemaic flimsy. So as will be seen in Chapter 13, although it was not until the appearance of the telescope in 1609 that any *observation* seriously embarrassed Ptolemaic astronomy, at least two decades earlier—by around 1590—supporters of Ptolemy were already on the defensive.

But even today, there is no clean line to be drawn between irresponsible fudging of data ("tuning the computer") and good use of craft and intuition in working out the details of a model. If we go back to the sixteenth century, there was not even a commonly understood way to discuss such problems. We see the difficulty in Copernicus' own treatment of the nested-sphere argument (*de Rev* I.10), where he clearly wants, as of course he must, to dismiss the purported remarkable fit as not to be taken seriously, but he does not quite know how to do it, and so settles for a few rhetorical flourishes.

Summing up: For a contemporary of Copernicus, the Ptolemaic system yielded the coherent and detailed and empirically powerful view of the world I have sketched, revealing that it has been contrived by God in a way whose beauty can be perceived by human beings even if the divine reasoning cannot. In terms of that Ptolemaic way of seeing the world, the sun must be what common sense, Aristotelian physics, and the Bible tell you it is: a heavenly body circling the earth. The earth and planets cannot be in orbit around the sun, for the nested-sphere coherence would become hideously unstable in such a world. The snugly fitting spheres would suddenly have huge amounts of room in which to rattle about. Improving Ptolemy—getting rid of the equant, for example—was a very plausible endeavor. But to suppose that the system might be fundamentally wrong would seem crazy. Getting rid of the major epicycles—which to a Copernican was the most striking single consequence of the heliocentric view—would seem like kicking the legs out from under a beautifully set table, for it was those very epicycles that gave the world its three-dimensional structure. Someone who was able to consider such things could find that they yield a system far more impressive than the Ptolemaic system. The puzzle is how anyone could (how Copernicus could) somehow have wandered, or been pushed, "uphill" [9.3] to the point where these crazy ideas could be seen to perhaps make sense.

11.10

I hope it is now clear why, on the account I have been sketching, it would be a gross mistake to treat the Ptolemaic centrality of the earth as an isolated belief, or to treat the Copernican discovery as the discovery of an alternative belief. We are dealing with a structure of belief in which even the most important single point (the earth is at the center) is not nearly as important as the pattern (gestalt, paradigm) within which that point fits. Given that way of seeing things, the Copernican alternative (the earth orbits the sun) is immediately incoherent, so of course it is hard to discover that way of seeing things, or to make sense of it if confronted by it.

Further, as I have already suggested—and as Kuhn has insisted—a deeply entrenched view like the nested-sphere view, or like the Copernican view when that came to be an entrenched view (at first only for Copernicus himself), cannot be reduced to an explicit set of claims. One can learn the Copernican argument by carefully going over the explicit points of the preceding sections. But in the cognitively crucial sense I want to stress, that alone is not enough to not make a person a Copernican. A person who is a Copernican is not just someone who knows the argument but someone who has come to have an all-at-once sense of the pattern of things within which that argument has a certain sense of inevitability, so that like Poincare with his theorem or like Mozart with his symphony [4.10], he has a sense of how the whole thing fits together. Similarly, the spelled-out version of the Ptolemaic view could not be a complete account of what it was to be a Ptolemaic astronomer. In the language of Chapter 4, I could only be giving a reasoning-why account of what a Ptolemaic astronomer knew in a more direct way by seeing-that [4.6]; and having that automatic, all-at-once, tacit sort of knowing the pattern of the thing is what it means to be a Ptolemaic astronomer.

If all that were *not* so, then the Ptolemaic commitment would have been far more vulnerable to explicit counterarguments than it was, once Copernicus had produced his alternative. It is essential to allow for the tacit (automatic, not subject to conscious control or awareness) way in which the established habits of mind are invoked, as illustrated by the scenario effects in Chapter 8. And while we have already considered many other examples of tacit mental habits during the course of this study, I want to conclude this chapter with an example tied to the substantive issues at hand.

11.11

The best opportunity for glimpsing the effects of entrenched habits of mind comes when we can find good reason to suppose that an entrenched intuition develops from experience in the world which the very person relying on the intuition would insist is irrelevant to his judgment [1.1]. Chapter 8 provided various "toy" examples. But the point would not be important if it was relevant *only* to toy problems. Since we do not have introspective access to the basis of our intuitions, if the basis of an intuition does not make sense in terms of other closely related intuitions we will just not conjecture those experiences as the basis of the intuition. In the hockey helmet example [7.5], we did not expect that a person would defend his aversion to helmets on the logically absurd, though cognitively plausible, grounds I suggested. The same holds for Nisbett & Ross's (1980) well-known example of ladies choosing hose. Instead, some conjecture not certifiably wrong, and not grossly inconsistent with other beliefs, will be offered. This does not mean that the reasoning-why we give for a judgment provides no evidence for a real connection with the cognitive basis of seeing-that. But our reasoning-why is only a conjecture, not a direct introspection that therefore cannot be wrong. If I say, "My tooth aches," I cannot be mistaken. But if I claim I believe *X* because I believe *Y*, that is some evidence for the claim but not at all a decisive proof. I might well be mistaken.

Indeed, every reasonably sophisticated person realizes that he holds inconsistent beliefs. He can see that all his friends are vulnerable to that failing, so unless a person has managed to convince himself of his unique wisdom, he will have to allow that like everyone else he too presumably has inconsistent beliefs. But that does not imply he can give an example. For if he found an example, he would somehow adjust those beliefs, whose incompatibility would now no longer be successfully tamed [7.7]. Similarly, I am sure that I (and needless to say, that you also) believe some things for reasons that, while sounding plausible, or at least not terribly implausible, in fact are false; but I can't give an example, anymore than you can. But in a particular such case, you might see it in me; and by showing me another reason, you might persuade me that I was wrong about some belief. For historians of science, I will try to do that now with respect to a belief which will be familiar to them.

Here is the example: one of the few major points in astronomy that was never in dispute, since everyone's intuition strongly supported the same conclusion, was that Mars, Jupiter, and Saturn are the "outer planets" (from the earth in Ptolemaic astronomy, from the sun in Copernican astronomy), coming in the inverse order to their periods: Mars (2 years), then Jupiter (12 years), then Saturn (30 years).

This universal intuition is usually explained on the grounds that, other things equal, the periods are naturally longer when the planet has further to go. But then, since Venus, Mercury, and the sun all have the same mean period, perhaps they also should be at the same mean distance. We will see in Chapter 12 how an astronomer who did consider that (correct) conjecture might rather easily have been led to the Copernican argument. No one, however, not even Copernicus, ever seems to have seen that argument, which suggests that in fact the experience in the world that underlies this strong, very widely shared period/distance intuition is not the one that has usually been asserted.

So we are led to consider what other basis for the intuition might exist. And once you have been prompted by the general argument to look for that, it is not hard to find. To see the point, you should try this experiment. Tie a small, dense object (such as a bolt) to the end of a piece of string, and tie another to a second string about twice as long. Take one string yourself and give the other to a friend and start them twirling around like planets in orbit. Which goes faster? (You could also do the experiment alone by counting aloud at each revolution of the first string, then repeating, counting aloud again, with the other string.)

What you will find—and you will not be surprised by it once you are actually in the situation—is that the bolt on the longer string seems to want to go slower. Physical experience in the world teaches us all that the longer the string, the slower the object on the end of it seems to want to go. From everyday experience in the world, Ptolemaic astronomers would have had this intuition as much as the rest of us.

This account of the basis in experience of the period/distance intuition escapes the problem posed by the usual "other things equal" explanation. If you let out some more string on a bolt tied to a length of string, it will seem to want to go round more slowly. But you can keep up the speed if you work at it. If you twirl both strings together, or even separately—one in the left hand, one in the right—they will tend to synchronize. So although experience with such things in the world would give a person some tacit sense that the period that feels most natural (which will be the minimum speed, hence the minimum effort, which keeps the string taut) slows as the string gets longer, you would also know in the same experience-based way that the object on the shorter string does not necessarily go faster. So if one object is in a slower orbit than another there is an intuitive basis for supposing that the slower one has the longer radius of rotation; but experience would not have taught you to expect that if the sun, Venus, and Mercury have the same mean period, they are necessarily at the same mean distance.

11.12

The bolt-and-string experiment suggests a simple explanation, based on physical experience with making things move in the world, for the universal intuition about the outer planets. But a Ptolemaic astronomer or Aristotelian philosopher could never have accepted that explanation, for he was explicitly committed to the view that the heavens consist of an ethereal element not subject to whatever physical laws governed sublunar elements. To explain the ordering of the outer planets by an analogy with objects on earth would be *logically* ridiculous. If accepted as a legitimate argument, it would undermine standard reasoning-why that gave answers to questions such as how the sphere of fixed stars could rotate at the millions of miles per hour implied by Ptolemaic astronomy (each star circling the universe in 24 hours) without being torn apart. So a Ptolemaic astronomer could not see the basis of the period/distance relation in terms of physical experience with swinging objects. But that did not imply that cognitively, in terms of the patterns of intuition that made some arrangements look right and others look implausible, the habits of mind acquired by physical experience were not powerfully in effect, as in the analysis of Chapter 8 the tacit scenarios exerted a powerful effect in contexts where the language of the problem would absolutely rule out their relevance.

Similarly, the nested-sphere intuitions that I have been describing in this chapter as at the cognitive core of Ptolemaic beliefs must be understood to exert their effect through the force of the tacit habits of mind that underlay Ptolemaic astronomy, not merely or even mainly through explicit counterarguments to seeing things as Copernicus saw them.

Twelve

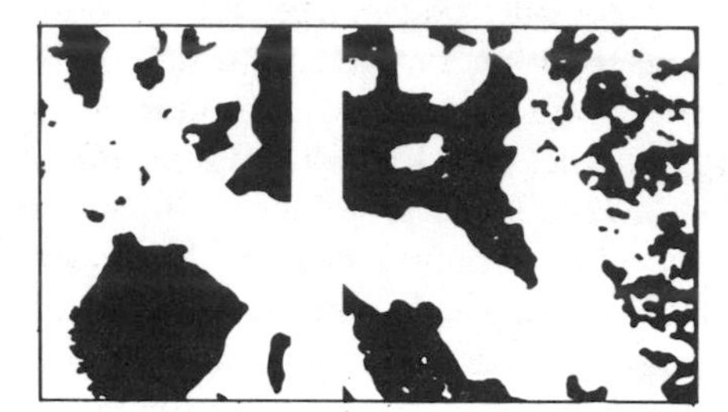

The Copernican Discovery

A route to the Copernican discovery that fits the *P*-cognitive argument very well was suggested by Copernicus himself. It starts from the commonplace observation that Venus and Mercury move in a regular pattern from being somewhat ahead of the sun (seen as morning stars just before sunrise) to somewhat behind the sun (seen as evening stars setting shortly after sunset). The possibility that these relations hold because Venus and Mercury orbit the sun was noticed at least a thousand years before Copernicus. In terms of the Kepler diagram (fig. 11.2), that would mean that the epicycles of Venus and Mercury would be moved to center on the sun rather than lying between the orbits of the sun and moon.

Any competent astronomer who took this possibility seriously was getting very close to the Copernican discovery (that is, to seeing the golden chain argument sketched out in Chap. 11). Copernicus points to that logic, though as a report about how the reader can reconstruct the discovery as part of coming to understand it, not as a report about how he himself came to make the discovery (Koyre 1973:41). On the other hand, many years earlier, in the *Commentariolus,* Copernicus said that he found his theory by looking for "a more reasonable arrangements of circles," prompted by his dissatisfaction with Ptolemy's use of equants (Rosen 1939:57), which is a reasonable description of how the path to the discovery I will sketch out starts.

In Rosen's translation (*de Rev* I.10): " . . . according to [Martianus Capella] Venus and Mercury revolve around the sun as their center. This is the reason . . . why these planets diverge no farther from the sun than is permitted by the curvature of their revolutions . . . If anyone seizes this opportunity to link Saturn, Jupiter and Mars also to [the sun] . . . he will not be mistaken, as is shown by the regular pattern of the motions." Martianus was a fourth century Roman schoolteacher or administrator who had compiled a summary of the wisdom of his times. Efforts to locate a source for his putting Venus and Mercury in orbit around the sun have not found an unambiguous precedent (Eastwood 1982). Certainly the idea played no role within astronomy before Copernicus.

And on the background sketched out in Chapter 11, it would indeed have been hard to see the Venus possibility, however obvious it seems today. Before the rise of Ptolemaic astronomy, all the heavenly bodies were assumed (as in Aristotle) to move on spheres centered on the earth, and (again as in Aristotle) all the planets were taken to lie beyond the sun. Further, in contrast to the other (outer) planets, within the limits of naked-eye observation, neither Venus nor Mercury exhibits the variation in brightness or size that would suggest variation in distance from the earth.

Hence, before Ptolemy, there was nothing to prompt the insight that perhaps Venus orbits the sun. After Ptolemy, astronomical theory did imply such a variation in Venus' distance [11.7], but only a well-trained astronomer would know that. And for that trained astronomer, the variation in distance was a theoretical consequence of the nested-sphere structure of the heavens, not something that could be directly seen. Until the telescope was available, there was no direct evidence that any such variation occurred. Psychologically, what was primary was the epicycle, which played a role in the Ptolemaic sense of the heavens that a leg does for the table you sit at [11.8]. Within the Ptolemaic framework, moving the epicycles of Venus and Mercury to center on the sun was a dreadful step. It moves the epicycle of Mercury inside that of Venus, and then moves the pair to center on the sun. But if that is done, then instead of neatly filling the space between the moon and sun, the spheres of Venus and Mercury would fill only one-third of the required space, and the snugly nested Ptolemaic world would collapse.

So the idea that the inner planets orbit the sun was logically easy. But that nevertheless no astronomer seems to have seized on that possibility (Eastwood 1982) shows that cognitively it must have been difficult, which makes sense in terms of the nested-sphere argument in Chapter 11. Martianus himself was not an astronomer and could have been merely misinterpreting something he had read. Certainly he shows no awareness that he is saying anything radical, or important, or controversial. And there is no evidence that any *astronomer* was interested in Martianus' remark until after Copernicus had occasion to draw attention to it by way of assuring his reader that he could find ancient precedents for his unorthodox views.

In a variant of the transmutation-of-species point of the Darwinian account of Chapter 10, there is a mismatch between the sort of person who could do anything with the Venus intuition if he could take it seriously even for a few moments, and the sort of person who might see such an intuition as making sense. The problem in the Copernican case, then, as in the Darwinian, is to see how it comes about that someone (Copernicus, Darwin) who is expert enough to exploit the idea somehow escapes the habits of mind that put the idea outside the range of ideas an expert might seriously consider.

12.2

However, *if* an expert could get started along the road Copernicus suggested (if he could consider the idea that perhaps Venus and Mercury orbit the sun), then a series of much easier further steps takes a person to the full Copernican argument. Here is one example of how such a sequence might go: (a) if the Venus/Mercury assumption were true, then it would be easy to calculate the relative distances and the heliocentric periods of these planets [11.4]; (b) the resulting periods and distances for Venus and Mercury (8 months and 88 days, respectively) now fit in with the universal intuition about a period/distance relation, but with the sun as the center of the orbit; and a person might now "take this opportunity"—as Copernicus says—to extend the idea to the outer planets; (c) if we do, we will find that a variant of the argument that leads to relative distances for Venus and Mercury yields the distances of Mars, Jupiter, and Saturn as well (fig. 11.5). So all the planetary distances are determined, relative to the sun/earth distances, where under the Ptolemaic scheme none were (without auxiliary assumptions); and the sequence that results looks intuitively compelling.

Continuing, (d) an immediate further consequence then is that the earth, with a period between those of Venus and Mars (8 months, 1 year, 2 years, respectively), fits between the two in terms of period and also fits between the two in terms of the independently calculated distances from the sun. The calculations also confirm—as Copernicus mentions—that indeed there is ample room not only for the earth but also for the space required for its orbiting moon to fit between Venus and Mars; and (e) with this much prompting you now imagine how things would look if the earth was actually in that space it seems to fit. If you do that, you can see (in the mind's eye) that looking toward the center from an orbiting planet, the inner planets must have the morning star/evening star relation with the sun that has always been observed, and how the planets might exhibit retrogressions without any actual looping in the heavens.

At this point, in terms of the discovery schema [9.3], a person would be at the climax of the uphill phase, where he isn't yet sure, but the radical idea looks too pretty to be *merely* wrong. Notice in particular that although the nested spheres logically collapse when Venus and Mercury are imagined to orbit the sun (call that Step 0), cognitively the situation is not so harsh, since it is only after the further relations that come when the later steps are in hand that a person might find himself tempted to *believe* the heliocentric idea. At Step 0, moving the epicycle can be rationalized as just trying out "arrangements of circles" (*Commentariolus*), with no active doubt about the usual view of the heavens.

On the *P*-cognitive argument, there should now be a striking story to tell of how an astronomer finally (after so many centuries) got past the immediate "looks crazy" barrier to considering the idea of moving the epicycle of Venus at all. Here it is important to emphasize the distinction between obtaining that first glimpse (Step 0), which entails no explicit rivalry, and reaching the point where the idea is seen as plausibly true (Step e), hence as a direct rival to the entrenched intuition that has to be overthrown. If we can get Copernicus to Step 0, the rest of the *uphill* phase follows easily.

Further, having reached that point of overt rivalry, *consolidation* can also come quickly—in contrast to Darwin's situation where it is very understandable that it came slowly. For Darwin, there was no sharp test available that could provide striking support for natural selection or against the intuition that good design implies a designer [10.2]. On the contrary, what more easily came to mind were what seemed overwhelming difficulties. Years later, Darwin told a correspondent (Asa Gray) that for a long time he was unable to think of the intricate structure of the eye without shuddering. But for Copernicus, the situation was very different.

A really striking test of the heliocentric idea was immediately available. For though the qualitative relations in (e) is very striking, it does not yet show that the retrogressions for each planet will actually match what is observed in the sky. But unless the calculated retrogressions from the heliocentric model match the observations, and do so for every planet, the beautiful relations in (a) through (e) can only be an illusion. There is a test, then, so salient that even a much more timid canon than Copernicus' career makes plausible could hardly resist doing the calculation, though perhaps mostly expecting that it would show that the striking but crazy idea is wrong.

Now, in fact, logically this is not a test at all. For the numbers come out right automatically (see Chap. 11, n. 5). But it was not easy to see that [13.2], so that discovering how everything works out exactly as it must if the world is really heliocentric is itself enough to make a man a Copernican, and we have very adequate grounds to suppose it did make Copernicus a Copernican.

12.3

Now, except for the last, the steps required for the discovery sequence are not likely to seem very difficult, even for patient readers with no background in this material. Step 0, in particular, is extremely simple, and we know that in fact it had been considered (by Martianus, at least) long before Copernicus; having—however tentatively—taken Step 0, the remaining steps look like they would be difficult for any technically competent astronomer to avoid. But cognitively, Step 0 must have been extremely difficult, because for more than a thousand years the sequence was logically available to every astronomer, but no one took it. Even after Copernicus pointed it out, nearly all other astronomers were at first blind to its power. Only after several decades (after a new generation had arisen which had learned its astronomy aware of the Copernican alternative) would the very same arguments come to be seen as hard to resist, so that (as will be reviewed in Chap. 13) even astronomers unable to believe Copernicus were still led to abandon Ptolemy.

In general, the more we examine the logic and history of the episode, the more reason we find to suppose that somehow the habits of mind that made a person a competent Ptolemaic astronomer must have blocked the initial insight (that Venus might orbit the sun) that would open the way to the rest of the Copernican argument. That would account for all three of the points just mentioned: the very long delay before anyone discovered the Copernican argument; the difficulty Copernicus' contemporaries—still deeply entrenched in the Ptolemaic habits of mind—had in making sense of the argument; and the power to persuade of that same argument that was revealed a few decades later once a new generation was in command which had learned astronomy aware of the Copernican alternative. The climb "uphill" (if you review steps a through e) is almost self-propelled, provided the potential discoverer can reach Step 0. The puzzle is to say how Copernicus might have done that, and in particular to give an account in which reaching that crucial initial step somehow grows out of Copernicus' logically irrelevant concern with the equant [11.2].

The first thing to look for [9.8] is some way in which Copernicus would be less entrenched in the nested-sphere way of seeing things than others who might have made the discovery but didn't. In fact, we can point to a number of such points, some tied to the partial outsider idea, others to technical details of his work.

Copernicus' career looks very little like that of a typical astronomer either of his or our time. He studied law, theology, and medicine as well as astronomy and mathematics. His doctorate was in law, and his subsequent official duties were entirely concerned with administration, law, and medicine. Astronomy was always an avocation. And he pursued it in isolation. The *Commentariolus* gives a very condensed argument that could only have been read with any comprehension by another astronomer (Swerdlow 1973:471). It presumably was circulated primarily among academics in Cracow, one of whom left a record (in the form of an inventory of his library) that he had a copy by May 1514. No hint of a response to the *Commentariolus* has survived. And that for the next several decades Copernicus had no further known contact with astronomers suggests that the response was not encouraging. As will be noted later, a remarkable range of people in those years managed to be on record as showing an interest (usually but not always favorable) in Copernicus' work prior to the 1540s, and several urged him to publish. But until the young Rheticus presented himself to Copernicus in 1539, no astronomer is on record as showing any interest in what he was doing.[1]

In general, Copernicus was free of all the usual connections and responsibilities of an astronomer of the period. He was never subject to the pressure to make astrological forecasts which anyone in that period who was a working astronomer could hardly escape. The premises of astrology fit in obvious ways with seeing the planets as satellites of the earth. But the heliocentric view makes the earth along with the planets mutually independent satellites of the sun. An important "partial outsider" advantage for Copernicus, therefore, may have been just that his independence let him escape entanglement in the particular habits of mind that would facilitate active work in astrology.

Copernicus (again reflecting his independence from the usual responsibilities of an astronomer of the time) had no students to teach or patrons to entertain. Here is a man familiar with the intricacies of Ptolemaic epicycles, equants, deferents, and the rest of that difficult apparatus, but not bound by any need to produce conventional astronomy (or astrology) or to teach others conventional astronomy. Like Darwin or Newton or Einstein—each for somewhat different reasons—Copernicus was in a position to acquire and develop the habits of mind needed for fluent use of the technical wherewithal of his field in a way that could keep salient rather than tame some anomaly that looks like it might be important. Unlike (for example) someone with responsibility for training students, or for carrying on other sorts of what we would today call mainstream activities, Copernicus (like Darwin or Newton or Einstein) was in a position to tolerate the clumsiness that would go with the transition from the usual way of seeing his problem to another radically different way. His isolation preempted the awkwardness that would otherwise go with seeing things in a way that made it difficult to communicate comfortably with professional colleagues.

12.4

These partial outsider effects would interact with the nested-sphere issue. For the technical problems (in particular, dispensing with Ptolemy's departures from uniform motion) that were his initial concern involved working in detail only on the model for one body at a time (the moon, Mercury, etc.). At the level of the overall structure of the heavens, the anomalies that equants resolved were subtle details which would scarcely be mentioned to any but the most advanced students, or discussed even with the most advanced students when the focus was on the overall structure of the heavens. So while training students or explaining the system of the heavens to a wealthy patron required fluent use of the nested-sphere notions, the problem that had caught Copernicus' interest did not *actively* engage those ideas at all, as someone today working with Keynesian macroeconomic models has almost never any occasion to give thought to the fine details of microeconomic theory.

Further, even within technical astronomy there was a related cognitively significant division that would work to Copernicus' advantage. In [11.3] I said that logically there was no incompatibility between the plane diagrams illustrated by figure 11.1, used to work out schemes for computing the position of heavenly bodies (the sort of work Ptolemy did in the Almagest and Copernicus in *de Rev*) versus the more intricate diagrams used for detailed illustration of nested-sphere configurations (fig. 11.3), such as Ptolemy gave in his "Planetary Hypothesis" and writers of Copernicus' time worked out under the label of "Theorica" (Swerdlow 1973). As with micro- and macroeconomics or statistical mechanics and thermodynamics today, a person competent in one of these things will always be knowledgeable and often expert in the other as well. But most people will work mainly in one or the other of any of the pairs mentioned.

But it was the nested sphere configurations, not the plane diagrams, that were seriously incompatible with the golden chain view (n. 5, Chap. 11). The diagrams used to work out predictive models show the deferent as the path of the center of the epicycle, as in the Kepler diagram, but the explicit nested-sphere models explicitly treat the deferents as snugly fitting shells within which the epicycles could turn (fig. 10.2 vs. fig. 10.3). It was to Copernicus' advantage that, as a side-effect of his absence of responsibility for training students, he would rarely or never have occasion to explicitly work with nested-sphere configurations. Rather, the added three-dimensional complexity of the equant-free models would have the cognitive effect of inhibiting thinking in nested-sphere terms (thinking about figs. like 10.3). A person learns not to think about what is clumsy to deal with, and especially so when very natural reasoning-why is available to justify treating that as a problem for someone else or some other occasion. Here there was a tradition going all the way back to Ptolemy himself of treating the two aspects of astronomy separately.

In sum, the absense of any need to continually use the nested-sphere intuitions in explaining things to students would be supported by the presense of a technical focus (on the equant) which made it clumsy to think in detail about nested spheres (C. Wilson 1975). And yet even the combined force of these could hardly account for how Copernicus broke free of constraints that had bound astronomers for so many centuries. Indeed, we know that Arab work in the thirteenth century, some of which seems to have been indirectly available to Copernicus (and which must have been available to a series of astronomers from the thirteenth century to the sixteenth), had produced equant-free models very similar and sometimes identical to those Copernicus would produce. But there is no hint of the heliocentric idea over those 200 years and more. There is nothing intrinsic to his situation or his focus on the equant to challenge the nested-sphere sense of the structure of the world that had governed the thought of astronomers for many centuries. What we have are some reasons for supposing that if some astronomer in the early sixteenth century was going to see the heliocentric insight, Copernicus—as someone who would be somewhat less deeply entrenched to the nested-sphere view than more typical contemporaries—was in a particularly favorable situation to be the person to do it. But nothing so far shows us how he was not merely more favorably placed than a more typical sixteenth-century astronomer but favorably enough placed to actually do what he did. However, a further technical detail of Copernicus' work supports the conjecture that he in fact made the discovery in the way suggested at the beginning of this chapter.

12.5

Copernicus seems to have achieved his first encouragement to attempt bold things in astronomy from success in resolving a gross anomaly in Ptolemy's account of the motions of the moon. For on Ptolemy's model, the distance of the moon varied by a factor of 2 between full moon and half moon. If that were true, then the half moon would have twice the apparent diameter of the full moon and twice its brightness (since the reflecting area goes with the square of the radius). Even casual naked eye observation shows that is absurd. Copernicus, either on his own or adapting untranslated Arab work that had somehow reached him, worked out a model in which this gross discrepancy disappeared.

The way Copernicus accomplished this—using what is now called a Tusi couple—turned on dispensing with a detail of Ptolemy's lunar model that violated uniform circular motion. If indeed Copernicus moved from success on the moon difficulty to work on some other important flaw in Ptolemy's astronomy, the salient case would be Venus. For Ptolemy's model for Venus also yields an observational discrepancy akin to that involved in Ptolemy's moon model, and in using an equant (fig. 11.1) it violated the axiom of uniform motion.

On Ptolemy's models (and in fact) both Venus and Mars must vary by a factor of about 7 between closest and furthest distance from the earth. Mars indeed showed a very marked difference in apparent brightness over this cycle; but—as Osiander mentions in his notorious preface to Copernicus—Venus did not.[2] So on the argument of Chapter 9, it makes sense that Copernicus would move from a confidence-building resolution of the moon puzzle by way of the Tusi couple to an attack on the Venus puzzle using the same technique. And indeed the same technical device (use of the so-called Tusi couple) could be adapted to work out equant-free models for the planets. Yet the result would be disappointing. The point of wanting to dispense with the equant was that the equant seemed to make no physical sense. Getting rid of it, one would hope, would give a model that would be more realistic and give better results. For the moon, a version of the Tusi couple removed a departure from uniform motion and at the same time produced a striking empirical improvement. For Venus, though, we get a more complicated model than Ptolemy's but one that does nothing at all to correct the gross anomaly. Somehow, something else needs to be done about Venus, and we have already seen what a sufficiently bold—though logically very simple—conjecture about Venus would lead to if a person were moved to consider (as Copernicus says he was) "alternative arrangements of circles." This piece of the story has an odd but cognitively significant resolution which I will take up at the end of this chapter. Here I only want to mention that although the heliocentric idea does indeed solve the Venus puzzle, Copernicus himself did not realize that. For how it did so was not discovered until Venus could be seen through a telescope. But as I tried to show via the sequence of steps at the beginning of the chapter, by the time he had gone far enough to meet that disappointment, Copernicus would have also gone far enough to see the golden chain argument that made the heliocentric idea capable of winning converts anyway.

So we have now gotten Copernicus to the point where he can be seen as peculiarly well-placed to do what he did (on the partial outsider arguments), and even how the logically irrelevant work on the equant could plausibly have made him peculiarly prone to exploring alternatives to the usual model of Venus. But we *still* do not adequately close the cognitive gap. For on the cognitive argument [7.9], we cannot expect that a person would be prompted to respond to a cognitive puzzle by overthrowing some deeply entrenched habit of mind unless that puzzle appears inescapably tied to intuitions as deeply entrenched as the intuitions that have to be displaced. Here, the Venus anomaly had been known for more than 1000 years, as had Ptolemy's use of the equant. It was something that astronomers had learned to live with. A failure of the latest attempt to work it out (repeating the Kuhnian point mentioned earlier) would be seen as a failure for the puzzle solver, not as a failure for astronomy.

Summing up: The point of trying to rid astronomy of the equant was to be rid of a physically implausible aspect of Ptolemy's work. It is hard, consequently, on the argument to this point to see how even for a moment a person would see moving the epicycle of Venus, hence collapsing the nested spheres, as a move that looked interesting as a response to the Venus anomaly. That would be like swatting a fly that had landed on a priceless painting. For 1300 years no astronomer showed that response to the Venus anomaly. What might have prompted Copernicus to a different response?

So we really would like, and indeed on the *P*-cognitive argument we really need, something that would help Copernicus get by the nested-sphere barrier that kept all other astronomers from the idea that perhaps Venus is always near the sun because it orbits the sun. But once we are led to look for a suitable cognitive shock—once, I would argue, we are thinking about the problem in terms of a view that tells us to look for such a shock, as Kuhn's argument does, and as in a more specific way the *P*-cognitive argument does—something sticks out very sharply. For Copernicus' insight followed curiously closely on another grand discovery; and in the light of the argument here it is obvious that the other discovery also gives evidence of a striking cognitive difficulty. The balance of the chapter will spell out the hunch that perhaps Copernicus' discovery had some link with Columbus' discovery. For this turns out to offer a detailed solution to the problem of filling in a cognitively plausible path to the Copernican discovery.[3]

12.6

That Columbus died believing he had discovered something far less important than what he did discover has usually been understood as showing how hard it is for a man to give up a belief he had expounded for many years. Columbus had committed his life to the "enterprise of the Indies," and he died still committed to it. But it turns out that he was far from alone in finding it hard to see that he had discovered a new world, not just a shorter route to the extremity of the known world. Columbus was not shaken in his belief even though he had known since 1494 (from the enormous outflow of fresh water from the estuary of the Orinoco River) that he had encountered something very big. Indeed, it remained very hard to believe that a new continental land mass had been discovered even after Vespucci reported an even vaster outflow of fresh water further south (the Amazon). Somehow, there was some cognitive barrier to seeing that there might be a new continent with no connection to the traditional inhabited world. An inability to see that was not a difficulty peculiar to Columbus. And the unthinkable character of the notion that there could be new continents in the West turns out to have an intimate link to the notion that the earth could orbit the sun: namely, they are both consequences of the nested-sphere structure of the world.

Logically, the inference required to appreciate what Columbus had discovered is trivial. The most sophisticated argument against Columbus' "enterprise of the Indies" had been that he greatly underestimated the extent of the ocean to be crossed (as indeed he had). So, logically, a perfectly obvious possibility when Columbus found land thousands of miles short of where Asia should have been would be that he might have reached something other than Asia. But no one seems to have suggested that until voyage after voyage brought back yet more word on the vast extent of the new discoveries, with no sign of things Chinese. No definite evidence has survived that anyone prior to the turn of the sixteenth century drew the inference that perhaps what Columbus had reached was not Asia. Somehow that seems to have been unthinkable, until the evidence for it became so gross that the thought could hardly be avoided any longer.

In an often-quoted passage, Herbert Butterfield (1965:29) remarks:

> . . . if you grant Copernicus a certain advantage in respect of geometrical simplicity, the sacrifice that had to be made for the sake of this was tremendous. You lost the whole cosmology associated with Aristotelianism—the whole intricately dovetailed system in which the nobility of the various elements and the hierarchical arrangement of these had been so beautifully interlocked.

But what if that intricately dovetailed system with its nobility, hierarchy, and the rest had already been disrupted?

12.7

Figure 12.1 reproduces two cosmological diagrams from a volume celebrating the five-hundredth anniversary of Copernicus' birth (Beer & Strand 1975:xxvii). Chosen by the editors to make an attractive display, not to make any point, the pair show two images of the universe printed 31 years apart (1493 and 1524). Both are labeled Aristotelian. Naturally both show a world of nested spheres. In the 1493 diagram, at the center of all is the sphere of earth; next the sphere of water; next air; next fire (hence the appearance high in the sky of meteors, lightning, comets, aurora). Surrounding these we have the nesting of heavenly spheres as in Kepler's diagram (fig. 11.2). All this is as it had been described by Aristotle close to 2000 years earlier. We have the classical scheme, invented by the Greeks, elaborated by the Arabs, and now developed yet further by European Christendom.

But the 1524 diagram is different. If you look closely, you will find that the way the world was seen had changed over the several decades following Columbus' voyage, and in a way that required—for the first time in the nearly 2000 years since it was propounded—a change in a basic point of the Aristotelian system of the world. A sphere has disappeared. There are no longer separate spheres of earth (the *oecumene*) and of water (the ocean), but now a mixed "terraqueous" sphere of water and land.

Endorsed and elaborated by the greatest intellects of the high middle ages (Roger Bacon, Dante, Aquinas), the nested-sphere idea had for several centuries been understood as continuing to work down to yet finer detail, so that the inhabited world (the Greek *oecumene*, the Roman *orbis terrarum*) became a continuation of the pattern of the heavenly world. For the Greeks, the center of the habitable world was the Eastern Mediterranean, from which spread out the three major components of the oecumene: Europe, Asia, Africa. But now this way of seeing the world was powerfully reinforced by Christian theology, with Jerusalem the center of all. In the center of the heavenly world was the earth, a mere point compared to the whole, but with a special importance and in a central location; in the center of the mundane world was Jerusalem, a mere point compared to the whole oecumene, but with a special importance and a central location. Figure 12.2 shows this oecumene as it was still conceived well into the age of discovery (1459). If the reproduction is adequate, you will see that the mapmaker superimposed the hub of a wheel at the location of Jerusalem.

When the bishop of Rome wished to proclaim his preeminence over the whole world, he called himself the bishop of the oecumene. He did not mean to leave anything out. Rather, the oecumene was part of the pattern of the universe, continuous (going inward) to the symbolic if not

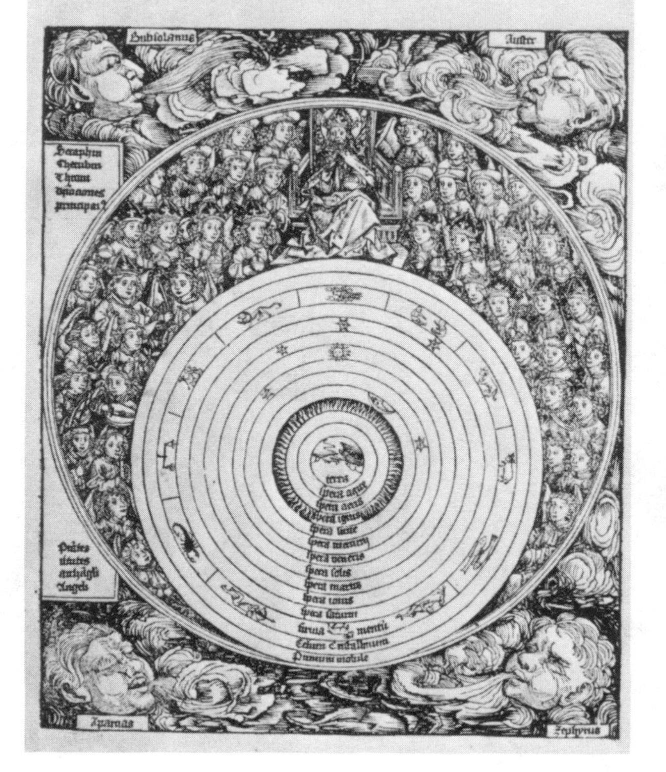

De sanctificatione septime diei

Consummato igitur mundo: per fabricam diuine solercie sex dierum. Creati enim dispositi et ornati tandem perfecti sunt celi et terra. Compleuit deus gloriosus opus suum: et requieuit die septimo ab operibus manuum suarum: postquam cunctum mundum: et omnia que in eo sunt creasset: non quasi operando lassus: sed nouam creaturam facere cessauit: cuius materia vel similitudo non precesserit. Opus enim propagationis operari non desinit. Et dominus eidem diei benedixit: et sanctificauit illum: vocauitque ipsum Sabatum quod nomen hebraica lingua requiem significat. Eo quod in ipso cessauerat ab omni opere quod patrarat. Unde et Judei eo die a laboribus proprijs vacare dignoscuntur. Quem et ante leges certe gentes celebrem obseruarunt. Jamque ad calcem ventum est operum diuinorum. Illum ergo timeamus: amemus: et veneremur. In quo sunt omnia siue visibilia siue inuisibilia. Et a domino celi: domino bonorum omnium. Cui data omnis potestas in celo et in terra. Et presentia bona: quatenus bona sint. Et veram eterne vite felicitatem queramus.

Figure 12.1. Two versions of the nested spheres as they were conceived in versions printed 31 years apart. Aristotle's basic scheme, after 2000 years, has changed: the distinct spheres of earth and water, still shown in 1493, have been merged by 1524 into a single "terraqueous" sphere. Reprinted from *Vistas in Astronomy* 17, *Copernicus: Yesterday and Today,* Arthur Beer and Peter Beer, general eds. With permission.

Prima pars. Col.6.

COELVM EMPIRRE VM HABITACV LVM DEI ET OMNIVM ELECTORVM

Decimum Coelu. Primu Mobile. Nonu Coelum. Et Cristallinu.

Coelum Saturni. Coelum Iouis. Coelum Martis. Coelum Solis. Coelu Veneris. Coelu Mercurij. Coelu Lune.

Figure 12.2. A world map as printed in 1459. The mapmaker has imposed the hub of a wheel at the location of Jerusalem. Courtesy of Art Resource, New York.

quite literal center at Jerusalem, and continuing out to the highest spheres of the heavens.

Of course, educated people understood that such maps were only schematic (though fig. 12.2 gives a very reasonable representation of details near the Mediterranean, for example, for Italy). The sophisticated view was, in geography as well as astronomy, provided by Ptolemy. His system of the heavens was very complicated compared to the simple schematics of a diagram like figure 12.1. Similarly, Ptolemy's map of the world—which was as familiar to scholars of Copernicus' time as the Ptolemaic nested-sphere diagrams—presents a much more complicated picture than the neatly circular oecumene of figure 12.2.

Figure 12.3. Inset (12.3a (above)) from the Waldseemuller world map of 1507 (figure 12.5) shows the world as envisioned at the opening of the sixteenth century. It remains basically the world of Ptolemy's *Geography*, as the iconography implies. The main innovation was that by now the Indian Ocean was shown as open, while Ptolemy had described it as a closed sea, symmetrical with the Mediterranean. Maps like this would have been entirely familiar to any educated contemporary of Copernicus. It showed what everyone "knew" was the land surface of the earth. Figure 12.3b (right), again an inset from the Waldseemuller map, then shows the new world, and its delineator Amerigo Vespucci. Columbus was credited with finding land in the West; but it is Vespucci that Waldseemuller credits with providing sufficient information to make it clear that what lay out in the Western ocean was a new continent, not merely the eastern extremity of Asia. The new continent, Waldseemuller decided, should therefore be called "America." Figures 12.3a and 12.3b reproduce two panels of the 12 which make up the map of figure 12.5. When the panels were assembled they formed a map very much larger (4′ × 8′) than any printed map published earlier. The rather intellectualized account of the shattering of the oecumene I have given in the text cannot capture the emotional impact this map would have had on a sixteenth-century viewer (such as Copernicus) who had come to maturity "knowing" that the land area conformed to the Ptolemaic picture of figure 12.3a, and for whom, until confronted by this monumental map, a new continent on the back side of the earth had been unimaginable. Courtesy Map Room, University of Chicago Libraries.

Figure 12.3a looks very much like other world maps printed in the early 1500s. It was still possible to see the world as Ptolemy had seen it almost 1400 years earlier: a complex version of the circle of land shown by figure 12.2 but still recognizably the oecumene, and except for some relatively small and nearby islands that could be understood as outlying elevations of the sphere of earth (Britain, Ceylon, Japan), topologically equivalent to it. The oecumene remains intact.

But eventually the evidence brought back from voyages to the West became grossly inconsistent with the traditional view. For anyone who came to maturity after 1492 the idea of new lands in the West would become familiar even before the notion of the oecumene could become entrenched. Only then did the notion arise that there existed—that there conceptually could exist—continental land masses across the high seas, remote and not assimilable to the traditional oecumene. As with pre-Copernican notions of a moving earth, isolated passages can be found which to our ears easily sound like they are or could be or ought to be references to new mainlands far out in the ocean. But nothing that holds up to a close reading has been found, either with respect to writing between Aristarchus and Copernicus regarding an orbiting earth, or with respect to new continents in the West before the end of the fifteenth century. Rather,

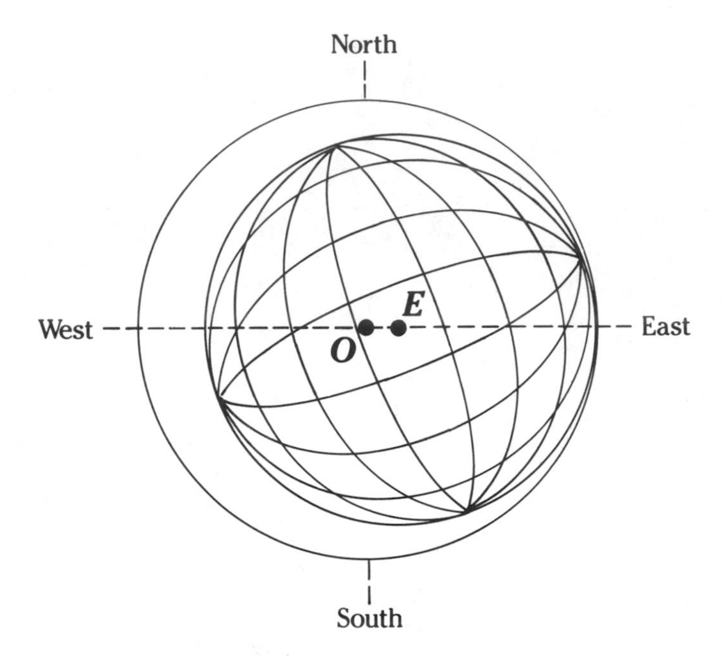

Figure 12.4. The *orbis terrarum* surrounded by the Aristotelian sphere of water seems to have been envisioned as shown: that is, as a globe tangent, from the inside to a larger globe, with centers at *E* and *O*, respectively. Slight irregularities (relative to the radius of the earth) would account for mountains, peninsulas, offshore islands, and so on. But continents of the far side of the globe would be unthinkable. If you imagine the sphere of earth as only *slightly* smaller than the sphere of water, then the ocean would be fairly narrow, and offshore islands might dot it. If you followed Ptolemy, the ocean was broad. In either interpretation, there was no place for the vast continents far from both Europe and Asia.

it looks like new continents that cannot be assimilated to the traditional oecumene, like the earth as a satellite of the sun, were cognitively impossible in the absense of some potential rival itself as deeply entrenched as the intuitions that had to be challenged. Absent such a shock to entrenched habits of mind, either would be like trying to see out the back of your head. At the center of the universe lay Jerusalem, surrounded by the sphere of earth (orbis terrarum: the oecumene), then the sphere of water, then the sphere of air . . . and so on out to the sphere of the fixed stars. How could there be a second orbis terrarum? It was like asking if a person could have a second head. Figure 12.4 illustrates how the surface of the oecumene surrounded by ocean could be envisioned as the Aristotelian three-dimensional spheres.

12.8

What is the earliest reasonable date for what might be termed the "shattering of the oecumene"? According to a sketch thought to be drawn to the specifications of his son, Columbus eventually came to believe that what he had encountered was an enormous extension of the Indochina peninsula. Early sixteenth-century maps hedge matters by failing to show clearly the eastern edge of Asia, or the western or southern edge of America. Such ambiguities were common through at least the 1520s. Through the balance of the century there are still rear-guard efforts to be found by defenders of the traditional oecumene (O'Gorman 1961; Randles 1980).

The first map that clearly showed both coasts of a new large (though still not nearly large enough) land mass was that of Waldseemuller in 1507 (fig. 12.5); and it was this map, published on (for its time) a colossal scale (4′ × 8′) which then gave the namé America to the new land. A fair estimate of when Copernicus (in his "darkest corner of Europe") might have become aware of the radical new idea would certainly go no earlier than 1507. But a very reasonable estimate of the latest date is only five years later, for in 1512 a map picturing the new world even more boldly than Waldseemullers', was published in Copernicus' own metropolis of Cracow. For reasons I will come to, we can be certain that Waldseemuller's map was closely studied by Copernicus.

When did Copernicus get his heliocentric insight? In his introduction he says that the book had lain among his papers into its "4th nine" before he was persuaded by others to publish it. So if his memory was correct (he was nearly 70 when the introduction was written), some sort of draft of the book was ready at least 27 years earlier, by 1516, but later than 1507. In fact, the key insight could hardly have come later than about 1512. For as mentioned earlier, a scholar in Cracow had a copy of the *Commentariolus* by (at the latest) May 1514, and the theory is sketched there in sufficient technical detail (Swerdlow & Neugebauer 1984:59–64) that Copernicus must have been working out the argument for some time. Indeed, the *Commentariolus* looks like the equivalent of Darwin's 1844 paper [10.5] marking the climax of the consolidation phase of the discovery. As Darwin was moved to immediately take steps to see that the 1844 draft would survive [10.5], Copernicus shows signs of moving quickly to make sure that the ideas would not be lost even if something happened to him. But, again as with Darwin, he settles into a long downhill [9.3] phase, which continues even after friends have begun urging him to publish, until circumstances deprive him of the option of delaying any longer.

The preface to a little book of translations Copernicus published in 1509, written by a friend (Corvinus), alludes to Copernicus' astronomical

Figure 12.5. Waldseemuller's map of 1507: the first printed map to clearly show the shattering of the oecumene. Earlier maps, even after it was realized that the new discoveries were vast, were drawn in a way that left open the possibility that beyond the margins the new discoveries were connected to Asia. Courtesy Map Room, University of Chicago Libraries.

work, especially by what seems to be a reference to improvement on Ptolemy's model for the orbit of the moon. This suggests that when the preface was written Copernicus had not yet made the much more striking heliocentric discovery. We then must allow time for the maturing of the idea from first glimpse to the *Commentariolus*' fairly detailed descriptions of equant-free models and still more time for that manuscript to be circulated and eventually turn up on an inventory dated mid-1514. Allowing for all that, when we could reasonably suppose Copernicus got his great idea must coincide very remarkably indeed with the period (after 1507) during which Copernicus could plausibly be first learning of the shattering of the oecumene.

So we find something very striking if we follow up a hunch that there might have been some cognitive difficulty in comprehending the discovery of the New World that would turn out to be connected with the cognitive difficulty in seeing the Copernican argument. It is very clear that indeed there was a severe cognitive difficulty for both, and the nature of the difficulty in both cases is essentially the same: each novelty violated the nested-sphere intuition, which was the very core of a universally shared sense of how the world is organized.

Further, what had appeared to be an interesting—but given how much was happening in the period, not intrinsically compelling—coincidence in time turns into something much more precise. There is a spread of only 50 years from Columbus' voyage in 1492—which would overturn Ptolemy's geography—to Copernicus' publication in 1543—which would overturn Ptolemy's astronomy. But looking closer, it turns out that the spread is much narrower. After almost 1400 years, we finally reach the earliest point (1507) when we could expect any astronomer to realize that the system of nested spheres is false with respect to Ptolemy's geography. Within six more years, at the very most, the heliocentric argument that will overturn Ptolemy's astronomy is finally noticed.

12.9

We can look for two sorts of supporting indications: (1) Biographical clues: Is there anything about Copernicus' life that would lend explicit support to the supposition that his thinking could be influenced by the shattering of the oecumene? (2) Textual clues: Is there anything in Copernicus' writings that suggests his argument was in fact somehow linked to the oecumene? The surviving record is comparable to that available for Shakespeare: references to Copernicus during his lifetime have all been printed in one thin book (*Regista Copernicana,* Biskup 1973: cited as *R*). The great majority of 500-odd items in *R* tell us nothing about Copernicus' serious work. Most are records of routine administration of Church funds, land, etc. We learn more detail about his affair with his housekeeper, which generated a good deal of correspondence with the presiding bishop, than about the heliocentric theory. There presumably is significant indirect evidence in archive materials revealing attitudes by astronomers other than Copernicus toward the nested spheres, the shattering of the oecumene, and so on. But the bits of supporting evidence I will rehearse are from readily available sources (mainly the *Regista*).

A striking point about the *Regista* is that there is almost no indication of contact with astronomers between the circulation of his *Commentariolus* and the arrival of Rheticus in 1538. A (disputed) exception is a reference to a debate about comets involving Copernicus (*R*:335). In particular, there is no evidence of contact with the astronomers in Cracow, though its university was an important center for astronomy. Copernicus certainly attempted to explain his view to his fellow astronomers by 1514, since the *Commentariolus* is too technical to have been intended for any audience but astronomers. The lack of any evidence of communication in either direction for many years thereafter tells us something about what sort of response he got. He seems to have worked in isolation from other astronomers until the arrival of Rheticus near the end of his life, which is hard to make sense of except in terms of an account in which (as in the nested-sphere argument we have been led to) there is some cognitive barrier—some Kuhnian incommensurability—between the way Copernicus had come to see things and the way other astronomers saw things.

On the other hand, since it is only an expert in astronomy who would be deeply entrenched in the nested-sphere sense of cohesion, we might hope to find evidence that nonastronomers found his work interesting, as laymen were interested in evolution before Darwin [10.1]. And, indeed, this provincial canon, in his self-described "darkest corner of Europe" seems to have become known everywhere. The *Regista* shows that even before Rheticus published his "Narratio Prima" (1540), popes, kings, and distant scholars have heard of him and his idea. Further, the reactions in the record are favorable, with the main exception of well-known remarks by Luther (*R*:421) and his colleague Melanchton (*R*:478).

Others who appear in the record seem favorably impressed—Pope Clement (*R*:339), Erasmus (*R*:348), the duke of Prussia (*R*:475), Cardinal Schoenberg (*R*:359), among others. It is apparent that such people could appreciate that Copernicus had worked out what today we would call a neat argument. They wanted to hear more of what he had to say.

If we examine the *Regista* for specific Copernican involvement with geographers, a remarkable number of connections turn up in this record. Copernicus' 1509 translation from the Hellenistic scholar Theophilactus (*R*:57) was provided with its preface by the man who gave the lectures in geography at Cracow which Copernicus would have attended as an undergraduate. These lectures had become the first printed geography text that was more than a translation of Ptolemy.

Copernicus' kinship to geography continued through his career. There are several references to involvement in map-making projects (*R*:299,472, etc.). Bernard Wapowski, secretary to the king of Poland, seems to have been (c. 1534) the first consequential figure to urge international attention to Copernicus' work (*R*:345). Wapowski was well-known as a cartographer (see *Dictionary of Scientific Biography*). Another earlier supporter was the internationally known scholar Gemma Frisius (*R*:469). Frisius, professor at the University of Lovain, was also known primarily as a geographer (*DSB*). And of course it makes sense, on the account of the discovery I have sketched, that not only should Copernicus have a lively interest in geography but that others with a lively interest in geography, much more than experts only in astronomy, should be especially open to his argument. Having been prompted to check into this, it seemed impressive to me how well the meager biographical record turned out to support that account.[4]

If we then turn from the biographical scraps to Copernicus' work we again find something reasonably remarkable, since Copernicus turns out to have discussed the discovery of the New World in his book (*de Rev*, I.3). The key sentence refers (in various translations) to "a new group of inhabited countries," "a new Mainland," "a new group of inhabited islands." Earlier in the same chapter, we have a remark translated as "What is a continent, and even the whole of the mainland, but a vast island?" Naturally, on the argument I have given we would suspect that the Latin for which the phrases I have mentioned are given as translations would be "orbis terrarum" (the Latin term for the oecumene), and that turns out to be so. Having referred at the beginning of his chapter to the orbis terrarum in a way that carries its usual sense as a singular thing ("What is the orbis terrarum but an island larger than the rest?"), in the later passage he calls the newly discovered lands "algun orbis terrarum," which as the term had always been used in such a context would be like saying that a person discovered he had another head.

Rosen (1943) tracked down the origin of this passage, motivated by concern that Copernicus was being blamed for promoting the naming of the New World after Amerigo Vespucci instead of after Columbus. For in the oecumene discussion, Copernicus specifically refers to "America." Rosen wished to absolve Copernicus of bad judgment on the matter. But the aspect of significance to us is how closely Copernicus kept to Waldseemuller's 1507 publication except on one detail. Rosen lists eight points on which Copernicus follows Waldseemuller so closely that he must have been writing with the Waldseemuller text (or detailed notes on it) at hand. Copernicus, Rosen reports, "departs from Waldseemuller in only one pronouncement, namely that America is thought to be a second orbis terrarum." For Waldseemuller had described America as a fourth part of the world (in addition to Africa, Asia, and Europe), and as a large island, but not as a second orbis terrarum.

Rosen traced the stronger Copernican language to a map of the world inserted in some copies of a 1507 edition of Ptolemy's *Geography.* The map also shows the discoveries as a new continent in the West, and carries the remark (Rosen's translation): "This region is believed by many to be a new orbis terrarum," while Copernicus writes: "They think it to be a second orbis terrarum." So Copernicus departs from Waldseemuller on only that single point of detail, and he departs in just the way we might expect if the shattering of the oecumene had played the role sketched out here. And it is significant that with great exactness, Copernicus repeats what was said in the earliest publications (both from 1507) that could have informed him of the shattering of the oecumene. It is as close to a direct proof as we could reasonably hope to find that that cognitive shock indeed made the deep impression on Copernicus that the account earlier requires.

12.10

The shattering of the oecumene demonstrated that Aristotle was not always right, but it hardly implied that the earth could therefore be thrown into the sky and made a planet. But no such "therefore" argument is meant here. I am only saying that when Aristotle turned out to be fundamentally wrong about the sphere of water, the cognitive basis was laid for someone recently startled by that to be open to the intuition that a violation of some other detail of the nested-sphere view might be thinkable. I want to close the chapter with a parallel illustration of how such a connection might have worked.

On the sketch of a cognitively plausible uphill sequence, a person thinking about Venus and the various familiar facts about Venus rehearsed earlier might easily be prompted to the thought that Venus' movements look the way they would look if Venus (and similarly, Mercury) was in orbit around the sun. But a Ptolemaic astronomer knew that was false, as you know that the upper arrow in figure 1.1 is not really longer than the lower one. If you look at the arrows, you do not have a sequence of intuitions: "The outward arrow is longer; no, that's only an illusion," so that if you didn't go through the second step you would be left believing the upper arrow is longer. Rather you know this situation so well that you have a single all-at-once impression: "The upper arrow looks longer although it isn't." Similarly, many a Ptolemaic astronomer must have had the compound perception: "Venus moves as if its epicycle could be centered on the sun, although it can't be," if only because occasionally some student not yet well-trained in seeing things in the nested-sphere way must have raised that possibility, as Martianus raised it [12.1].

All that is required for the argument here is that Copernicus should have been capable of seeing that usual presumption as conceivably too strong. The proximity between the shattering of the oecumene and the Copernican insight (in particular, Step 0, opening the door to that insight) is just the sort of thing that would fit that requirement on the general *P*-cognitive argument. If you had recently become aware of an extraordinary piece of news (that not just a route to the East but entirely new and vast land masses had been discovered in a place that was hitherto unthinkable), which had the consequence that two of Aristotle's spheres had to be collapsed into one sphere, then you might be prompted to a double-take at the ordinarily automatic dismissal of the possibility that the doctrine of nested spheres might be violated in the heavens. Where until the shattering of the oecumene a well-entrenched habit of mind would make that idea look utterly wrong to a competent astronomer, recent striking experience in the world now makes available the intuition that perhaps it isn't utterly wrong. After a while, this cognitive shock is tamed, as the Aristarchan shock about the immense size of the sun was tamed (Chap. 11). The new terraqueous sphere comes to be seen as a clarification of Aristotle, not a contradiction (as discussed earlier). But there is a narrow window within which a favorably situated early sixteenth century astronomer could have discovered what we know that a favorably situated astronomer named Copernicus in fact did discover.

12.11

As it happens, yet another Copernican puzzle, also involving Venus, illustrates the kind of relationship I have been suggesting between the shattering of the oecumene and the possibility of being prompted to look closer at the Venus intuition. Here the issue is an oddity in how Venus was seen by the earliest Copernicans. For until a year after the invention of the telescope, no astronomer, Copernican or otherwise, realized that if the world is heliocentric a famous anomaly in astronomy could be resolved by supposing that Venus goes through phases like those of the moon. It was not until shortly before Galileo first saw Venus in its "new moon" phase as a thin crescent, and nearly a year after he had been using the telescope, that he (and, independently, his protégé Castelli) had that idea (Drake 1984).

Logically, Copernicus himself, and all subsequent Copernicans, should have foreseen the phases of Venus. For a Copernican logically should see that Venus is probably opaque, like the earth and moon; and if so it must go through phases like the moon as it goes around the sun. But Copernicus did not think of that; even 70 years later, Kepler was taken by surprise by the news. And Galileo barely avoided being taken by surprise.

So why were Copernicus and Kepler blind to this obvious and obviously important inference, with Galileo only marginally better? On the *P*-cognitive account, this is a "scenario" effect, of the sort discussed in connection with the Kahneman & Tversky and the Wason illusions of judgment in Chapter 8. We all know very well from experience that if we see something in dim light or in the distance, when we see it more clearly we may find we have suffered a misperception. But far more common are the cases where perception is sound, even under unfavorable conditions [2.9]. For usually the things we see under unfavorable conditions are things we are familiar with, so that even when weak cues prompt a pattern that goes well "beyond the information given," the imputed pattern is generally a pattern that works. Experience in the world discourages doubts, for if we were prompted to look closer every time we went beyond the information given we would be crippled by constant doubting. Consequently, a universally well-entrenched habit of mind is to be prompted to doubt our perceptions only when we have encountered positive cues which prompt a closer look.

But seen with the naked eye, Venus *looks* round, even when a telescope will show it as a thin crescent; and until the invention of the telescope there was no experience that would associate things seen in the heavens with things that sometimes had a misleading appearance. So there was nothing in experience to prompt a person to second thoughts about a possibility that Venus (more exactly, the lighted face of Venus) might not be round. Explicit beliefs logically should have required a Copernican to guess at what the telescope would show was a crucial truth, but by the time a telescope existed, the logic of the situation was a tamed anomaly. That Galileo (and his student) finally did see the point is consistent with that argument. For they did not do so until they had about a year's intense experience with startling disparities between how heavenly objects look to the naked eye versus how they look through the telescope. There the sharp edge of the terminator on the moon becomes ragged, the lunar sunrise reveals peaks and valleys of mountains, Jupiter turns out to be accompanied by moons, and so on for corresponding remarkable new views of the Milky Way, the planets, and the sun.

Therefore, what Galileo did when he finally thought of the phases of Venus is in a sense—an incomplete but meaningful, not trivial, sense—

what I have suggested the macaque did when she discovered how to wash rice [6.7]. A pattern that worked in one context is seen as looking like (shares a sufficient range of cues with) a later experience. Having found over and over that things in the heavens look very different when seen through the telescope, and coming into a period in which—after some months of bad viewing opportunities for Venus—he was turning his attention to that planet, Galileo was prompted to think of how Venus would look as it moved from around the sun. Having gotten to the point of thinking of that question, the answer is obvious, so it is not surprising that his student Castelli also thought of the same idea at essentially the same time.

For Galileo, as in an entirely different way for Copernicus on the account I have sketched, being prompted to ask the right question about Venus is the crucial step in reaching the discovery. This occurs in a simple and I should think reasonably uncontroversial way for Galileo's prediction of the phases of Venus, and in a more complicated and no doubt highly controversial way in the account of the Copernican discovery. In both these cases, what is involved at that critical step is seeing a question to ask: how Venus might look different if we could see it better; how Venus would look if its epicycle was centered on the sun. In both cases, this requires being prompted to doubt a habitually automatic blindness to considering such a possibility. In both cases, we can point to vivid recent experiences that could prompt such doubts, and we can also specify circumstances that can account for how the intuition that would prompt doubt would itself be prompted: for Galileo, the imminence of the opportunity for close study of Venus, and for Copernicus (here of necessity, only conjecturally) a focus on Venus tied to his interest in the equant.

Thirteen

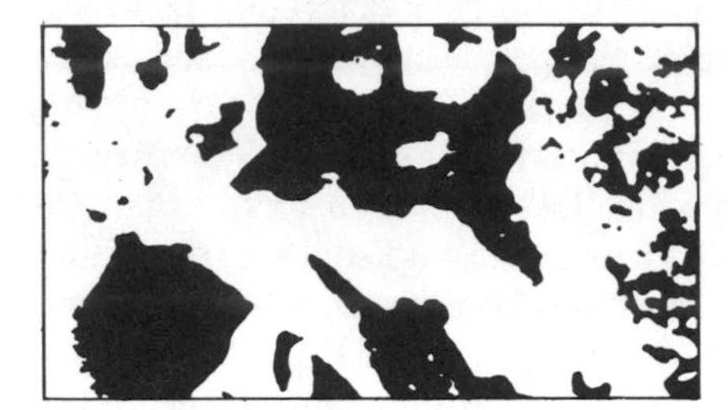

The Copernican Contagion

Sometimes an idea that initially seems hard to believe comes to seem hard to doubt. A *P*-cognitive analysis of the contagion of such an idea should require no different sort of analysis from what is required for an account of discovery. In philosophy-of-science language, but in contrast to a common philosophy-of-science claim, the kind of analysis that can account for what goes on in justification should also account for what goes on in discovery. But the main *P*-cognitive puzzle for justification (meaning here, contagion, spread of belief)[1] is intrinsically different. In the discovery context we need to uncover a path "uphill" [9.3] that plausibly meets the requirement of cognitive continuity. But in the contagion context, potential believers are given what the original discoverer had to work out for himself. Both the key insight, and supporting reasoning-why intended to make that insight understandable and believable, will be presented as forcefully as the discoverer can manage. What needs to be explained is why the initial reaction to the new idea tends to be far less favorable than seems reasonable in hindsight, and what happens to change that.

For an account of discovery, Pasteur's maxim ("chance favors the prepared mind") aptly summarizes what is required: we seek an account of how a particular mind came to be prepared for a radical discovery when the opportunity arose. The sequence-dependent character of *P*-cognition [4.3] then plays an essential role. It makes us look for an uphill path such that the most deeply entrenched habits of mind are not directly challenged until late enough in the process to allow the radical idea to survive reasonably untamed. Chapters 10 and 12 sketched out how such an account can be constructed for the Darwinian and Copernican discoveries.

But the audience the discoverer then must persuade will not have had that preparation. It will be confronted abruptly—relative to the experience of the discoverer and despite all efforts to soften the shock—with a radical idea that then will tend to look immediately wrong. The taming response to a conflict between a deeply entrenched habit of mind and a radical intruding intuition [7.9] will tend then to be such as to entrench seeing the new idea as illusion, not insight [7.10]. We want an account of how that changes and of the way the community changes once the cognitive barriers have been broken.

When contagion really takes hold, it does so as an interaction among social (not atomistic) individuals who talk to each other, and among whom the novel idea has emerged as something people have become interested in talking about. Contagion takes on an essentially social character, and we effectively are moved from retail (individual by individual) shifts in belief to wholesale (classes of individuals, such as Jesuit astronomers). Then, even if the arguments for the new view were only what had been provided from the start, what had been important cognitive barriers will still lose much of their strength [9.11]. But by the time social contagion takes hold, additional or improved arguments produced by the first converts will have strengthened the case the original discoverer was able to make. The two effects will be mutually reinforcing. Eventually it requires Kuhn's (1977) "hermaneutical" effort to recapture how anyone found the old way plausible, or the new way implausible.

The transition from acquaintance to belief, therefore, is not difficult to understand once the stage of social contagion has been reached. Hence the psychologically most interesting part of a contagion story turns on the early converts. All of this applies, of course, most strikingly to the case of radical discovery, where the new idea conflicts with deeply entrenched habits of mind which are shared across the entire relevant community. For less radical novelties, contagion is much easier [9.9].

13.2

In recent years, examination of sixteenth century manuscripts has revealed the recurrent, apparently independent, efforts of a number of leading astronomers born around the time Copernicus' book was published to satisfy themselves about just how the Copernican scheme related to the Ptolemaic. Astronomers of the generation active when *de Revolu-*

tionibus was published mostly ignored the heliocentric arguments despite their interest in Copernicus' equant-free models (Westman 1975a). By the 1570s, however, there was much interest in working out ways of clarifying the Ptolemaic/Copernican observational equivalence through diagrams that can be interpreted in Ptolemaic terms, but which with some ingenious alternative labeling of points and lines yields a Copernican equivalent (Westman 1975b; C. Wilson 1975; see also fig. 11.1 and n.5, chap. 11).

The equivalence diagrams revealed that the difference between the Ptolemaic and the Copernican models was like the difference between seeing the well-known gestalt drawing as an old hag versus as a young woman. In terms of computing the future location of the planets, as seen by an observer on earth, there is no essential difference between Copernican and Ptolemaic astronomy, just as it makes no difference for the purpose of locating the coordinates of some point in the gestalt drawing whether we see the picture as an old hag or as a young woman. But in terms of what the world is taken to be really like, hence (aside from natural curiosity about that) what a promising agenda for further work would be like, Copernican and Ptolemaic astronomy are radically different.

13.3

An even closer equivalence relation holds between the Copernican scheme and the Tychonic compromise which appeared and rapidly took hold around 1590. Once the Ptolemaic/Copernican relations had been painstakingly worked out, it became relatively easy (not absolutely easy, but relatively easy) to notice how the Copernican system could be "inverted" to yield the "Tychonic" system, which combines stability of the earth with the Copernican golden chain relations.

In figure 13.1 you can see Tycho's sun located on its orbit around the earth at the point due north of the earth. The orbit of the moon also goes around the earth. But the orbits of the planets go around the sun. Suppose the sun continues on its orbit for three months. What would this Tychonic system now look like? The answer is shown in figure 13.2. The sun, having moved through a quarter of its yearly cycle, is now a quarter turn to the right, due east of the earth. But it is no closer to intersecting the orbit of Mars than it was in figure 13.1. The sun can never collide with Mars or be anything but the usual distance from Mars. For the sun carries the orbits of all the planets around with it, as in the Copernican system the earth carries the orbit of the moon around with it; or as in the Ptolemaic system, the sun keeps the epicycles of Mercury and Venus between it and the earth. The Tychonic system looks more puzzling than either the Copernican or Ptolemaic, and I will comment on the cognitive basis of that later. The essential point for the moment is only that the key to the Tychonic system is that the orbits of the planets move with the sun, which in turn carries the whole system around the earth.

Another way of thinking about the Tychonic system is to imagine that the sun has six spokes, like a clock with so many hands. In the Copernican scheme, the hands go around in the usual way, each carrying a planet around at its own pace. For five of the hands, exactly the same thing happens in the Tychonic scheme. But the earth, at the end of the sixth hand, is fixed in space at the center of the world. So when the hand connecting the sun and the earth turns, it is the sun (carrying the five rotating planets along with it) that is turned around the motionless earth rather than the reverse. Hence, in figure 13.2, each planet has moved on its orbit, while the sun has moved carrying all the orbits (Venus moving from *V* to *V**, and so on).

All the technical properties that made the Tychonic scheme an interesting alternative to Copernicus depend on the point that the Tychonic orbits of the planets are carried around with the sun as the sun itself moves around the motionless earth. The moving orbits (as they are all carried around by the sun) combines with the movement of each planet along its individual orbit to produce the looping retrogressions that occur when a planet is on the opposite side of the earth from the sun. All other interesting features of the Tychonic system also depend on this interaction. If you imagine the sun moving through figure 13.1 to S′, then you have misunderstood how the Tychonic system works. It is impossible for the sun to move without the orbits of the planets moving with it. The whole diagram must turn, giving figure 13.2. This point needs emphasis, since until you are well-accustomed to it, trying to envision this double motion (in the mind's eye) is almost painful. There is something deeply counterintuitive about the Tychonic motions, and that will play a role later in this chapter, and again in the account of Galileo's confrontation with the Church in Chapter 14.

What I will try to show is that on the *P*-cognitive analysis the crucial turning point for the Copernican contagion comes with the appearance of the Tychonic system around 1590. In Kuhnian terms, if we want to mark the transition from the Ptolemaic to the Copernican paradigm, 1590 is about the right date. Or, in language used earlier in this chapter, the transition from individual to social—"retail" to "wholesale"—contagion of the Copernican view (I will try to show) comes with the appearance at the end of the 1580s of the Tychonic system just reviewed.

This claim varies sharply from what seems to be the prevailing view. It turns on the cognitive centrality of the nested-sphere versus golden-chain gestalts. For so long as the central issue is seen as whether a person accepts the Copernican orbiting earth, then the Tychonic system is obviously best understood as a way of assimilating the Copernican advantages

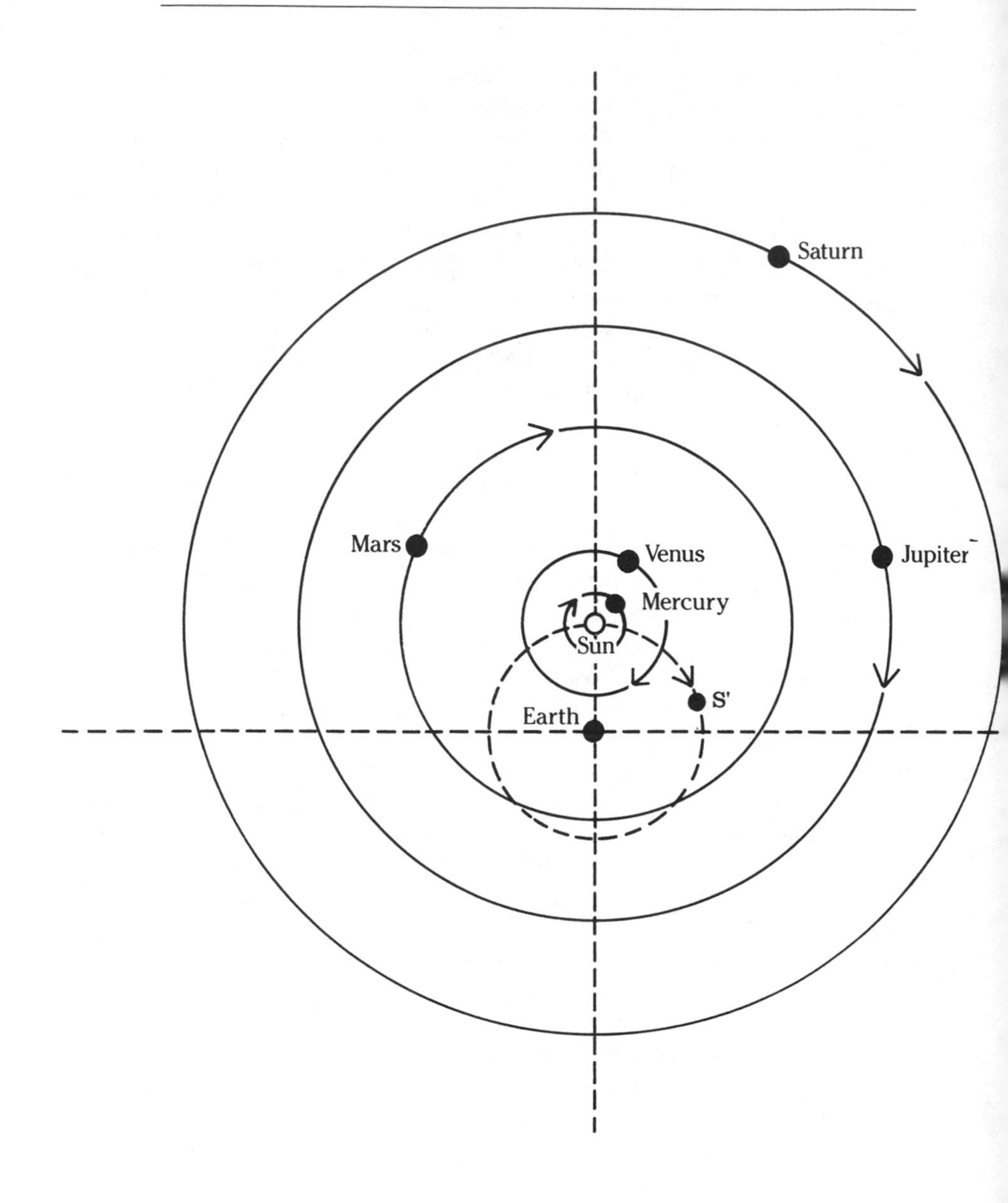

Figure 13.1. The Tychonic system. This can be derived from the Copernican by a shift in origin (leaving all relative distances and angles of observation unchanged), as described in the text using a clock analogy.

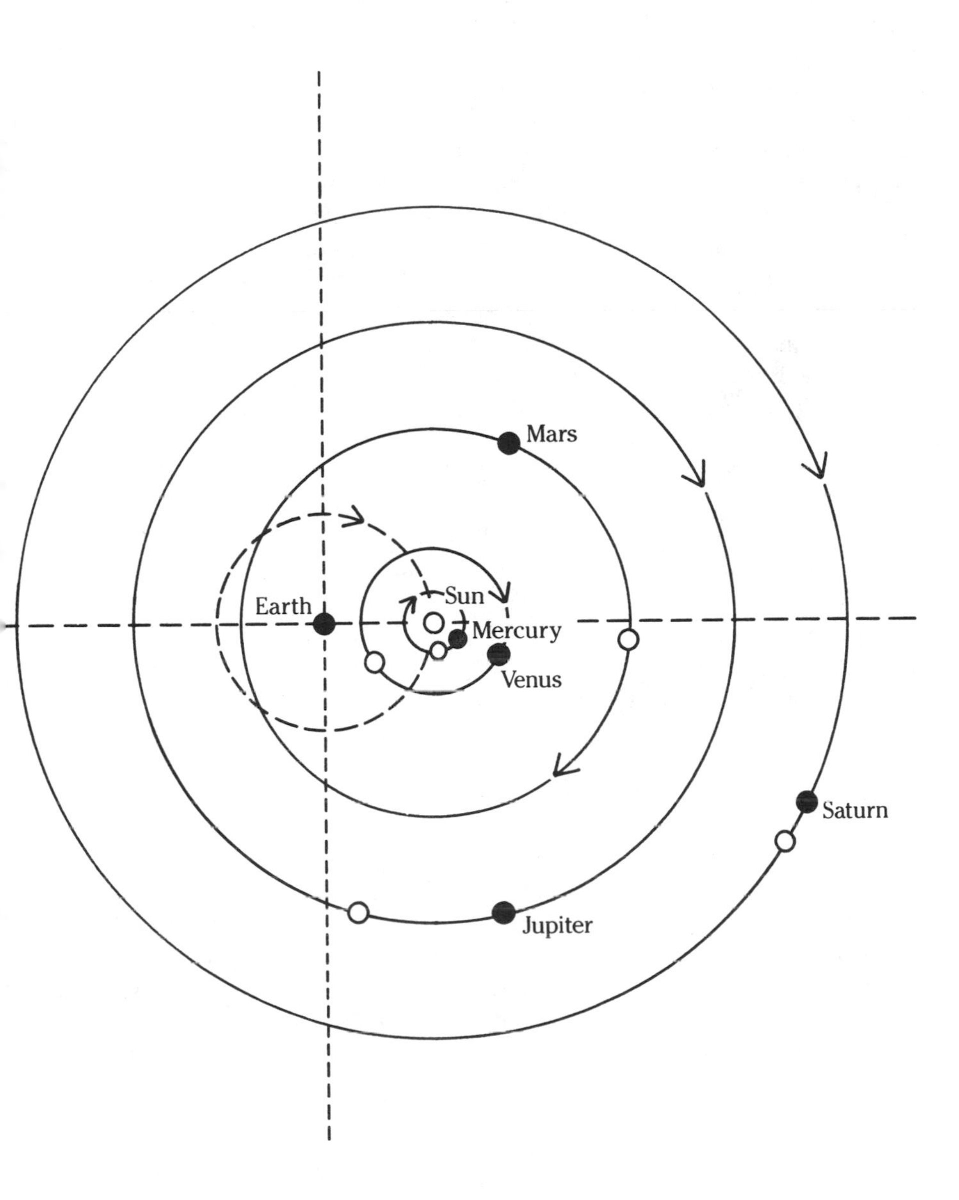

Figure 13.2. The Tychonic system of fig. 13.1 three months later. The sun has moved 90 degrees, carrying the orbits of the planets with it, while each planet has moved three months along its orbit.

to Ptolemy. But on the argument I will make, a person could only make sense of the Tychonic system if the habit of seeing things in terms of the nested-sphere cohesion had already eroded. Once that has occurred, a person might not yet see the Copernican system as right, but he would no longer see it as absurd or incomprehensible. In the contagion version of the discovery schema [9.3], the situation has moved from "It's absurd" to no worse than "It's wrong." In terms of the cognitive argument, whether the Tychonic system is essentially Copernican or essentially Ptolemaic is more than a "half-empty versus half-full" sort of point. On the contrary, what I will try to show is that various striking details of the episode make sense if we suppose that the critical cognitive transition is marked by (*not* caused by) the appearance of the Tychonic system. Those same details do not make sense if we suppose (as in Grant 1984, and much other writing on the episode) that Tycho and other supporters of his or similar systems are still holding onto an essentially Aristotelian or even essentially Ptolemaic viewpoint.[2]

13.4

How did it happen that in the late 1580s not only Tycho but several other astronomers of the period almost simultaneously abandoned Ptolemy and proposed some version of the scheme in figure 13.1? There is a "usual story" about that, turning on how Tycho's observations of the great comet of 1577 demonstrated that there were no heavenly spheres, hence that Ptolemaic cosmology must be wrong. On the other hand (continuing this "usual story"), if there are no spheres in the heavens, then a major objection to the Tychonic scheme is preempted: if there are no spheres, then the intersection of the orbits of Mars and the sun (fig. 13.1) is not the impossibility it would be in a world in which each orbit marked a path along an impenetrable sphere. So on one side, observational evidence assertedly makes the Ptolemaic system untenable (there are no impenetrable spheres), and on the other hand the same evidence shows there is no insuperable bar to the Tychonic alternative. So the way is cleared for this new system, which simultaneously avoids all the usual physical and theological objections to the Copernican scheme, yet preserves its golden chain advantages.

However, although sketches of this usual story appear routinely in writing on the period,[3] you will not find a detailed story, fitting all the pieces together in a coherent way. For an attempt to produce a neatly rational reconstruction of the episode is bound to be either factually wrong or logically incoherent. Rather, the Tychonic episode (I will be arguing) makes sense in terms of a cognitive argument in which nothing happens between 1543 and the 1580s which is nearly as important as the mere passing of time. With that passing of time, astronomers deeply entrenched in the nested-sphere view (so that they find a golden chain argument bewildering) are replaced by a generation that has come to maturity always aware of the Copernican possibility, and finds that the Copernican argument explains so much so neatly that it can hardly be wholly wrong.

13.5

Clearly, around 1590 the Tychonic idea was somehow one whose time had come. Not only do we have multiple discoveries, but the contagion is remarkably rapid. When Kepler (1596) came to publish his outspoken Copernican first book, he and his teacher (Maestlin) assured the authorities that there was nothing scandalous in that since virtually all astronomers accepted the Copernican arguments (Aiton 1981:22). A few years later (1601), when Kepler was working for Tycho, he was more careful about treating the Tychonic scheme as merely an implausible inversion of the Copernican scheme. He was more precise, writing that "today there is practically no one who would doubt what is common to the Copernican and Tychonic hypotheses, namely that the sun is the centre of motion of the five planets" (Jardine 1984:147).

Such remarks have to be understood as comments on the state of expert opinion. For the wider community, the question was always only whether the sun went around the earth or the reverse, so that what was a radical change in expert opinion (from Ptolemaic to Tychonic views) could go virtually unnoticed outside the expert community. It was only within the expert community that the question of whether the planets (other than the earth, of course) went around the sun could be a terribly important question, since it was only within the expert community that the nested-sphere versus golden chain cohesion could be either cognitively important (governing what looked reasonable) or logically important (governing the explicit arguments a person would make in accounting for the structure of the world). About the last serious contribution to Ptolemaic astronomy was Praetorius' manuscript (described in Westman 1975b:194 ff.), which itself was (significantly) an attempt to work Copernican results into a Ptolemaic framework. By that date, there is much evidence that astronomers as a community are seeing the world in a new way: more likely the Tychonic version than the Copernican, and often no doubt with no firm commitment either way, but in any case seeing things in a way that made no sense in terms of the pre-Copernican snugly nested spheres. Even someone trying very hard to be a Ptolemaic astronomer, like Praetorius, reveals that he has lost that vision of the world: you can graft the golden chain onto a Ptolemaic model, but it is like dressing up a corpse: it cannot be long before everyone sees that the thing is no longer alive.[4]

On this account, there is a reversal of causality (relative to what has usually been said) with respect to Tycho's argument that the Ptolemaic heavenly spheres do not exist. Astronomers do not abandon Ptolemy because the Ptolemaic nested spheres have been killed by empirical evidence; rather, astronomers who learned their craft aware of the Copernican argument find it easy to believe that the nested spheres do not exist. From a generation for whom the Copernican golden chain argument was puzzling and uncomfortable to deal with (Westman 1975a), we come to a generation of astronomers who have grown up aware of the Copernican alternative and to whom it starts to look irresistibly right.

Being aware that the nested-sphere cohesion was not the only way to understand the world—that is, being aware that if it went away there was something that might take its place—weakened the cognitive pressure to tame doubts about the Ptolemaic story. The golden chain gained in comprehensibility and credibility with the passing of pre-Copernican astronomers deeply entrenched in the nested spheres (not because the Copernican argument was getting any better); similarly, the nested-spheres view lost credibility—not because the argument against it was getting much better but mainly because the new generation could see what had logically always been there to be seen. We find an instructive clue to what was going on in Tycho's complaints in the mid-1570s (before the crucial comet had even appeared) about astronomers who were improperly mixing up Copernican and Ptolemaic arguments (Moesgard 1972:32), as would occur if you continued to treat physical arrangements as Ptolemaic but nevertheless used the Copernican golden chain harmonies.

13.6

Copernicus himself uses (and other times ignores) the heavenly spheres in a manner that reflects the combination of the taken-for-granted reality of the spheres that characterized astronomy in his time, qualified by the way that Copernican cosmology undermines the significance of belief in the spheres. From the cognitive relations among habits of mind, use, and entrenchment of belief, we would expect that the loss of the significance of the spheres' solidity for giving structure to the world would have the consequence that the spheres would become marginal. A person who had acquired the habit of mind of seeing the world in terms of the Copernican golden chain might still believe in such spheres but not in a way that would make it wrenching to consider the alternative.[5]

Therefore—not abruptly but over time—astronomers for whom it would be painful to imagine the world without nested spheres to give it structure (astronomers for whom reduction of retrogressions to an illusion of parallax had meant eliminating an essential element of the structure of the universe) would be replaced by new people for whom the elimination of the major epicycles could come to be seen as striking evidence for the plausibility of the Copernican view.

It is only against that background that we can make reasonable sense of Tycho's argument for his system.[6] We want to consider how far the sort of argument Tycho actually produced could plausibly be supposed to carry the burden of persuading the community of astronomers to abandon the nested-sphere foundation of cosmology. The answer is going to be, "Not very far at all." What is left after an even moderately critical scrutiny of Tycho's argument is only a particularly grand example of how far beyond the instructive but substantively modest cases usually discussed (Nisbett & Ross 1980) the possibility goes that seeing-that sometimes governs reasoning-why more than the reverse [4.8].

I will push this claim very far, arguing that despite all that has been written to the contrary, not only was Tycho's argument much too weak to convince someone still seeing things in the Ptolemaic way but too weak to have plausibly convinced even Tycho himself. Rather, the situation is akin to the one I sketched of the distance/period intuition [11.11] where what makes cognitive sense of an intuition—what patterns of experience plausibly could have prompted the intuition—has nothing in particular to do with the reasoning with which people who have the intuitions try to explain them. As we all have habits of movement or speech that we would never notice unless someone points them out to us, so we have habits of mind that we ordinarily have no way of noticing but that yield intuitive "looks right/looks wrong" judgments about things in the world, such as the arrangement of a room or of an argument. What we believe is not linked in a necessary way to why we think we believe it. I want to suggest how what Tycho came to see as the basis for belief in his system could hardly have been essential to why he actually came to believe it (could not have had much to do with the patterns of experience available to Tycho which give us a coherent account of the emergence of the Tychonic system). In particular, until a pattern of experience underlying an intuition has been explicitly noticed, it cannot have been linguistically described, hence could not be available in the repertoire of patterns of argument available for explicit reasoning-why. A person would then necessarily come to believe some artifactual reasoning-why to justify beliefs that are rooted in that intuition.

13.7

In 1572, a nova (new star) appeared—the first in several centuries. This provided evidence that Aristotle's claim that the heavens were unchangeable was wrong. The measurements required to demonstrate that the new object showed no parallax (hence was in the heavens beyond the moon) are extremely simple: for a person who knew what he was doing, a piece of thread was the only instrument required. Somehow, however, neither ancient astronomers nor the highly skilled Arab astronomers who were active during Europe's Middle Ages saw what was obvious to post-Copernican astronomers like Tycho and Maestlin. What had been available all along, then, was now actually noticed. Apparently something had changed about the way astronomers saw their phenomena (Kuhn 1957:208).

By itself, however, the nova could not have had any substantial effect on astronomy: it was generally taken to be a portent of some momentous event on earth (like the star of Bethlehem); and it could be understood as a temporary brightening of a star that was previously and afterward invisible. In any case, whatever it might imply about a detail of Aristotle's cosmology, it implied nothing about the nested spheres or any other salient feature of Ptolemy's astronomy, which—in the use of eccentrics, epicycles, and equants—had always allowed departures from Aristotle.

The new star was shortly followed by the unusually vivid comet of 1577. Tycho, Maestlin, and others made observations that convinced them that comets too were (contrary to Aristotle) beyond the moon. So perhaps did Copernicus (*R*:335, 451). The significance of this view of comets is partly only the same as the new star: providing yet more evidence (on top of the nova and the terraqueous sphere) that Aristotle was sometimes wrong but implying nothing in particular about Ptolemy's astronomy. Tycho and Maestlin, however, noticed that over the two months it was visible, the comet's angular movement across the sky slowed as its brightness dimmed. The simplest explanation was that the comet was moving away from the earth. Both Tycho and Maestlin concluded that it orbited the sun in the vicinity of the orbit of Venus, which would account for both the apparent variation in distance and for the fact that over its 10 weeks of visibility the comet moved through about a quarter of the sky, giving it a presumed period about the same as Venus. Figure 13.3 is Tycho's diagram.

However, no comet, in fact, has an orbit near Venus, as Tycho and Maestlin supposed. Indeed, if they had correctly guessed (as Kepler did) that the comet passed completely beyond the planets, Tycho's argument against the spheres would have been stronger. But the logical force of his claim, of course, has to be judged by the argument he made, not by the better argument we would be tempted to have him make. Further, there was great controversy at the time about every aspect of the comet, including whether even the basic claim that it was beyond the moon was right. The belief that there was any closed orbit at all was open to question; in fact, both Kepler and Galileo later did question it, and Ptolemaic astronomers—who had much more reason than Copernicans to question it—certainly had grounds for doing so. A comet was a transient appearance in the sky, which (for all anyone at the time could say) may or may not have been an actual object rather than a mere optical effect of some sort, like a rainbow. Giving it particular properties and supposing it had a particular orbit was a speculative, not a deductive, enterprise.

Suppose, however, that Tycho and Maestlin were right in saying that the comet orbited the sun at roughly the sun/Venus distance. Then *if* the world were Ptolemaic, the comet would be passing through the supposedly solid spheres of the planets. That is the totality of Tycho's argument that the spheres do not exist. But Tycho's conclusion doesn't follow, which accounts for why Maestlin, who had computed essentially the same

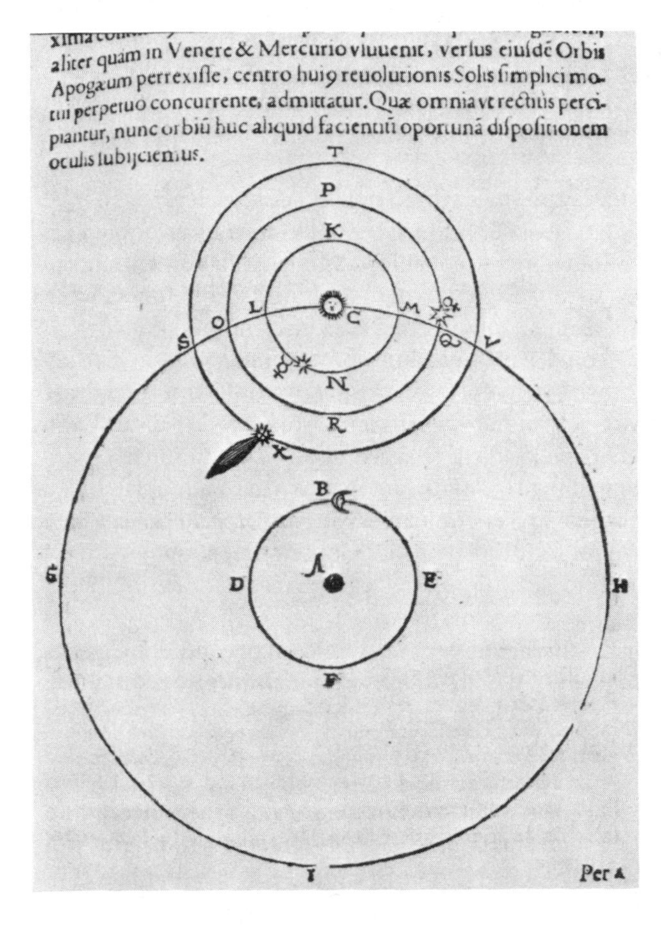

Figure 13.3. The orbit of the comet of 1577 as shown by Tycho in the publication (1588) which announced the Tychonic system. Reprinted from *Vistas in Astronomy* 17, *Copernicus: Yesterday and Today,* Arthur Beer and Peter Beer, general eds. With permission.

orbit as Tycho, did not come to doubt the reality of the spheres from the evidence of the comet; and for a decade, neither did Tycho. He seems to have reached that conclusion only in 1587, although he had all the essential information he would ever have ten years earlier (Christianson 1979). On the face of things, then, the connection between the evidence of the comet and disbelief in the solid spheres is not something simple or unambiguous.

In fact, if you were a Copernican, and hence also if you favored the Tychonic system with its identical solar distances, there could be nothing in the evidence of the comet that would strike you as an argument against the existence of solid spheres. For with the disappearance of the major epicycles (the circles that take up most of the space in the Kepler diagram, fig. 12.1), there is plenty of room for special regions for comets (Field 1983). So the comet argument was irrelevant to a choice between the Tychonic and Copernican systems. *Only* if you were still faithful to Ptolemaic nested spheres, where the greatest distance of each planet brushed up against the least distance of the next planet, would you face a problem of explaining how there was room in your heavens for an orbit varying widely in distance from the earth like the one Tycho and Maestlin claimed for the comet.

But a Ptolemaic astronomer would not feel defenseless, even if he accepted the Tycho/Maestlin observations. He could start by asking why a Ptolemaic astronomer should even consider comets as falling within the domain of astronomical theory. Right down to Newton's time, comets were seen as transient portents of earthly events, and since no one knew what comets were, no one could say anything convincing about how they would interact, if at all, with heavenly spheres. How could you then draw any conclusion at all about planetary astronomy from Tycho's data? Even if you accepted Tycho's orbit (and there were other ways to interpret his data), there was no compelling reason to see that as anything more than a showing that the spheres could be penetrable by comets, as obviously they were perfectly penetrable by other transient phenomena, such as light rays. Nothing in this contradicts the planetary spheres' mutual nonpenetrability.

In general, it is not hard, but very easy, to concoct additional interpretations of the Tycho/Maestlin data which are consistent with sustained belief in the traditional nested-sphere view. For examples: (1) There is no cometary orbit: the partial path that was observed is all there was; and it is misinterpreted as showing the comet moving away from the earth: rather, the changes in appearance are a clue to the meaning of the portent (the motion is really slowing, and the comet is really getting less bright, not merely appearing to do so because it is receding). (2) Tycho and Maestlin had only weak reasons for supposing that the comet moved between the earth and the sun. It could have been always somewhere in the region of the outer planets, for example, between Saturn and the fixed stars, so that even if we accept the argument that the comet is receding it would not follow that it was passing through the planetary spheres: it could have been entirely beyond them all the while. (3) A bolder Ptolemaic astronomer could have taken a position like the one that Kepler (1596) himself took with respect to his own platonic solids account of the world. Kepler's vision was of a world in which the spheres are mutually noninterpenetrable, just as in the traditional view, but transparent to comets and to physical objects in general, since they are not themselves solid things but something like ideas in the mind of God.

We can extend this list, but all that is important here is that there was no reason for a Ptolemaic astronomer to feel defenseless against Tycho's claim. On the contrary, he could construct an elaborate defense.

This possibility has to be considered in the light of the sort of anomalies that Ptolemaic astronomers managed to live with for upward of a thousand years: for examples, the moon anomaly [12.5]; the Venus anomaly [12.11]; the tendentiousness of the reasoning required to produce the crucial sun/moon ratio [11.9]; and the puzzle of the physical meaning of Ptolemy's equants [11.2]. On that record, it is hardly either psychologically or logically plausible that the core idea of traditional astronomy (the snugly nested spheres) would be convincingly overthrown by a transient event like the comet. All the anomalies mentioned (and others) were somehow tamed, and indeed tamed even though no one could give any plausible suggestion at all of how to make sense of them. Astronomers simply learned to ignore them. But the evidence of the comet did not even have to be ignored; it could be explained away in any number of ways (as suggested above), none conspicuously less plausible than Tycho's claim that it showed that there were no heavenly spheres.

It then follows that the rapid spread of the Tychonic idea could not have been driven by Tycho's proof; rather, the repeated claims that Tycho had any such proof makes little sense except as a symptom of a change that had already occurred (which made it easy to believe an argument against the spheres), not as the cause of a change that was contingent on Tycho's proof.

In sum, then, the logic of Tycho's proof is worse than weak. It scarcely goes beyond a bit of handwaving. Rather, Tycho's argument could—at most—show no more than that *either* the world is not Ptolemaic, *or* (if it is Ptolemaic) then it cannot be built out of spheres impenetrable to comets. The argument gives no strong claim against the existence of spheres impenetrable to each other (whether or not the world is Ptolemaic) and no argument at all about spheres if the world is not Ptolemaic. So to accept Tycho's claims that his comet argument somehow removed a physical impediment to the Tychonic system, you have to believe that Tycho somehow is entitled to have something more than both sides of this disjunction at once, which would be truly a wonderful case of having your cake and eating it twice.

I have gone out of my way to emphasize the weakness of Tycho's argument, since it leads to an important but perverse effect which would not be expected in terms of a more Whiggish view—in which the move to Tycho is seen as an empirically driven step away from Ptolemy toward Copernicus rather than as a step back from being an overt Copernican, taken by people who are already far more committed cognitively to Copernicus than they perhaps are aware. On the *P*-cognitive argument, astronomers came to the Tychonic view by way of the Copernican golden chain cohesion, which had undermined the habits of mind that were the cognitive basis of firm belief in impenetrable spheres. Naturally, Tycho produced reasoning-why in support of his new views, but we are scarcely surprised to find that the logic of that reasoning-why is inadequate. We are alert to the possibility that reasoning-why can easily diverge from a plausible approximation of what patterns of experience are the actual framework for intuitive seeing-that [11.11], and we are prepared to find that even an outstanding individual like Tycho is quite capable, like the rest of us, of failing to notice flaws in his own argument.

Moreover, if there are aspects of the Copernican golden chain cohesion that are lost in the Tychonic world, then someone committed to Tycho would become habituated to those losses [7.7]. Such people would tend to see arguments of the sort they have learned to put out of their minds as not really important or relevant or convincing. They will acquire a certain trained insensitivity to new evidence that involves that sort of argument. And (leaving a more detailed discussion to Chap. 14), this effect can be expected to be especially strong when they become actively involved in fighting for (in this case) the Tychonic view against Copernican adversaries.

13.8

Consider, then, how the Tychonic system in fact fell short of a completely satisfying package of Copernican harmonies minus Copernican heresy. Even Kuhn (1957:203–4) writes that under Tycho's scheme "all the main arguments against Copernicus' proposal vanish . . . and this reconciliation is achieved without sacrificing any of Copernicus' major mathematical harmonies. . . . The relative motions of the planets are the same in both systems, and the harmonies are therefore preserved." Comparable statements can be found in virtually any account of the situation. Nevertheless, the positive arguments for Tycho's system consist of only a subset of the Copernican arguments. In particular:

1. The looping motions of the planets remain just as real as in Ptolemy, but they are produced in Tycho's scheme [13.3] by the way that the planetary orbits are carried around the earth by the sun rather than by epicycles synchronized to the movement of the sun. So the Tychonic retrogressions lost their Ptolemaic significance for the structure of the world without gaining the startling simplification of that structure that Copernicus offers. The radical simplification of the heavenly machinery—the reduction of the looping to a mere illusion of parallax—is completely missed in the Tychonic scheme.

2. The intuitively neat Copernican ordering of the planets (Chap. 12) becomes only a puzzling coincidence under Tycho. In the Copernican system the earth (with a period of one year) turns out to be—as Copernicus (*de Rev* I.10) emphasizes—in just the place it ought to be if indeed it is a planet (between Venus, with a period of eight months, and Mars, with a period of two years). But under Tycho we get the odd result that the sun turns out to exhibit a combination of period and distance from the earth that would make sense if the earth were a planet turning around the sun, though Tycho's earth is not a planet at all but a unique object at the center of the universe. When Kepler (1621:914) came to this point in his list of 18 reasons for rejecting the Tychonic system, he commented, "Does not the nature of things shout out loud that the world is Copernican?"

So Tycho could offer only a degraded version of the Copernican golden chain. But in addition there was a further problem of another sort:

3. There are the clumsy (not at all golden) dynamic problems that come with the Tychonic system, which account for the difficulty of "seeing" the Tychonic system go through its motions in the mind's eye. That can easily be done by anyone familiar with the Copernican system. But even if you have studied it carefully, it is usually impossible to do it for the Tychonic system.

Here, as Kuhn (1957:204) points out, "It is difficult to imagine any physical mechanism that would produce planetary motions even approximately" like Tycho's. Yet the problem (I want to argue) is more than a logical one of that sort. There is something about the Tychonic scheme that seems to stubbornly conflict with our intuitions about how things should go.

The *P*-cognitive explanation of those stubborn intuitions is akin to the explanation of the distance/period relations of the planets. There is something powerfully counterintuitive about the Tychonic scheme. It is also the case that *if* heavenly spheres were mutually noninterpenetrable, then the Tychonic system would be logically impossible. But that Tycho's system would then *be* logically impossible need not make it *look* cognitively impossible (look definitely wrong in an intuitive sense); and the converse also holds: fixing things so that the system is no longer logically impossible does not necessarily make it then look cognitively plausible. If the problem were the logical one (so that although the system looked workable in an intuitive way, something needed to be fixed up if it was to be logically coherent), then Tycho and others would have been prompted to look hard for some line of reasoning-why that might remove the logical difficulty. And that would not be hard to do, since obviously on the Tychonic scheme the sphere that carried the sun around the earth could easily be a qualitatively different kind of sphere than the spheres that carry planets around. Just as a Ptolemaic astronomer could have dismissed Tycho's argument on the ground that no one had ever claimed that the planetary spheres were nonpenetrable by comets, only by other planetary spheres, a Tychonic astronomer (if the only problem with the system was really the confounding of spheres) could have made a similar distinction.

But Tycho was sure that both the Ptolemaic and Copernican alternatives could be rejected, and we have no trouble understanding both these views: the former on the golden chain argument I have been stressing, and the latter on an argument perhaps even more readily understandable. For we have in Tycho a man very conscious of his claim to be the greatest observer astronomy had ever produced. After careful effort, he found no hint of the stellar parallax that must exist if the Copernican system were correct. Hence, for Tycho, either his observations were not good enough or Copernicus must be wrong in putting the earth in orbit. But he also would have had, as everyone who looks at the Tychonic diagram has, a strong intuition that something is wrong with this scheme. If, indeed, he had proved there were no heavenly spheres, then he would have a complete story: the Tychonic scheme looks wrong, and that is because the spheres overlap; but the spheres are not real; therefore, although the system continues to look counterintuitive, at least until you have worked with it for a long time and become habituated to its peculiarities, in fact it is logically OK. Despite its flaws [13.7], this argument provided a result very hard to resist for Tycho and others who could neither give up the golden chain nor accept the whole Copernican story.

But now look at figure 13.4. This is yet another version of the Tychonic diagram already used in figures 13.1 and 13.2. But in figure 13.4 the orbit of Mars has been deleted, so now there is no collision of spheres. If it were true that the reason the Tychonic system looks implausible is because of the collision of spheres, then figure 13.4 should look intuitively reasonable (there is no collision). I do not expect that many readers will

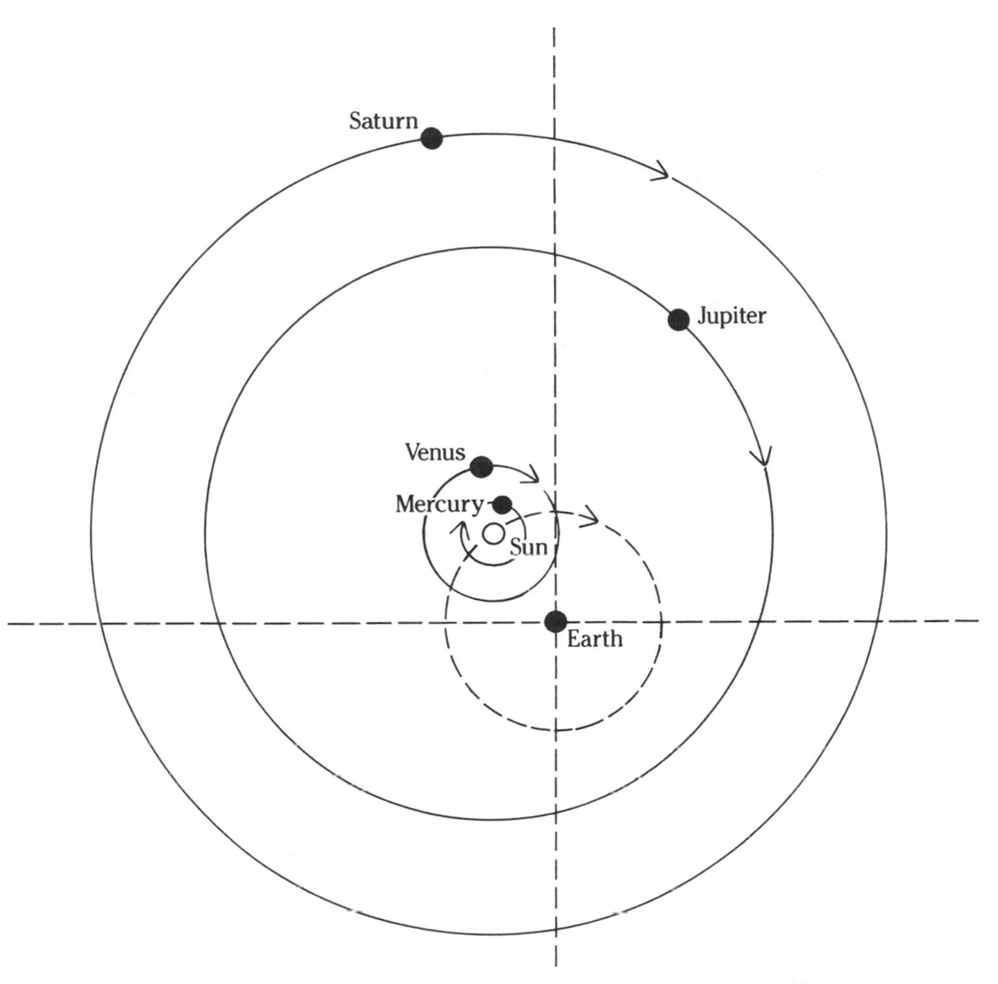

Figure 13.4. The Tychonic system again about nine months later than fig. 13.1 and with the orbit of Mars deleted.

find that erasing Mars' orbit "cures" the Tychonic malady. For it goes very much against our experience with things in the world to have huge things swirling around a center which is itself being swirled around another center (item 4 in Kepler's (1621:915) list of 18 anti-Tychonic arguments). It is this double center of rotation, combined with the large size of the secondary rotations (the planets going around the sun) relative to the primary (the sun going around the earth), that violates our tacit intuitions of how things work in the world, quite aside from whether the orbit of Mars intersects the orbit of the sun.

This holds for us today, and we can notice that it held for Tycho's contemporaries; and no doubt it held for Tycho himself though we would hardly expect to find him discussing it. On the *P*-cognitive argument, we would suppose that Tycho saw something looked wrong about the scheme and came to believe that the thing that looked wrong was the thing particularly easy to notice (the collision of the spheres), which Tycho eventually persuaded himself he could take care of. This is not so strange as it may sound, and a thoughtful reader will, I think, realize how vulnerable we all are to such slips. After all, Tycho was convinced that Ptolemy was wrong; and he had reasonable grounds (his own failure to detect parallax) for believing that Copernicus also must be wrong. From his situation (though not necessarily from some other person's, such as Maestlin or Kepler), it looked like his own system was the only viable alternative, hence it *must* be right.

13.9

The transition in habits of mind for the Copernican case has parallels in the Darwinian context 400 years later. Copernicus' contemporaries, like Darwin's, admired his detailed work, and above all (for Copernicus) his escape from the Ptolemaic equant [11.2] and (for Darwin) his richly detailed natural history that made the idea that evolution had *somehow* occurred almost irresistible. So Darwin rather quickly persuaded enough of his contemporaries that evolution had actually occurred to almost immediately put that aspect of his work into the phase of social contagion, as Copernicus' technical work almost immediately established him as the modern successor of Ptolemy, guaranteeing that any ambitious young astronomer would study his book. But for each, the more radical aspect of their work did not penetrate the expert community sufficiently to lead to major new results for some decades. Natural selection could not be compelling so long as no-design-without-a-designer remained deeply entrenched [10.2], as the Copernican golden chain could not be compelling so long as the nested-sphere sense of the world remained deeply entrenched.

But with the passage of more than a generation, the more radical aspects became more thinkable, though at first a tamed version won most adherents. As Tycho wanted the Copernican harmonies to be rescued from the Copernican heresies, biologists wanted evolution and even a role for natural selection, but in a way that at least muted the claim about design-without-a-designer.

With hindsight, of course—but only with hindsight—we know that the appeals of equant-free astronomy and then of the Tychonic alternative were intellectual dead ends, while the basic idea of evolution and then the

Mendelian idea would prove to be marvelously fruitful. But that is looking at history backward. For the people who were living that history, the two sets of moves have cognitive roots that are quite exactly parallel. Darwin's audience was luckier than Copernicus' audience, not cognitively different.

The roughly 40-year lapse in both cases before anything substantial was done with the new view meant that the next step was deferred until a generation was reached that was taught by people who themselves learned of the new ideas (though as a problematical novelty) early in their careers. It was only after a complete turning of generations—like the biblical 40 years in the desert (Walzer 1985)—that we reach people free enough of the older habits of mind for the contagion to reach the stage of social knowledge.

Finally, in both cases (to emphasize that fundamental point), what is in the head of the 40-years-after generation is *not* merely some particular controversial fact now accepted. Rather, in the Darwinian case as in the Copernican, what was crucial was the tacit entrenchment of a new sort of pattern. As Tycho insistently saw himself as proposing an independent invention, not merely a variation on Copernicus, so the early twentieth century Mendelians saw themselves as offering an alternative to Darwinism. Prior to Darwin, a claim about an invisible hand process that produced good biological design without a designer would lie somewhere between the incomprehensible and the absurd. By 1900, however, the new generation was full of people who had that sense of things, including the new Mendelians, even if they saw themselves as proposing a superior alternative to Darwinism. Indeed, they were so imbued with Darwinism as to not even be self-consciously aware of it, as an astronomer like Magini was in a significant sense a Copernican in spite of his determination to uphold Ptolemy; or as many people today are deeply imbued with Freudian and Marxian (and also, of course, Adam Smithian) ideas in spite of their professed contempt for them. In between, there was a generation that had learned the ideas but for whom it was rarely what Polanyi called "tacit knowledge," or what I have been calling an entrenched pattern. It was not (for that intermediate generation) something that did not need to be explicitly thought about to be used, and it therefore could not be used with the facility that comes when a move has become so automatic that it need not be done self-consciously.

13.10

We can now briefly cover two further points that will be needed for the discussion of Galileo's confrontation with the Church in Chapter 14:

1. How, specifically, did the Copernican golden chain view influence Kepler and Galileo?

2. What persuaded the rest of the scientific world not severely constrained by extrascientific influences (essentially, everyone knowledgeable enough to make a serious judgment other than Jesuits loyal to Rome and Scandinavians loyal to Tycho) to be won over to Copernicus decades before Newton finally provided an unambiguously overwhelming argument?

Newton himself never had occasion to explicitly make any argument for Copernicus. He simply took it for granted that anyone reading his book was already a Copernican (Westfall 1980). Even Descartes—in a book written by 1630 but then suppressed after the condemnation of Galileo—tried only to add detail to the Copernican view: he did not treat Copernicanism as a controversial novelty that needed advocacy.

The key to (1) lies in noticing how entrenchment in the Copernican view would make salient the issues which, in fact, caught the interest of Galileo and Kepler. If, in fact, the earth moved, but birds and cannonballs and so on were not left behind by the onrushing earth, then a radical reshaping of Aristotle's account of motion is required. Even a non-Copernican *might* see the problem of motion as a topic where a major improvement on Aristotle perhaps could be worked out. But for a Copernican that was not a possibility but a necessity. If the Copernican view was sound, then Aristotle's physics, and in particular what Aristotle said about motion, had to be radically wrong. From one among many interesting problems, motion became, for a Copernican, a crucial problem. This salience could hardly have been achieved on the basis of an implausible but not provably false conjecture (the mere idea that the earth might orbit the sun); what it came from was a sense of the world within which the movement of the earth was not an academic possibility but the only way to really make sense of things.

On the other hand, another salient problem arose not from Aristotle but from the unfinished business in technical astronomy that Copernicus left to his followers. As Kepler noticed at the start of his career, if the earth is a planet then why should the center of the earth's orbit be allowed a privileged position? Exploring that showed Kepler that some of the minor epicycles that remained after Copernicus eliminated the major epicycles were also only an illusion: the apparent motions in latitude disappeared once the planes of the planetary orbits were focused on the sun. At this point, doubts that any epicycles at all were really needed could hardly be avoided. The epicycles lost the intrinsic plausibility they had so long as direct observation of the looping motions of the planets made it hard to doubt that the major epicycles, at least, existed in *some* physical sense. The planets, after all, did loop as you tracked their movements from night to night against the background of the zodiac. But for a Copernican the looping was reduced to an illusion of parallax; and pushing the Copernican view a step further, Kepler found that some of the remaining minor epicycles (those governing variations in latitude) also disappeared. A Tychonic astronomer would not have found it easy to follow the same route, since terrible complications arise of the sort Kepler (1621) mentions in item 2 of his list of 18 reasons for rejecting the Tychonic system, and which Galileo works out in his sunspot argument (*Dialog:* 345 ff.). Further, if a Tychonic astronomer as able as Kepler existed, he would face other urgent problems, growing out of the logical flaws in Tycho's disproof of the heavenly spheres [13.7], and out of the puzzle of how to conceive some mechanisms that could account for the Tychonic double motions.

So the problems that Galileo and Kepler, respectively, had taken on in the 1590s were peculiarly salient Copernican problems. At the peak of their careers, both explicitly present themselves as followers of Copernicus, and they had good reason to do so. They did not make great discoveries which persuaded them to become Copernicans. They were already Copernican at a time when not many other people were, and as Copernicans made the discoveries that would make everyone Copernican.

13.11

However, given that a few astronomers could become Copernicans even before 1600 (and many more could be followers of Tycho, or agnostics no longer committed to Ptolemy), it is not hard to understand how the rest of the scientific world could be won over by the work that Kepler and Galileo did, long before Newton had produced arguments that would have made further scientific resistance pointless. The odd coincidence (on Tycho's scheme) that the earth and sun have a relationship that would make sense if the earth were a planet grows starker as Kepler's astronomy adds further detail to the ways in which the earth/sun connection is just what would make sense if the earth were a planet. Galileo's discoveries complement that most strikingly. Even the major empirical embarrassment for the Copernican side—the absence of detectable parallax of the fixed stars—became ambiguous (rather than an unqualified embarrassment) when it turned out that the telescope did not enlarge the stars. The planets, seen through a telescope, were disks, revealing the shape that would be expected from seeing how the telescope enlarged distant objects on earth. But the stars became points of light, just as occurred (Galileo showed) when seen by the naked eye if care was taken to avoid the glittering effect of the atmosphere. If it had been possible to see the stars as disks (possible to see them as having distinct cross-sections), then it would be impossible for a Copernican to plausibly explain why no parallax could be detected. But that stellar disks couldn't be seen supported the Copernican prediction that they were not, as the anti-Copernicans claimed, just little further off than Saturn but very far beyond the planets. So even this strongest anti-Copernican argument turned out to have a pro-Copernican aspect.

On other points, the new evidence was routinely favorable to Copernicus. The world kept looking more and more the way a Copernican might suppose it would look, and less and less as if the earth was not really a planet but a unique body at the center of the system.

The phases of Venus must occur in either a Copernican or Tychonic world and cannot occur in a Ptolemaic world. But they make a kind of intrinsic sense in a Copernican world that does not hold for a Tychonic world. And discovery of the phases of Venus did not come by itself but as part of a series of discoveries, no one alone terribly embarrassing to the Tychonic view but all of which together revealed a world that almost insistently seemed to be trying to show that the earth is a planet. If the earth is a planet, given that it has a moon, it makes sense that other planets could have satellites—as Jupiter and Saturn did. Similarly, it makes Copernican sense that the moon would have mountains, like the earth. And it did. Similarly, if the earth rotates, then it makes sense that other bodies might rotate; and observations of sunspots showed the sun does so. Similarly, if the moon noticeably lights the earth, the earth should light the moon, and Galileo showed that earthshine could be detected on the moon.

Similarly, since if the earth is a planet Aristotle's physics must be wrong, commonplace observations take on a new significance. If you pour a glass of wine inside the cabin of a smoothly sailing ship, you cannot tell which direction the ship is moving by seeing the wine pour at a slant: it goes straight down, just as if the ship were standing still. You could not detect the motion of the ship that way, and similarly you could not disprove the earth's motion by the fact that rocks dropped from towers fell straight down.

Of course even all of this together could not prove in a mathematical sense that Copernicus was right and Tycho wrong. There is nothing so clear-cut as the way the phases of Venus make it quite impossible to doubt that Venus orbits the sun, so that the Ptolemaic nested-sphere structure collapses. But even Newton's work gives no mathematical proof that Tycho—or even Ptolemy—*must* be wrong. Newton added whole new strands to the Copernican golden chain, but by the time he came along the chain had already been strong enough for decades to pull the whole scientific world its way.

A pattern-recognition analogy—on the argument of this study, more than an analogy—is the following. Imagine that a "pixel" version of a well-known image is being made, like the Lincoln grid [2.9]. Gradually decrease the size of the pixels (increase the fineness of the grid). There is no unique point where we can say it has been proved that the image is (say) Lincoln. Some people will see Lincoln vaguely, while others see only noise; as the process continues, some will feel sure it is Lincoln, while others are not sure, and some continue to insist there is no clear pattern at all. Eventually, we will get to a point where, although we still have no mathematical proof, no one capable of recognizing an image of Lincoln any longer doubts that as the fineness of the pixels continues to increase we are going to see ever more clearly an image of Lincoln. In fact, if later on it turns out that it isn't quite Lincoln after all, as later on it turned out that the Newtonian image was not quite right after all, that comes as a shock.

13.12

I mentioned earlier [13.7] that there is a perverse corollary of this account of the appearance of the Tychonic system, which stresses that a competent astronomer could only become a supporter of the Tychonic system if he were already won over to the Copernican golden chain sense of cohesion (hence had already lost the entrenched nested-sphere sense within which the Ptolemaic system could look plausible, given the Copernican alternative). A person who became actively involved in defending the Tychonic position as against the Copernican could be expected to become habituated to the loss of those features of the Copernican golden chain [13.8] that are incompatible with Tycho's system. A person would come to be insensitive to those features whose importance has to be denied to feel comfortable with the Tychonic system. What I will try to show in Chapter 14 is how that tendency seems to play a crucial role in explaining how the Church became rigidly intolerant of the Copernican view just at the time when the case for it was becoming wholly irresistible.

Fourteen

Political Judgment: Galileo and the Pope

Sometimes I can choose *A* even though you choose *B* and Smith chooses *C*. Politics enters when one choice must be made for all of us, provided that no single actor can simply do whatever he pleases.[1] This last is empirically a very weak condition; even in the most rigidly totalitarian societies, power is shared (unevenly of course) in intricate ways. Once we are concerned with common rather than individual choice, everyone will have to accommodate to the decision, hence everyone's interest is engaged. Among other things, I will have to care about what you think, if what you think may affect the common social choices we will all have to live with. Since what you know will influence what you think, I will be interested in influencing what you know. Since people who work together to influence the social choice can be more effective than people who don't, I have an incentive to find a basis for cooperating with other people who more or less share my sense of how things ought to go. In the second half of this chapter, I will sketch the beginnings of an abstract analysis of all this, concerned mainly (though it is only part of the full story) with what I will call "political polarization." The part of the story I will develop immediately will be the part that interacts most directly with the cognitive themes of this study, and in particular with what I will call cognitive polarization. So we will be considering both cognitive and political "twoness," and the way they interact.

Of this twofold *twoness*, cognitive polarization is the ability to see a field of perception in two ways, but rarely in more than two ways. One can give counterexamples, but it is strikingly the case that all the familiar gestalt figures elicit just two ways of seeing the image. Cognitive polarization is probably at least as significant in thinking as in perception, though less simple to demonstrate.[2]

Cognitive twoness appears especially in the context of larger-than-life social judgments, where the issues go far beyond the range of things we can know as familiar experience. It is hard (for example) to see the efficiency and equity sides of an economic issue as simultaneously in focus. Rather, the tendency is to see the situation in terms of equity, or in terms of efficiency, as we see the gestalt figure as a rabbit or a duck, but not both at once. We can arrive at what we feel is a balanced judgment by switching back and forth between the two gestalts, but it is hard to merge the two into a single gestalt.

Some other examples are seeing "profits" as greed, or as incentives; or seeing the US/Soviet nuclear balance in terms of a common interest in averting disaster versus each side's interest in deterrence of the adversary. Most generally, perhaps by anchoring on salient features around which other things are cognitively organized, a person often comes to see costs versus benefits, or pros versus cons of a controversial issue in the polarized way I have been sketching. On all these matters, a person (once alerted to the tendency) will sometimes be able to notice switches in the way she or he is seeing an issue, akin to the sensation of a switching gestalt figure.

We can expect, then, that sometimes a person might have trouble seeing one or the other gestalt, or might exhibit a bias that makes it easier to see one gestalt rather than the other. Then one gestalt is dominant, the other subordinate. It is this possibility for dominance to appear, or to be accentuated, that makes cognitive twoness an important aspect of social judgment. For on the earlier argument, strong dominance of one gestalt could be expected to arise rather easily when the general tendency to tame anomalies is accentuated by the "choosing-up-sides" incentives of a political context.

So in a political context, twoness of another kind (political polarization) will interact with cognitive polarization. The choosing-up-sides pressure of political choice implies competitive incentives to build a winning coalition. When one choice must be made for all, those who behave as if only an individual choice is being made will do badly compared to those who coordinate their efforts. Often, the social result will be a compromise in which everyone is reasonably satisfied; but the tendency toward political polarization will still be important in the process that eventually produces the compromise. On the other hand, sometimes the outcome will be so one-sided that it is eventually seen as excessive even by its original supporters, as when conflicts (wars, strikes, revolts) intensify beyond what the rational interests of either party could possibly justify. In the general analysis later in this chapter, I will try to show how either situation could arise.

14.2

A convenient test for the polarization argument is provided by continuing the analysis of the Copernican discovery and contagion of Chapters 11–13 into the clearly *political* context of Galileo's confrontation with his Church. What was at stake was the Church's policy with respect to Copernicus; and power was shared over this choice in the intricate, often indirect, ways characteristic of politics in complex societies. In principle, the pope could choose as he wished, but as a practical matter he was multiply constrained.

Even within the Church, commands that could not be overtly contested still could be quietly neglected or sabotaged; even commands that must be obeyed may be resented with consequences for resistance on other matters. Beyond the Church, indeed within the Church but beyond the Vatican, opposition need not even be covert. The pope was therefore necessarily enmeshed in politics in many ways: in civil life in Italy, in the internal politics of the Church, in European politics in the midst of the Thirty Years War, in the counterreformation rivalry of Catholics and Protestants for the allegiance of princes in various parts of the continent. Urban VIII was the most important pope in a turbulent century, entirely aside from his role in the trial of Galileo. His handling of policy with respect to advocacy of the Copernican view turned on many matters besides theology. And even his personal stake in the issue could never be reduced to merely private pleasure or annoyance with Galileo: his effectiveness in his many political roles, and in particular his effectiveness in being able to navigate his way among these many, often conflicting, roles depended on his being seen as authoritative, shrewd, effective, not vaccilating, confused, blundering.

That the Galileo affair occurred at a sensitive moment in Urban's reign, when these conflicting pressures were particularly acute, has been noted in earlier accounts of the trial (Santillana 1955:196–97; Geymonat 1965:137–39). But as in Chapters 11–13, the story I will tell is different from the story as it has usually been told. I will try, in particular, to make sense of two striking points that have never been plausibly accounted for:

1. Galileo began work on the book after a series of private meetings with the pope in 1624, consulted again with the pope when he

brought the finished manuscript to Rome in 1630, and obtained the imprimatur of the Vatican from one of the pope's closest associates. How, then, did a book in which the pope had taken a personal interest, and which was published with the explicit approval of the Vatican, become heretical six months later?

2. Galileo's *Dialog* appeared in 1632, 40 years after astronomers began to openly abandon Ptolemy (Chap. 13), and 20 years after the evidence of the telescope made that tendency irresistible, so that the only coherent choice available to a competent astronomer became Copernicus versus Tycho. After the telescope revealed the phases of Venus, the Ptolemaic snugly nested spheres could hardly be defended by anyone even minimally competent at astronomy. So far as I have been able to discover, by the time Galileo's *Dialog* appeared there were no Ptolemaic astronomers left; Ptolemaic astronomy survived only in outdated textbooks. Yet, in the nearly 500 pages of the *Dialog,* Galileo argues always against Ptolemaic astronomy and never once mentions the Tychonic system.

All other matters aside, then, how could a masterful debater like Galileo fail to say a single word to guard himself against the obvious counterattack that poor Galileo, in his old age, was still beating the dead horse of Ptolemaic astronomy, when he should have known that the real alternative to Copernicus was not Ptolemy but Tycho.

These questions invite straightforward answers. What I will argue is that:

1. The simplest way—so far as I can see, the only way—to account for why Galileo never mentioned the Tychonic system is to look for some reason why he may have supposed that his book would be acceptable to the pope if he limited his overt discussion to an attack on Ptolemy's system, and not so if he explicitly attacked the Tychonic system as well. I will try to show how that situation would arise.

2. Similarly, the simplest way to account for why the adverse reaction came only after the book had been in circulation for six months would be to find some way in which an unanticipated reaction of readers to the book made it intolerable. For since the book was acceptable when published, but then *became* unacceptable, it seems that the heart of the difficulty lay not in the book itself but in something about the effect the book was having on readers.

The discussion in Chapter 13 of cognitive and logical peculiarities of the Tychonic system [13.2, 13.8] suggests how both points can be resolved. I will try to show (a) why an attack on Ptolemy leaving the Tychonic possibility open could have served the pope's interest, but (b) why Galileo's silence about the Tychonic system was worthless, so that there could be a gross mismatch between what the pope believed he had approved and what he eventually realized he had gotten. I will try to say just enough about the politics of the situation to suggest how the pope could very plausibly have made the choices he did, both in authorizing Galileo to go ahead, and then authorizing the Inquisition to go ahead. Somehow (and I will try to say how) the book came to be intolerable and had to be disowned, but in a way that made Galileo guilty of deceiving the pope, not the pope guilty of incompetence in failing to foresee what he was setting loose in approving the project in the first place.

14.3

The participants in Galileo's *Dialog* debate the pros and cons of Copernican view for nearly 500 pages. In the course of this Tycho's name appears some 30 times. But the Tychonic system is treated only with absolute silence. On the face of things, Galileo did not incidentally neglect to discuss the Tychonic system; he very deliberately avoided discussing it. Deliberate intent is suggested by the title page (fig. 14.1), which says the *Dialog* will deal with "the two GREATEST SYSTEMS OF THE WORLD: PTOLEMAIC AND COPERNICAN." The title seems to have been suggested by the pope himself (Santillana 1955:183). A reader who wondered whether the title meant that the third greatest system (the Tychonic) was going to be left out would then notice that Galileo suggests as much again in the dedication to the Grand Duke, again in preliminary remarks addressed "to the discerning reader," and yet again in the opening speech of the actual *Dialog*.[3] In the first, second, and fourth of these Galileo specified that the *Dialog* will concern the Copernican versus Ptolemaic systems. In the third, he says he will "strive by every artifice to represent [the Copernican view] as superior to supposing the earth motionless—not, indeed, absolutely, but as against the arguments of some professed Peripatetics" (followers of Aristotle). But Tycho was almost as contemptuous of the Peripatetics as Galileo.

When the book was finished, Galileo brought the manuscript to Rome, conferred with the pope, and left it in the hands of the Master of the Sacred Palace for censorship. There the manuscript remained for about a year before it was finally released. It was finally printed in Florence in February 1632. Galileo made various changes at the request of the censors or the pope. Even after the storm broke, he was not charged with departing in any serious way from the instructions of the censors. The censors, in turn, knew they were dealing with a manuscript by a famous and controversial writer, concerning a subject so sensitive that the Church had 15 years earlier (1616) acted to silence public discussion. Therefore, no one involved could be unaware that the book was going to be very widely read, or that its appearance under an imprimatur would inevitably be interpreted as signaling a change in the Church's position. Indeed, in the carefully reviewed introduction, Galileo specifically reminds readers of the events of 1616, referring to the "seasonable silence" then imposed. Obviously, the season for silence was now over.

Further, from very early on, the pope knew essentially what Galileo's book would say; for what amounted to preliminary sketches of the argument had been circulated in the Vatican by 1624.[4] As to what the pope required of Galileo as conditions for his approval of the project, what is known consists of one simple point and another subtler one: (1) Galileo was to leave explicitly open the possibility that his pro-Copernican arguments could be wrong: so in the introduction he labels the Copernican the-

DIALOGO

DI

GALILEO GALILEI LINCEO

MATEMATICO SOPRAORDINARIO

DELLO STVDIO DI PISA.

E Filoſofo, e Matematico primario del

SERENISSIMO

GR.DVCA DI TOSCANA.

Doue ne i congreſſi di quattro giornate ſi diſcorre
ſopra i due

MASSIMI SISTEMI DEL MONDO
TOLEMAICO, E COPERNICANO;

Proponendo indeterminatamente le ragioni Filoſofiche, e Naturali tanto per l'vna, quanto per l'altra parte.

CON PRI  VILEGI.

IN FIORENZA, Per Gio: Batiſta Landini MDCXXXII.

CON LICENZA DE' SVPERIORI.

Figure 14.1. Title page of Galileo's *Dialog* specifying that the Dialogo would deal with the Ptolemaic vs. the Copernican systems. In the 500 pages that follow, Galileo never mentions the Tychonic system.

ory a "mathematical fantasy," and throughout the text there are many similar disavowels of any claim to literal truth. (2) In addition, there was a more specific requirement about the tides argument that concludes the book. Galileo attached great importance to this argument, though it turned out to be the only important claim in the book that was wrong. His argument was that the tides are caused by a sort of sloshing due to the complicated motions of the earth. He even planned to call the book "On the Flux and Reflux of the Tides," but the pope apparently preferred the title already quoted. The "tides" argument, however, remained as the closing section of the book, along with a concluding remark, which followed an explicit instruction to Galileo, to the effect that no such argument could be decisive, since God could have arranged the tides without being restricted to some scheme that men would understand.

Obviously, a great deal more than the bits we know about went on in the meetings between Galileo and the pope. They talked for a long time. On the other hand, they were old friends, and there is no reason to suppose that the talk was all business or that it ever had the character of a negotiation. No one (prior to the account here) has read the arrangement as one that would leave the Tychonic system as an undiscussed, hence also unchallenged, alternative to Copernicus. Yet that would make sense of an otherwise bewildering situation. From the pope's side, that arrangement would have the virtue of distancing the Church from the rigorously anti-Copernican position it had appeared to take through the Edict of 1616 (forbidding Catholics to "hold or defend" the Copernican view). But that would be done in a way that did not conflict with a strict construal of the Edict of 1616, hence raised no real issue of whether the Church had erred in issuing the edict at all.

Some aspects of this can be dealt with more clearly later in the chapter (after I have spelled out the argument about cognitive polarization). But the most important general point is to notice that the Church had never taken a position in favor of Ptolemy or Aristotle but only against the Copernican claims that the earth moved and that the sun stood still. But unless a person had a fair knowledge of astronomy, this distinction was not easily noticed. Outside the expert community the issue was seen [13.5] as simply whether the earth went round the sun or the reverse; so if you supported the first you were a supporter of Copernicus, and if you supported the second you were a supporter of Ptolemy. To a considerable extent, the same simplistic dichotomy occurs in writing about the episode today.

But a simple dichotomy between Copernicans and anti-Copernicans is fundamentally wrong, for reasons I hope I have adequately emphasized in Chapters 11–13. A Tychonic astronomer (on the nested-sphere versus golden chain argument I have presented as cognitively the heart of the story) would feel intellectually closer to a Copernican than to someone holding onto the shattered remains of Ptolemaic astronomy and Aristotelian cosmology after the evidence of the telescope was in hand. Even Tycho himself, who was dead before the telescope appeared, had for decades taken a dismissive attitude toward Aristotelian cosmology and had always treated Copernicus as a great figure who advanced astronomy beyond Ptolemy.

The difference between the way the Copernican controversy tended to be seen by experts versus nonexperts would have substantial consequences both in terms of internal management of the Church's affairs and in terms of the pope's political interests in the conflicts north of the Alps. These consequences, I want to argue, were striking enough to provide a resolution to the central puzzle at hand. For it is strictly inconceivable that the sophisticated and tough-minded head of the most sophisticated organization on earth could allow Galileo's sensational (in the context of the Edict of 1616) project to proceed without thinking about its effect on the interests of the Church and on his position as leader of the Church. To suppose otherwise would be like supposing that a concert violinist might perform on a badly out-of-tune instrument because he didn't realize it was out of tune. It makes no more sense to suppose that Urban could be insensitive to the political implications of approving Galileo's project than to suppose that a professional violinist would be insensitive to the badly mistuned violin. In each case, the person could not be in the position he is in if he were insensitive to such things.

Three mutually reinforcing possibilities now become evident, all linked to the nonexpert's tendency to miss the distinction between opposition to Copernicus and endorsement of Aristotle and Ptolemy. The most obvious point is that a person aware that knowledgeable astronomers had abandoned the ancient tradition, and had moved at least as far as Tycho, suggested that yet more evidence might appear that would make it hard to avoid going further, to Copernicus. A concern with this possibility argued for making it clear somehow that the Church was at least willing to permit hypothetical discussion of the Copernican view, so that an eventual victory for the Copernicans would not necessarily be a humiliating defeat for the Church. But this argument would not explain why Galileo was permitted to write a book obviously intended to entertain and inform a wide audience, not a technical treatise readable only by experts.

There were, however, at least two fairly obvious reasons why the pope could have seen a more popular kind of book as consistent with his interests and responsibilities. One concerns a delicate situation within the Church, and the other concerns a delicate situation in the Church's role in European politics.

Within the Church, the way the Copernican issue tended to be seen by nonexperts (the identification of an anti-Copernican stance with a pro-Ptolemy/Aristotle stance) meant that even men who adhered strictly to the Edict of 1616 would be vulnerable to attack. The Jesuit astronomers of Collegio Romano had openly abandoned the Ptolemaic system in favor of the Tychonic 20 years before the *Dialog* was published. They moved remarkably soon after Galileo announced his discoveries with the telescope. But energetic pursuit of even Tychonic arguments, or of work with the telescope which involved no explicit assumption about Copernicus versus Tycho, was intrinsically work that would offend defenders of Aristotle (Van Helden 1974). Such people would look to traditional theologians and philosophers as covert Copernicans—all the more so since many of them in fact *were* covert Copernicans.

North of the Alps, that such attitudes extended even to supporters of Tycho looked like what would today be called cultural imperialism. If the Edict of 1616 was understood as support for Ptolemy and Aristotle, it was evidence of both arrogance and stupidity. Rome was insisting on adherence to its own tradition in the face of new ideas coming from north of the Alps (from Copernicus, Kepler, and Tycho) that persuaded all reasonable men to abandon Aristotle's cosmology and Ptolemy's astronomy. On top of all the other recurring resentments of Roman pretensions to authority, here was yet another source of difficulty.

Authorizing Galileo to publish what he did publish responded to both these points. It made it very clear—even brutally clear—that the Edict of 1616 could not be interpreted (in either of the two contexts just sketched) as endorsement of Aristotelian orthodoxy. Yet, at the same time, defenders of orthodoxy could not effectively claim that the pope was ignoring the Edict of 1616; nor could enemies of the Church claim that the Church was admitting it erred in 1616. For the Edict did not actually forbid discussion of the Copernican idea. Rather, a defender of the Church could point out that discussion of the Copernican view as a hypothesis had never been forbidden. Strictly interpreted, the Edict of 1616 required only that Catholics not "hold or defend" the Copernican claims as proven facts. The ban on Copernicus' own book had been removed in 1620, requiring only a handful of essentially trivial changes. The only substantive remark deleted in the whole book was a sentence in which Copernicus warns theologians not to meddle in the matter lest they make fools of themselves (*de Rev*: Introduction). Although it seems to have been a close thing, the Church had stopped short of fully committing itself against Copernicus in 1616, leaving open the politically convenient possibility that a mere signal—allowing Galileo to publish—would dissociate the Church from any presumed or implicit commitment to the pre-Copernican traditions.

So it is not hard, in fact, to see why the pope had grounds to justify what in any case was certainly his own inclinations on the matter, that is, to make it clear that the Church's position was in opposition to treating the Copernican ideas as demonstrably true, *not* in favor of enforcement of Aristotelian and Ptolemaic orthodoxy. It then is easy to make sense of some understanding that left Galileo free to attack Ptolemaic and Aristotelian ideas but not to attack the Tychonic alternative. And absent some such understanding, it is hard to make sense either of how the pope could have approved Galileo's project, or of how Galileo could have failed to deal with the Tychonic alternative. The contrast between Galileo's treatment of the Tychonic system and Kepler's is striking and instructive. Kepler (1621:913–15), writing under the patronage of the Catholic emperor but beyond the reach of the Roman Inquisition, cannot hammer enough nails in the Tychonic coffin. He lists 18 reasons for rejecting the Tychonic system. But Galileo maintains absolute silence. Yet, as will be seen, various oddities about the way he handles his argument can be cleared up very simply if we consider the possibility that Galileo was perfectly aware that it was pointless to offer expert-level arguments against Ptolemy, since expert adversaries were supporters of Tycho and not defenders of Ptolemy.

14.4

When the *Dialog* came to look intolerable to the Vatican, six months after it finally appeared in print, Galileo was bewildered as to what he had done wrong; but the pope was convinced that he had been betrayed by Galileo (Santillana 1955:193). On nearly all accounts, the pope's anger was aroused because Galileo's enemies were able to persuade him that Galileo had formulated the pope's response to the argument on the tides in a way that was *intended* to make the pope look ridiculous. There was no reason, though, why Galileo would have wanted to insult the man who until that time had been both a personal friend and the most powerful of allies, nor would such a thing have plausibly escaped the particularly close reading we know the censors gave to the opening and closing pages of the book. Even after releasing the bulk of the text for final approval in Florence, the opening and closing pages were kept by the Master of the Holy Palace for further review. So the supposedly offending passages had in fact been reviewed with exceptional care, and by an official who had immediate access to the pope, and who indeed repeatedly in his communications about the book said he was making sure the pope's instructions were being followed.[5]

But the peculiarities of the Tychonic system [13.8], together with the tendency to cognitive polarization [14.1], provide a cognitive explanation of how Galileo's book could be seen as acceptable in manuscript but *become* intolerable six months after appearing in print. What angered the pope, I want to argue, and left him feeling betrayed by Galileo, is then not Galileo's book, and certainly not some easily amended detail of the book, but how Galileo's book came to be seen in the light of the response of the public. In contrast to more usual accounts, which require us to believe that the crisis had its indispensable roots in some intrinsically trivial misunderstanding, we find a serious problem with which the reaction to Galileo's book confronted the Church, and which Urban could have believed would not have occurred if Galileo had been more candid with him.

The easiest way to see the situation is to notice how the Tychonic versus Copernican arguments can be divided into complementary pieces in two different ways (see fig. 14.2). (A) We can divide the anti-Tycho arguments between those that can arise incidental to a discussion of the choice between Copernicus and Ptolemy and those that would only come up in a discussion explicitly about a choice between Copernicus and Tycho. On the argument I have been sketching, those in the first category would appear in the *Dialog,* but those in the second category would be forbidden. (B) We can also divide the arguments into rival cognitive gestalts of the sort discussed in [14.1]: (1) a quantitative gestalt, which sees the choice in terms of traditional arguments of positional astronomy; and (2) a qualitative gestalt, which concerns whether the world *looks* more Copernican than Tychonic, especially given all the new things that could now be seen through a telescope.

In the rows, in figure 14.2, part A.1 includes all the arguments relevant to a choice between the Copernican system and *any* alternative (not necessarily Tycho's) which assumes a stationary earth. In particular, this would include all arguments favorable or unfavorable to the basic Copernican notion that the earth is a planet—for examples, the stellar *parallax* argument, or Galileo's argument that ocean *tides* are a sloshing effect due to the motions of the earth, or arguments about ways in which the earth is *like* a planet or heavenly bodies are like the earth. Part A.2 then contains whatever other arguments might be brought to bear which turn on particular merits or weaknesses of the Tychonic scheme as against the Copernican. Within the Ptolemaic versus Copernican limits advertised by his title, Galileo could deal only with arguments that fit within A.1.

Galileo could not touch any arguments in A.2 (arguments relevant to Tycho but not to Ptolemy). But the significance of that restriction then turns on what is left to be said after everything in A.1 had been covered. Galileo could not explicitly mention the relevance of his points to Tycho. But of the three main ways in which Tycho's system is vulnerable to attack [13.8], Galileo could and did develop in detail the arguments about the way

	B. Gestalt	
	1. Quantitative	2. Qualitative
A. Relevant to:		
1. Ptolemy and Tycho	parallax	tides sunspots sequence retrogressions likeness
2. Tycho only		motions

Figure 14.2. Taxonomy of arguments relevant to a judgment on Copernicus vs. Tycho. The "sunspot" argument (*Dialog:* 345–57) turns on the point that if the earth is motionless, an extremely complex set of motions must be postulated to account for the observed movements of sunspots. We would need to account *simultaneously* for rotation of the sun on its own axis, daily rotation of the sun around the earth, and annual movement of the sun through the zodiac. If the daily rotation is transferred to the earth, the difficulty is resolved. So the argument is not directly one for the Copernican system but only for daily rotation of the earth. However, anyone accepting this "semi-Tychonic" compromise gives up the main physical and theological objections to the Copernican system [11.5].

that the Copernican scheme elegantly solved the puzzle of the looping *retrogressions* of the planets; and in a particularly neat bit of rhetoric, the Aristotelian spokesman (*Dialog*: 321–26) is led into putting the earth in its Copernican *sequence* as the most intuitively reasonable interpretation of what had been seen with the telescope. Galileo could not explicitly point out how much less satisfactory the Tychonic scheme is on these two points, but the stage was fully set for the points to be instantly available whenever anyone responded to his argument by suggesting the Tychonic alternative.

What was left out was any explicit discussion of the "monstrous" (Kepler 1601) *motions* required in the Tychonic system. Yet, what really needed to said about that? The problem for the Tychonic advocate was that the scheme looked implausible on its face, not that it seemed reasonable unless arguments to the contrary were brought to bear [13.3]. In short, if we ask what Galileo left out that was important to be in if the book were to be an effective argument against Tycho as well as against Ptolemy, it is hard to find anything of consequence.

A striking bit of tacit evidence for the account here is that despite the heavily advertised Copernican versus Ptolemaic focus of the book, Galileo never bothers to attack the snugly nested spheres, which were the fundamental idea of Ptolemaic cosmology [11.8]. Here was an easy (given the discovery of the phases of Venus) and important target—if Galileo's concern really was with Ptolemy—which Galileo ignores. It is a bizarre omission, unless Galileo was not really motivated to expend effort on material that would be irrelevant to a choice between Copernicus and Tycho.[6] He introduces the main argument against Ptolemy with a piece of dialogue that suggests that Aristotle himself would have preferred the Copernican system to Ptolemy's (*Dialog*: 320 ff.). The language is double-edged: in its context it applies to Copernicus versus Ptolemy; but as with several other passages (for example, *Dialog:* 326), if someone suggested the Tychonic alternative the language would be immediately interpretable as equally relevant to Copernicus versus Tycho.

A final example is the often-noted absense in Galileo's discussion of any reference to Kepler's laws, or even to the more general success of Kepler's sun-centered models in both improving the accuracy of astronomical predictions and completely eliminating the Ptolemaic epicycles. This has usually been given some invidious explanation, tied to Galileo's unwillingness to share the Copernican glory. And, indeed, if Galileo saw his task as demonstrating the superiority of Copernicus to Ptolemy, then why else would he ignore the powerful arguments Kepler made available? But if the Tychonic alternative is the real alternative, then Kepler's laws are far less relevant, since they could be interpreted to fit the Tychonic scheme. In fact, if you look to Kepler's list of 18 reasons for rejecting Tycho [14.3], you will find that Kepler himself does not invoke his own laws.

14.5

In general, then, and despite Galileo's repeated statements in the front matter that the book would consider only Ptolemy versus Copernicus, it in fact ignores powerful arguments against Ptolemy when they are not important against Tycho as well; on the other hand, it gives a nearly complete, highly adverse discussion of what really needed to be said to provide a powerful piece of advocacy for Copernicus over Tycho as well as over Ptolemy. If what the pope thought he was authorizing was a book that would leave the Tychonic system as a plausible alternative consistent with the anti-Copernican Edict of 1616, then he had some reason to feel betrayed.

But while we know that the pope indeed saw himself as betrayed by Galileo, Galileo could feel (as all the evidence indicates he did feel) bewildered by the pope's anger. From Galileo's view, as from Kepler's, it was puzzling how anyone who understood the Copernican arguments and was aware of what the telescope had revealed could take seriously the Tychonic scheme. Church discipline might make a man–even a pope—reluctant to openly say that, but if he was competent it was hard to believe that he was serious. Leaving Tycho unattacked was like the pro forma disavowals of conviction at the beginning and end of the book: it was something that perhaps had to be done, but not something subject to a serious test of whether a sophisticated reader would believe the disavowals. Rather, making the case for Copernicus, which Galileo was allowed to do, *entailed* setting a stage on which Tycho's alternative would seem implausible. Consequently, there is no reason to suppose that Galileo (despite an occasional twist of the knife [14.4]) saw himself as doing anything seriously beyond what the pope had authorized. In any case he had published nothing that had not been submitted for censorship and given the Church's imprimatur.

But the patterns of experience, hence the habits of mind entrenched by that experience, would be different for the pope and for the astronomers of Collegio Romano than for Galileo. For the pope astronomy, obviously, could be a minor interest but never a major focus of his life. For the Jesuit astronomers, working under the discipline of the Edict would train them (that is, what is involved is not some conscious decision: they would *become* trained, given their circumstances) to see the Tychonic system as an attractive alternative to Copernicus. Galileo, in contrast, had for many years been an energetic Copernican. This divergence would be aggravated by years of public debate between various Jesuits and Galileo on other matters which did not *overtly* turn on the choice between Copernicus and Tycho and in which the Jesuits generally got the worst of it. And that, of course, would promote a belief that Galileo was a tricky fellow in debate. For of course it is far more congenial to believe you are being taken advantage of by tricky methods than to believe you are vulnerable because your arguments are weak.

So even before we come to the fine details of the cognitive divergence that could account for the mismatch between Galileo's and the pope's perceptions of what he had done, it is easy enough to see that if the *P*-cognitive argument developed throughout the earlier chapters has any merit at all, then men like Kepler, Descartes, and Galileo could find the evidence for Copernicus completely convincing at the same time as the Jesuit astronomers and the pope saw the same evidence as no better than ambiguous. And for the Churchmen, if the technical evidence was ambiguous, obviously it was pernicious, given the theological difficulties, to persuade people that the Copernican system is true.

Reviewing: We know the pope authorized Galileo's project. The pope, therefore, must have supposed there was a viable stopping point that undercut extreme interpretations of the Edict, but without making the Church look incompetent either in its decision of 1616 or in failing to formally reverse that decision. The political situation facing the pope [14.3] makes it comprehensible why he would choose to walk this tightrope rather than either refuse Galileo completely or accept his proposal to rescind the Edict. In any case, the delicate choice (authorizing Galileo to write, but refusing to lift the restraint on open advocacy) is a matter of record. Since no one has ever accused Urban of gross incompetence, it follows that he had some basis for believing his choice was sensible. It is surely hard to imagine what that could be other than an expectation that supporters of Tycho could respond to Galileo's book with a reasonable showing that Galileo's arguments, insofar as they were clearly right, could be accommodated by the Tychonic system; and where they couldn't be accommodated, they were not clearly right. For if that could not be done, how could the Church be following a sensible policy?

Provided the Tychonic alternative was viable, the Church's policy could be defended as wisely imposing restraint on agressive advocacy of a position so unsettling to traditional learning and traditional understanding of scripture, and which in fact was confronted by a plausible alternative. But if, instead, the Tychonic response seemed futile, then the Church would look like it had tried to suppress the truth in 1616 and had now moved to vacillation.

The argument I want to make, then, is that what lies behind the pope's feeling of betrayal and the resulting trial of Galileo most plausibly was a realization that an attempt to defend the Church's position by pointing to the Tychonic alternative was hopeless. What I will try to suggest next is a cognitive account of how such a gross misjudgment (by the Vatican) could have occured.

14.6

What we require is some explanation of how the pope and his advisors could have failed to foresee how difficult it would be to present Tycho as a viable alternative substantially immune to Galileo's arguments against Ptolemy. Seeing how that could occur turns on returning now to consider the second, gestalt-oriented, way of dividing the arguments bearing on Copernicus versus Tycho [14.4].

In terms of *quantitative* gestalt [14.4], the Tychonic and Copernican systems are observationally equivalent [13.3], except on the point of stellar parallax. Given the absense of detectable stellar parallax, in terms of the quantitative gestalt, there was a clear advantage for Tycho, though less so than before the evidence of the telescope provided an observational argument that the stars must indeed be far beyond the most distant planet.

But in terms of *qualitative* features [14.4] that cannot be deduced from mathematical astronomy—in terms of what *looks* intuitively reasonable, the advantage goes overwhelmingly to Copernicus. The Copernican case was strong enough to convert men like Kepler and Galileo even before the evidence of the telescope. But then what the telescope had revealed—not just to the specialists but in a way that everyone could see with their own eyes—was a whole array of planet-like qualitative features of the earth and the earth-like features of heavenly bodies [13.11]. Not one of these features was logically ruled out in Tycho's world. But intuitively, they make no sense in such a world, where the earth is not at all like the heavenly bodies but is a unique, physically different sort of creation located at the center of the universe. To a Copernican like Kepler [13.8], "nature cries out loud that the world is Copernican." Was it plausible that God created the world to look like the Copernican system is true, even though it isn't? Paraphrasing Einstein, we can believe God is subtle, but do we have to believe he is malicious? Or paraphrasing a nineteenth-century theologian responding to the argument of one of his colleagues that God had created the rocks with fossils already in them: Is it possible to believe that God has filled the heavens with one enormous lie?

It was qualitative ("looks like") intuitions of this sort—not formal proofs—that made Newton and the rest of the scientific world see Copernicus as making sense in a way that Tycho could not. This conviction was well-settled before Newton got his first glimpse of what would eventually develop into the *Principia;* for Descartes (1630) and many others, the issue was settled before Galileo's book even appeared. But as I have sketched for the pope and for the Jesuit astronomers, if your circumstances were such as to make it uncomfortable to see things this way, a person could be well-informed and still a defender of Tycho: in particular, because the very process of trying to hold back from Copernicus and defend Tycho would promote what I called [13.7] a trained insensitivity to just those differences between the Tychonic and Copernican systems that represented what was lost to advocates of the Tychonic scheme.

In terms of cognitive polarization [14.1], there was a particularly natural twoness in the way that astronomical issues tended to be seen. Even back to the Greeks, there had always been a tendency to see cosmology separate from positional astronomy: one being concerned with the nature of the world, the second with models capturing the movements of the heavenly bodies. The telescope sharpened that dichotomy, since now there were two essentially different aspects of empirical astronomy. One could construct mathematical models tuned to positional observations and extract predictions of future positions from those models, or one could see what the heavens looked like with the aid of the telescope. The two were not mutually exclusive or intrinsically incompatible. But we have a salient pair of foci for rival gestalts. A Copernican would see both sets of arguments as mutually reinforcing in terms of the moving earth. But for a supporter of Tycho, things made much more sense if you saw the deductive set of arguments as what really counts in astronomy and the arguments about what the world looks like as a different, less scientific, inferior, intrinsically indecisive sort of thing.

So on this polarization argument, men committed to Tycho would come to see the qualitative arguments as secondary, and the appeal of the Copernican case on such arguments as disgraceful: Galileo was—yet again, it would seem to the Jesuits—exploiting the gullibility of the wider public, which could not be expected to appreciate the subtleties of what really counted in astronomy. Not only the academic philosophers the *Dialog* openly attacked but also Tychonic astronomers and the pope and the Church itself were being put into a difficult position by unfair methods leading to a conclusion that was at least doubtful.

That the pope and the Jesuits could see things this way is hardly open to dispute, since even today, long after the issue has been settled, the same view is still common. Except on something like the cognitive polarization account here, that is difficult to comprehend. After all, the people won over by essentially the arguments available by 1630—for nothing of great importance appeared in the half-century between then and Newton's *Principia*—included every major scientific figure alive in that span, and nearly all the minor ones as well other than Catholics sufficiently loyal to the Church or Scandinavians sufficiently loyal to Tycho. Even the ablest of the Jesuit defenders of Tycho (Riccioli) managed to leave readers in doubt about whether he really was a secret Copernican (Grant 1984:14–15). There are very reasonable grounds, therefore, for supposing that some odd cognitive blindness is revealed by claims—such as those of Feyerabend (1975) and many others—that imply that prior to Newton's discoveries a person would be a Copernican—Newton himself, for example, could be the Copernican he was—only out of irrational zealotry, or because he was deceived by the rhetoric or misrepresentations of a zealot.

So how easily cognitive polarization can come to make one gestalt dominant is illustrated by how commonplace it is for contemporary scholars writing on the issue to still see the issue of Tycho versus Copernicus in terms of a dominant quantitative gestalt. If such views can be held today, a fortiori they could easily be held by the Vatican astronomers circa 1630. Galileo's book is then seen as skillful propaganda, deliberately depending for its power to persuade on arguments that would not convince a properly informed rational chooser.

The rest of the story concerns the way that political polarization interacted with the cognitive polarization I have now described. That requires attention in far more detail than would be appropriate here to the intricate political alignments within the Church, within Italy, and within Europe in the midst of the Thirty Years' War and the counterreformation, all considering not only the varied sensibilities emphasized earlier of different actors to the Copernican issue but also the pope's particular concern that whatever he does is done in a way that limits the damage to his credibility as an effective leader. At a number of points, the story is further complicated (and aggravated) by chance factors, such as an outbreak of plague that delayed arrival of Galileo's book in Rome until it had already been the subject of lively discussion for some months in the great cities of the north. But on this *P*-cognitive account we can see how trivial things came to seem important, given the deeper and not at all trivial roots of the crisis. We are not asked to believe that if a few lines on the last page of Galileo's book had been rearranged (so the pope's argument was reported by Sagredo instead of Simplicio), the crisis would never have occurred.

14.7

In the Appendix to this chapter, I give a simple account of political choice in terms of the distribution of power across a large number of choosers who share responsibility for a choice. An excessive outcome most easily occurs when the controversy has become politically polarized (making compromise difficult) *and* there is an overwhelming preponderance of power on one side (making compromise unnecessary). Here, the pope's preference counted vastly more than any other, so that if his preference coincided with an even arguably predominant view of the Church establishment (the cardinals, the curia, the ecclesiastical orders—especially the Jesuits and Dominicans), the balance would be lopsidedly in favor of that view. If, on the other hand, his view went contrary to that establishment, then even the pope had to move, if at all, cautiously.

That prior to 1632 Urban handled the matter in the personal way he did (dealing directly and privately with Galileo) suggests that he expected that much of the Church establishment would be opposed to his judgment. The way he proceeded suggested that his intent was to have the book become a matter of debate only after it was a fait accompli. When it turned out to be not what was intended, the pope was put in a difficult position. Intending to be perceived as the wise manager of a delicate situation, he now discovered he was likely to be seen as a blundering mismanager of that situation. In that context, we can make sense of the pope's move away from tolerance for Galileo. And once that occurred, of course, the balance of power shifted very decisively to the conservative side.

At that point, the dynamics of political polarization (as sketched in the Appendix) would encourage the anti-Copernican forces to push hard for their view. The pendulum could someday swing the other way. For those strongly on the anti-Copernican side, the time to move decisively was while the pope was at least tactically on their side. Grounds were found—in politics grounds for almost anything can be found—for silencing Galileo without admitting that the pope was reversing his earlier judgment. The Church had not erred. Galileo had cheated. For—repeating now the central point—a person who saw the issue in the cognitively polarized way I have been sketching would be blind to the effect that qualitative arguments would *reasonably* have on Galileo's audience. Then he could easily feel *tricked* when that became evident.

It is essential here to see that the argument is not that the Church acted to suppress an argument that it knew was right. That would have made no sense. Rome's authority in this matter was rather narrow. The Protestant half of Europe would treat the condemnation of Copernican doctrine by Rome as an argument in its favor; and even within Catholic Europe, the authority of Rome was limited. Hearing that Galileo had been summoned to Rome by the Inquisition, friends invited him to return to Venice, where the Roman Inquisitors could be safely ignored. North of the Alps, Kepler had carried on his advocacy of Copernicus as chief mathematician to the Holy Roman Emperor, and his Catholic hosts did not seem to be bothered in the least that he ignored the Edict of 1616. So it made no sense whatever for Rome to commit itself against the Copernicans unless the case for the Copernican view was indeed much weaker than Galileo had made it sound, so that there was some plausible hope that over time the Church's judgment would come to look increasingly reasonable, not increasingly bad.

Summing up once again: Both on the cognitive analysis I have sketched out, and on an appraisal of the politics of the situation (in particular, allowing that Rome had to be powerfully concerned with the effect of any decision it would make on the prestige and authority of the Vatican), we would expect that the pope and others involved came to believe that Galileo had indeed cheated, though not necessarily in the way the formal proceedings claimed. For in terms of the cognitive polarization I have sketched, Galileo's adversaries could come to see him as intrinsically cheating to appeal to qualitative intuitions which (in their view) his lay audience was not sophisticated enough to realize had no probative weight. If the Tychonic system could not be appreciated by the wider public, but instead its defenders were to be subjected to ridicule, then what that showed was that the topic was not an appropriate matter for public debate.

Almost a century earlier, Luther's colleague Melanchthon, on the other side of the religious divide, remarked about Copernicus: "Wise managers should be able to stop the impertinence of such talents" (*R*:478). Although we can see the situation as a case of raw power being used to suppress an argument that those in power cannot effectively answer, there is no reason to doubt that the pope and others involved saw themselves as acting prudently to prevent a difficult matter from being overwhelmed by vulgar debate. Men well-trained in the doctrinal disputes of the Church—and the need to keep such doctrinal disputes from tearing society apart, or leaving untrained opinion to make commonsense judgment about such delicate matters as the mysteries of the trinity—would not have difficulty persuading themselves that what the embarrassment the book was causing showed was that the merits of the Tychonic system are not something appropriately dealt with by Galileo's style of argument.

We get an ironic result. Starting from encouragement of Galileo that would make no sense at all except as reflecting an intent to distance the Church from excessively strong interpretations of the anti-Copernican Edict of 1616, Urban committed the Church more strongly than ever against Copernicus. We come to a situation in which the Church becomes not only fully committed to one side—but to the *losing* side. And this happens even though all the major evidence and arguments that would carry the day for Copernicus were available to the Vatican when it acted. The pope produces an outcome which today, after 350 years, is still a lingering embarrassment to the Church, and which surely contributed to the radical weakening of the Church as a major force in European politics during the seventeenth century. It is an outcome that makes little sense in terms of actors rationally seeing the logic of the situation and rationally pursuing their long-run interests, given that logic. But it is an outcome of which we can make sense in terms of the cognitive polarization treated in detail in the body of this chapter, as that interacted with the dynamics of political polarization.

14.8

The overall argument of the last seven chapters might now be summarized. Consider a spectrum of judgments stretching from the narrow and artificial contexts of puzzles and experiments, continuing through ordinary personal experience, and reaching finally the broad contexts of large social issues (fig. 14.3). At the narrow end, we have limited experiences in impoverished environments. At the broad end, we have judgments that far transcend individual experience, so that the consequences of acts will often be so diffuse or delayed or ambiguous that close tuning of choices to experience cannot occur. Near the middle of that spectrum, on the scale of normal human experience, we can expect reliable tuning of acts to consequences. So we can expect cognitive anomalies to

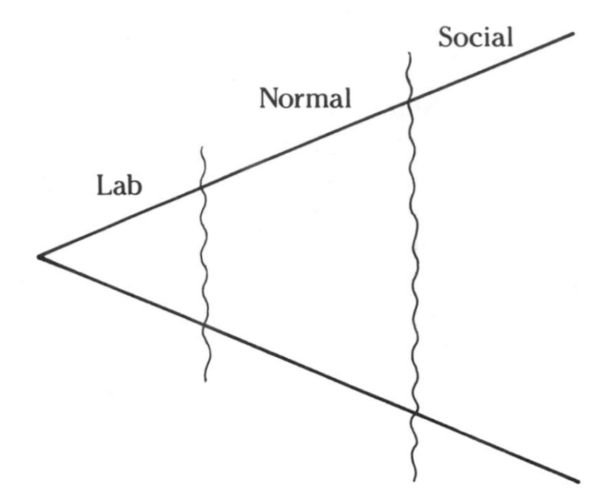

Figure 14.3. Spectrum of experience, ranging from narrow and impoverished situations characteristic of psychological experiments, through normal situations, to situations beyond the first-hand experience of an individual. Cognitive illusions should be rare in the normal range, where there should be good tuning of patterns to cues; but they can occur fairly easily both at the impoverished narrow end of set-piece puzzles and laboratory experiments, or in the larger-than-life situations at the broad end of the spectrum, as with theory choice in science, or large-scale political judgments.

be least conspicuous in the middle of the spectrum and most conspicuous at the ends of the spectrum. Hence, it is in the realm of social judgment (at the broad end of the spectrum) that we can expect to find large consequences linked to the anomalies revealed by the artificial little puzzles that give us the narrow end of the spectrum.

What I have tried to show in the last half of the study is that the *P*-cognitive argument can really be put to use, applying it first to several of the most thoroughly studied cases at the narrow (laboratory) end of the spectrum of experience and then to the general problem of theory change in science, with detailed application to a particularly famous and well-studied case in the history of science.

Across this wide range of material, we have been led to results that diverge strikingly from what can be found in the extensive specialist literature on these topics. The results are, I think, specific enough to invite further work which will either support or embarrass the *P*-cognitive argument. The variants of the Wason and the Kahneman/Tversky experiments in Chapter 8 are easily subject to replication and extension of the results. Further, the line of analysis (if sound) should produce interesting results across the whole range of cognitive anomalies reported in that rapidly growing literature. Similarly, the historical cases invite further work by researchers more expert in the relevant languages and cultures than I am. Such work, again, will eventually either embarrass or support the *P*-cognitive arguments. If the theory is basically sound, it must produce new results on many other cases of scientific innovation.

Finally, we have been able to reach the more complicated domain of political choice by way of Galileo's confrontation with the Inquisition. A nearly final version of the theoretical argument (the opening paragraphs of this chapter, plus its Appendix) was presented to a political science conference many months before the case of Galileo appeared as the occasion for an application of the polarization argument. Application of the analysis of Chapter 9 to the Copernican discovery began as a pure conjecture. If the argument were sound, something interesting (I thought) was likely to be found. When what was found turned out to be (it seemed to me) extremely interesting, I was encouraged to continue the analysis to cover the Copernican contagion (in particular, emergence of the Tychonic alternative). When that too worked out well, it seemed almost obligatory to see if the Galileo case—which obviously was no longer a matter of contagion within an expert community—could be accounted for in terms of the already written account of political judgment, turning on the interaction of cognitive and political polarization. This chapter is the result. On the general argument about cognition that I have been sketching, you should expect I would come to believe that a theory that working experience reveals to be so fruitful must be in some interesting sense right.

Appendix: Political Polarization

Social choice (other than on a very small scale) is never strictly and rarely even approximately determined on a one-man, one-vote basis. Even in contexts where decisions are nominally governed by a one-man, one-vote rule, asymmetries of power are routinely important. In a legislature, for example, there will always be some members more equal than others: for example, because of special expertise or authority on the issue immediately at hand, or because of special authority of a more general nature growing out of seniority or party office. But even in the case of a general election, there are important asymmetries. The agenda will be very limited (yes or no on a particular referendum question, or choose one from a small list of candidates), so a large component of the process that produces an outcome has taken place earlier, when the limited selection of the options that will actually appear on the ballot was determined. Later in the process, some actors have, but most do not have, significant power to influence the way others will cast their votes (by campaign contributions, popular appeal, etc.). So even in a nominal one-man, one-vote context, the simple model I will sketch requires interpretation to fit concrete cases, and a fortiori, the model will need interpretation to consider how it applies to situations (as in the Galileo case) where the formal decision rule is at the other extreme, so that a single actor—here, the pope—has in principle complete power of decision.

Nevertheless, the simple one-man, one-vote model can capture the basic nature of political polarization; and even the polar opposite situation can be interpreted in terms of the scheme I will sketch by thinking of "votes" as units of power. A predominance of power is needed to make a choice, analogous to the one-man, one-vote majority. The pope commanded enormous power, but in practice not absolute power. On the other hand, Galileo, even after he had surrendered himself to the Inquisition, was not wholly powerless. He was too well-known and admired a figure to be treated other than very carefully. He had many admirers outside, and even inside, the Church. Even the Church as a whole was not free as a practical matter to do whatever it wanted—there were things it could do in an absolute sense but could not afford to do considering the consequences for its future situation. And similar remarks apply to other actors within and beyond the corporate entity called "the Church." So although I will do no more here than sketch the simplest (one-man, one-vote) case, the dynamics of social choice captured by this model are, in fact, of very broad applicability.

Imagine that voters (or legislators) face a choice that involves just two considerations, labeled *A*-considerations and *B*-considerations in figure A.1. *A* versus *B* might be dollar costs versus benefits to health, or near-term versus long-run considerations, or efficiency considerations versus equity considerations; or any other pair that participants can easily see in terms of incommensurable scales of value. We will suppose that voters are not yet sharply polarized in the cognitive sense of [14.1]. *A*-ness and *B*-ness are not (yet) names for rival gestalts, one of which might come to be seen as dominant by some actors. An opinion survey, we will suppose, would find a distribution of judgments like that sketched by the dotted curve in figure A.1.

The curve is drawn with its peak a little off to one side to emphasize that we need not assume that the "middle of the road" position comes literally at the midpoint of the spectrum, only that it seems reasonable to expect that opinion would tend to clump into a "normal" sort of curve, reflecting prevailing knowledge, values, and sentiments. If there were no further complications, we would expect a social choice near *M* in the fig-

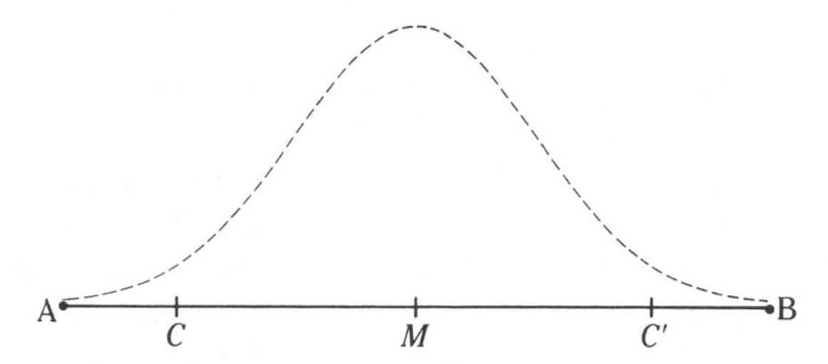

Figure A.1. A "normal" distribution of opinion about the relative importance of *A*-ness *vs.* *B*-ness.

ure. However, real social choices (even if the distribution of opinion should start in this way suggested by fig. A.1) often fail to conform to this "normal" pattern. What we want to go over is how the dynamics of social choice—the "choosing-up-sides" pressures of social choice—can account for that, especially as the political polarization ("twoness") interacts with the potential for cognitive polarization.

For an issue even to get on the agenda of choice is the outcome of a process. The narrowing of the range of salient choices to a few—often to just two—salient positions involves a process. Further, initially undecided or uncertain or only tentatively committed choosers do not merely passively respond from a fixed view but are partly targets of attempts at persuasion, partly seekers of information.

Naturally, it will be people with a strong prior disposition to a particular kind of choice (here a choice emphasizing *A*-considerations vs. *B*-considerations, or the reverse) who will be moved to commit resources (time, money, bargaining position on other issues) to an effort to shape this choice. For more typical political issues, these prior dispositions are often strongly motivated by economic interests; but it is also common to find activists motivated primarily by some sense of social values remote from such private interests, as is apparent in such cases as the environmental and right-to-life activism.

Whatever the balance and range of motivations, we expect to see efforts to organize this prior (or at least susceptible or latent) constituency. The preferences of voters will be ultimately controlling, which gives managers of a faction (or would-be managers who hope to organize a significant faction) an incentive to find a proposal that will gain enough support to carry the day. This argues for a middle-of-the-road proposal (near *M* in fig. A.1). But partially offsetting this incentive will be the need to develop a core of committed and active supporters. It would be hard to do that if the objective were merely to organize in support of *M*, or of some outcome only a shade away from *M*. Rather, it is the prospect of making a substantial impact (moving the decision definitely in one direction or the other) that is likely to arouse supporters, and indeed that is likely to initiate the effort at coalition building. But once a faction has been organized to promote (say) *C*, an incentive exists for those who dislike a choice dominated by *A*-considerations to develop a counterproposal (say *C′*), around which a countervailing coalition might be organized. Choice thus tends to become polarized in a tactical sense, reflecting the dynamics of a choice process whose outcome depends on putting together a winning coalition.

Suppose that initially this polarization was indeed merely tactical, with the activists still seeing things in terms of an unavoidable tradeoff between *A*- and *B*-considerations. The dynamics of the choice process will nevertheless push each faction to stake out a commitment to particular positions. In figure A.1, *C* and *C′* represent positions sufficiently far from *M* that active support can be aroused from those with strong *A*- or *B*-inclinations; but they also represent bargaining positions from which an eventual compromise closer to *M* can be managed. Even individuals who would be quite happy with *M* as an outcome will have an incentive to support a coalition focused on *C′*, once a coalition supporting *C* has been formed. *C′* may be a sensible counteroffer for someone who expects and perhaps even wants the eventual outcome close to *M*.

Once coalitions form around *C* and *C′*, each side must work to win over the less-committed participants in the process, that is, they will try to win over people who will influence the decision but for whom this particular issue is not (yet) a central concern. For this purpose (among others), each side is motivated to look for what in various contexts would be called a theory of the case, or a view of the world, which supports one side or the other of *C* versus *C′* (Chap. 7). Cognitive propensities then come more prominently into play. Even people who began with some sense of a reasonable balance between the *A*- and *B*-considerations may then come to see things in terms of a choice of right versus wrong, or justice versus privilege, good sense versus hysteria. In the context of a gestalt drawing, where values are not at stake and choices (acts) are not being faced, it is easy to notice that there are two radically different ways of interpreting the picture; with experience a person gains facility in routinely and at will shifting from one way of seeing the image to the other.

But in the context of social controversy, that is vastly more difficult. Facility at seeing things both ways is a cognitive burden in such a context, where what is immediately useful is to be able to quickly see how a piece of information or an argument fits into the view to which you are committed, and to quickly see something wrong with arguments or information claims that do not fit that view. Consequently, political polarization stimulates cognitive polarization, and in particular encourages the tendency for one view or the other to become dominant (easily seen, comfortably worked with). This is not, of course, a claim about what we consciously choose to do but about what we can be observed to do. Naturally, we find it far harder to notice that tendency in our friends than in our adversaries, and most difficult of all to notice it in ourselves.

So the consequence of political polarization is that active partisans will be doing their best to "sell" one view or the other. Among those they most commonly sell are themselves. On the *P*-cognitive argument, this occurs because just such "selling" is what will produce the many rehearsals of a pattern of argument that makes that way of seeing the issue habitual. What may have begun as a tactical (bargaining) commitment to *C* slips into a hardened view that sees *C* as a sensible choice and *C′* as a choice reflecting not merely a less-attractive or less well-informed or less reasonable judgment on tradeoffs but in the far harsher light reserved for advocates of some absurd, perverse, corrupt, or incomprehensible position. A more detailed account of this process than I can attempt here would pay close attention to the way that innate cognitive cues, culturally conditioned symbols, prevailing sentiments, and so on are recruited in this effort, and by no means always in a consciously manipulative manner. As has been long noted, man is a political animal, and although we now have specialists in image manipulation, everyone of us has some natural talent in that direction.

Out of the processes I have been describing, we could arrive at (among other possible ways the social controversy could evolve) the situation of figure A.2 or that of figure A.3. In both figures (in contrast to fig. A.1), the situation has become polarized. I have now sketched the curves showing the distribution of preferences as solid curves (rather than as dotted in fig. A.1) to suggest that not only has the shape of the distribution changed in a way that reflects at least a tactical polarization, but also the

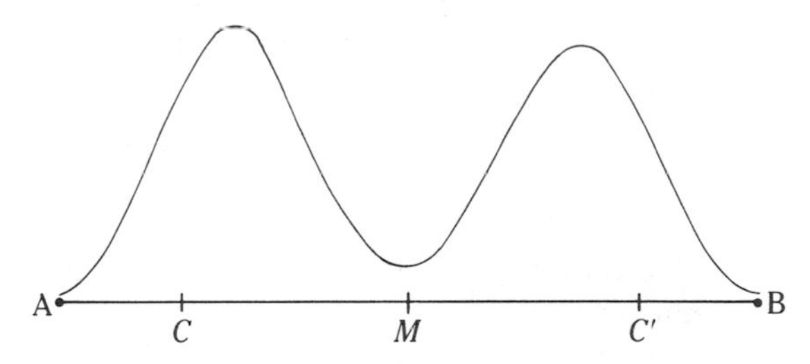

Figure A.2. Polarized judgments after a period of controversy but with the mean left essentially unchanged.

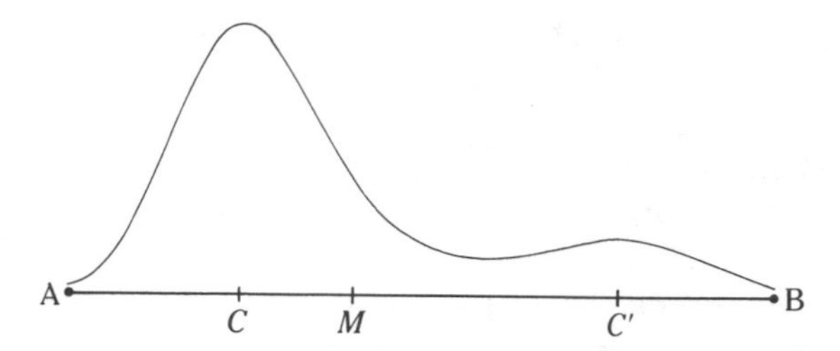

Figure A.3. Polarized judgments after a period of controversy but with the balance of opinion strongly favoring the partisans of *A*-ness.

positions are firmer, harder to shift. Instead of a general sense that a choice must be made that somehow balances *A*- and *B*-considerations, but with a good deal of openness about just where a reasonable balance might lie, individuals have now become committed to a relatively narrow range of alternatives clustered around either *C* or *C′*. Further, these choices now easily come to be seen (in the manner of a gestalt drawing) not as different choices along the spectrum of tradeoffs but as a discontinuous choice between different views of the world.

The intensity of the polarization is likely to depend a good deal on whether the situation that emerges is that of figure A.2 or figure A.3. In figure A.2 the strength of the two views is about equal; in figure A.3, position *C* is emphatically dominant. In general, though not inevitably, the situation of figure A.2 will lead to less-intense long-term polarization in the kinds of situations we are concerned with here. If the balance is close (and so perceived by both sides), proponents of position *C* will be moved to make their position seem reasonable to supporters of *C′* who might be tempted to defect, and vice versa. Further, if neither side is dominant (or if both are sufficiently uncertain about who is dominant as to be wary of an either/or choice), then there will be an incentive to return to the bargaining mode of working out a tradeoff along a spectrum of considerations instead of choosing between the polarized either/or alternatives. But once strong polarization has been established, this may be difficult to achieve, even if the balance of forces becomes more equal. Even further complications will be those that arise from misinterpreting the existing or prospective balance.

If the balance is one-sided, however, supporters of the dominant view have little incentive to see things in a balanced way. On the contrary, they will easily see an incentive to make the most of the opportunity by building into the choice as deep a commitment as possible to *C*, so that at a later point it will be as difficult as possible for supporters of *C′* to reverse the commitment that has been made. So we can get an extreme outcome as the result of a process that involves the reinforcing interaction between (a) the dynamics of building a winning coalition, and on the other hand (b) the way human beings process information and organize the way we see the world. The former lends itself to analysis with roots in the economic analysis of markets (focusing on incentives, competition, bargaining, and so on), while the latter turns on cognitive psychology and, in particular, the psychology of judgment and persuasion. Both these aspects will have to be integrated with a theory of leadership, which takes account of the importance (stressed here in connection with the pope) of the credibility of political leaders. But spelling that out will require another book.

Notes

Chapter 1

1. It would be a mistake to read this qualification merely as showing that mathematicians know better than to make this mistake. Something more specific is involved, which will play a large role in the overall argument. For most of us, the way experience would prompt us (tacitly) to "see" this question is in terms of looking down at our feet and seeing the string 36 inches too long. So we see the bit of extra string relative to our sense of the circumference of the earth. But for a mathematician, far more routinized habits in dealing with circles and their radii give priority to seeing the problem in terms of the fixed ratio between any circle and its radius. For both the nonmathematician and the mathematician, it is (intuitively) obvious that the extra height of the stretched string must be trivial. The difference is in the reference to which experience in the world makes a person tacitly sensitive. For the nonmathematician, the problem prompts "looking down" to the circumference of the earth from a normal position on the surface. For the mathematician, the intuitive reference is to the radius as a whole, as if "looking up" from the center. From one perspective, the increment seems microscopically small; from the other the increment will also be trivial, but relative to the radius of the earth (4000 miles). Six inches seems astonishingly large "looking down," but immediately plausible "looking up." This is the first of many examples of how experience in the world tunes cognition (establishes habits of mind), which then provides the anchor (point of departure, prototype) which—ordinarily tacitly, unself-consciously—guides judgment.

2. Readers familiar with the Wason work will notice that I am using playing cards in place of the usual cards labeled with letters on one side and numbers on the other. This change by itself has no effect on the base case results. But in conjunction with another by itself ineffective change it yields one of the novel remedial versions to be described in Chapter 8.

3. Occasionally claims are made that the Wason test really is very difficult, and these are sometimes supported by a painstaking spelling-out of a complete logical solution of the problem, which indeed looks forbidding (see, for example, Finocchiaro 1980). Often this is contrasted with versions of the test which prove to be easier. But a variant using a rule of the form "If *A* then not-*B*" (rather than "If *A* then *B*") would be at least as complicated in terms of spelled-out logic. Yet, as Wason found very early in his work on the problem, such a version is quite easy (see the discussion in Chap. 8). In general, if hard and easy versions are treated comparably, no difference in logical difficulty holds up across the range of variations which elicit the illusory response. A formal proof of 2 + 2 = 4 in Russell & Whitehead fashion would also be highly forbidding. But that would hardly support a claim that adding 2 + 2 is a very difficult logical challenge. On the argument of this study, what such demonstrations show is that indeed we all would be in terrible trouble if we could draw inferences only by going through an explicit and rigorously logical process. But that is not how the mind works. Rather, we become attuned to patterns of relationships, and see the inferences in the subjectively all-at-once way discussed in connection with Wason's and Rumelhart's comments on the Wason problem (Chap. 8). But on the detailed argument here, this is done with much more competence at generalized (non–task-specific) patterns of inference than Wason and Rumelhart seem to allow, offset by more vulnerability to misframing than they consider.

Chapter 2

1. In terms of what nonbiologists might think of as a big step, the step between any two generations that the advocates of "punctuated evolution" have in mind would be very small. We would not see an egg from what is clearly a reptile hatching into another sort of thing which is clearly a bird, or anything remotely like that. Indeed, the best nontechnical account of the conservative aspect of the Darwinian process was written by the best-known advocate of punctuated evolution (see Gould 1980).

2. In one of the rival views, hill-jumping is itself the result of a particularly fortuitous bit of hill-climbing. Macroevolution is a special case of microevolution. A species, climbing its local hill, gains what turns out to be a "preadaptation" for climbing some very different hill. If, in then getting to the top of some newly available niche, the creature tunes up its preadaptation into a striking novel capability, such as the ability to fly, then it will find itself with an extraordinary capability for hill-jumping, giving access to many niches which dominate those already occupied. Hence many areas become available for colonizing.

The alternative to this preadaptationist view involves some tidied up version of the "hopeful monster" argument put forward by Gouldschmit just before

World War II (Gould 1980:186 ff.). An updated though simplified version of this would imagine a creature at least conditionally viable (it can survive absent competition from its own species), and which happens to be in a situation that in fact gives, for a time at least, protection from strong conspecific competition. Imagine that this creature is misshapen in some way, such as having some part duplicated (an extra finger) or enlarged, or abnormal in some other way that would no doubt be disadvantageous in the short run.

In such circumstances, the ordinarily finely balanced genetic endowment of the creature is now somewhat out of balance, so that its descendants, competing among themselves, are much more likely to profit from further mutations or odd recombinations of genes than would the mainline species. A new capacity (a new function or a novel extension of an old function) has a relatively easy time getting started from the misshapen genotype, and once started ordinary hill-climbing can take it a long way. So if the initial disadvantage is not too strong, our misshapen but hopeful monster has a chance to develop some strikingly novel capability. Of a large number of such opportunities, a few will actually succeed. And once a new functional capability has appeared, then, as in the alternative formulation, hill-jumping becomes possible.

Chapter 4

1. My usage here implies that the only difference between a metaphor and an analogy lies in how hard it seems reasonable to push one thing as a model for another. But if that were all that was involved, then a metaphor would be just a weak analogy, and the idea that some metaphors can be powerful would be almost a contradiction in terms. What makes a powerful metaphor, however, is the tension between the model working very well on some particular points but in general being wholly inapt.

2. The neat phrasing here is taken from a remark of Alain Enthoven's with respect to policy analysis.

Chapter 5

1. Although I cannot develop the point very far, it will be apparent here, and in the discussion of learning to come, that I want to emphasize reflexivity in accounting for the special character of human (vs. nonhuman, including to date computerized) intelligence. Human beings make judgments about judgments, use tools to make tools, use language to define new language, and so on. I want to stress the value of seeing all these disparate sorts of reflexivity as of a piece, and as linked to consciousness.

2. I have inquired, casually but persistently, among people familiar with odd languages (anthropologists, linguists) for a counterexample. Perhaps a reader will be able to supply one.

3. On this account of knowledge versus belief, "know" is better characterized as an assessment-verb rather than as Ayer's (1956) achievement-verb, where the assessment is *always* made by the user of the verb. If I say I know *p*, my

belief state with respect to *p*, given my information, is $I^+ C^+$. But if I say that Willy knows *p*, then my belief state is $I^+ C^+$ with respect to *my* appraisal of *Willy's* information. For belief states are intuitive responses, and of course the intuition that comes to me looking at a situation is my intuition and could not possibly be Willy's intuition. On this account of what prompts such intuitions, there is no puzzle about such matters as unconscious knowledge ("He knows he's wrong, even if he won't admit it even to himself"). Similarly, the Gettier (1963) problem yields to straightforward analysis, so long as "know" is treated as an assessment-verb with respect to states in the belief matrix, and so long as we notice that (empirically, not normatively) when *A* uses "know" with respect to *B*'s judgment it is an assessment by *A* of what *A* takes to be *B*'s information.

Chapter 8

1. "Bayesian" is used to label a school of statistical practice; and it is also used to label adjustment of probabilities to take account of the background probability with which various possible cases are encountered. The term comes from "Bayes's theorem," which provides a formal basis for doing the adjustment calculations, and which plays an especially prominent role in the work of Bayesian statisticians. Bayes's theorem itself is an elementary relation in probability theory. So there is no more room for reasonable discussion about whether a person should accept Bayes's theorem than about whether a person should accept the Pythagorean theorem. Competent arguments are always about whether the empirical situation is one that requires a Bayesian adjustment, never about the uncontroversial point that in many empirical situations a probability estimate will be in error unless a Bayesian adjustment is made. In particular, Kahneman and Tversky's "taxi" problem has nothing to do with Bayesian statistics (as a school to which a person might or might not adhere), or with how to be a good Bayesian. Any competent statistician, whatever his views about Bayesian statistics, will agree with Kahneman and Tversky on the correct response to the taxi problem. Any set-piece puzzle gives an indeterminate answer if the solver is free to imagine additional conditions and complications rather than "play it as it lies." There is nothing "Bayesian" about the point. The only reason there is anything controversial about this note is that a considerable number of philosophers and psychologists who (it seems to me) should know better have made contrary claims. The puzzle of how that could happen will be taken up later in the chapter.

2. Many natural examples akin to the "negligible head injury" must be available. But of course we seldom notice them.

3. For a more usual view, see Hempel's (1973) article on "confirmation."

4. "Main" because aside from pragmatic ambiguities connected with if/then language, other unintended readings turn up, as is common in such experiments. Subjects will sometimes be uncertain about whether a violation could consist of failing to follow what were intended to be the preconditions of the problem, such as leaving blank the side of the card on which there should be a number; or subjects override the plain sense of the language and understand the problem as about which cards need to be turned over to find out what is on the other side, tak-

ing the rule as given. But these odd effects are too transient or rare to be of much interest, or they come when subjects are trying to rationalize a choice already made. But on the argument here a patient inquiry could track down some combination of semantic and scenario effects that accounts for them.

5. Wason (1983:67) rightly rejects claims that the problem is basically changed if instead of being asked to be sure that the rule is true or false for the cards shown, subjects are asked only whether it is ever violated (or, alternatively, whether it is always obeyed). The claim is that subjects are no longer being asked to assign logical truth values to each card, hence can get by without grasping material implication. Wason points out that the results that have made the problem famous do not depend on this, since they are generated about equally by either version. I want to add that a "true or false" framing could be in fact a significant source of noise in the results, in the following way: in formal logic, asking whether a proposition is false and asking whether it is true are complementary: answering one automatically tells you what the answer to the other must be. But that simple relation rarely holds outside of formal logic, where a proposition that is not false might be indeterminate rather than true. If the problem was actually intended to test whether subjects would use material implication, it would be of no interest, psychological or otherwise. Aristotle and everyone else who lived before that convention was introduced into formal logic around the turn of the century would be exposed as incompetent at logic by that standard, as well as every contemporary reader who does not interpret the problem as a technical problem in formal logic but merely takes it as the ordinary language problem it appears to be. See the earlier remarks on the irrelevance of material implication to ordinary language reasoning [5.5].

6. Examples of the two remedial forms:

A. Four cards have been picked from a mixed pack (some red backs, some blue). The person who chose was told to obey the following rule:

Rule: Pick any four cards . . . except that if a pick has a Red back, it must be at least a 6.

You see the cards lying on the table, with two face down and two face up.

Red back Blue back 7 of hearts 5 of clubs

Circle each card that must be turned over in order to be sure whether it violates the rule.

B. Attached is a packet of cards. Each has a letter (*A* or *D*) on one side, and a number (2 or 3) on the other side.

Rule: If the letter is *A*, the number must be 2.
Question: Do the cards obey the rule?

Without actually turning over any cards, sort them into the two categories shown: (1) cards that would need to be turned over to check for violations; (2) cards that don't need to be turned over. If you are uncertain how to sort a card, move it to the bottom of the stack. Continue until all the cards have been sorted.

The remainder of the instruction sheet is divided into two areas ("need to check," "don't need to check"). The final bit of instruction is important when in fact there are only four cards in the packet, to preempt the subject from spreading the cards out and effectively converting this version back into something close to the original version. The problem doesn't seem to arise if the packet is thicker (say, 10 cards).

7. Consider a string of 100 accidents. The expected fraction of Blue reports that are correct will be: (no. correct Blue reports) ÷ (no. correct Blue reports + *in*correct Green reports) = $(15 \times .8) \div (15 \times .8 + 85 \times .2) = 12/27 = .41$. To get an intuitive sense of what is going on, it may help to consider the case where first *no* Blue, and then *no* Green cab is working on the night of the accident. Next, consider what the probabilities must be if there is just *one* Green (but 99 Blue) or just one Blue (but 99 Green) cabs operating.

Here of course just one accident is involved, not the string of 100 postulated for the calculation. But the arithmetic works out in the same way for the single case. After all, if I give the probability of some event—whether tossing a coin one time, or the accident postulated here—as $p/100$, I must mean that if I considered 100 cases, each unique, in which I estimated the probability as $p/100$, then I would expect the number of cases in which the postulated event occurs to be p. What else could I (coherently) mean by such a statement?

8. My argument here was first prompted by a remark in conversation from Gilbert Harman. The comment about what was available before the appearance of mathematical probability in the mid-seventeenth century is related to but not the same as Hacking's (1985). His argument interprets the older notion as not really tied to experience (evidence, induction) at all. As I understand him, he treats the older notions of probability as limited to meaning something like "approvable by respected authority" and seems to exclude "plausible on the basis of what can from experience be expected to be encountered," which is not excluded in the sense of the older notion I will propose in [8.11]. On Hacking's account, treating this "probability$_b$" as a version of the prequantitative notion would need some cleaning up. On the argument of this study, however, we would expect that experience in the world would give everyone plausibility intuitions (probability$_b$) tied to what today we would call inductive evidence. What would be lacking would be conscious articulation of the basis of such intuitions (see, for a parallel situation, the argument about intuitions with respect to radial motion that concludes Chap. 12). So while I don't doubt that Hacking is correct in saying that *probabilitas* as used by medieval scholars was a notion explicitly tied to authority and not to evidence, my conjecture would be that very often a review of the context will find that the judgment often is tied to evidence anyway: for example, by treating authority as the *basis* for taking the judgment as believable (as indeed we often do today), or by mentioning what we would regard as inductive evidence as illustrative of the soundness of the judgment.

9. Wason found that many of his subjects have great difficulty in coming to see the point of the four-card test. But the experiments that revealed that were oriented to seeing how quickly a person could discover an error in judgment, under circumstances in which he or she was put into the position of defending the judgment. This turned out to be a remarkably slow process. Thus, an important

point about stubborn anomalies is that even simple anomalies can become stubborn when circumstances are such that the person digs himself into a commitment to the judgment by the exercise of trying to defend it. On a larger scale, this will play an important role later on, especially in Chapter 14. But the selection problem is not at all intrinsically stubborn. A subject who is informed that an answer was incorrect, and offered a coherent argument, will almost always quickly see the point. What is then interesting is that if given the problem again at some later time—even a few minutes later is often enough—the same mistake will be made again. And again, the person can immediately follow the logic of a correction. This makes sense if people are perfectly capable of the reasoning required to solve the problem but somehow see the problem in a way that yields a response of another kind. The scenario account here explains how that might happen.

10. The probability the witness is right when he reports Green is: $(.8 \times .85) \div (.8 \times .85 + .2 \times .15) = .96$. Across a large sample of such cases, we would expect the witness to report Green 71% of the time $(.8 \times .85 + .2 \times .15)$ and Blue 29% of the time $(.2 \times .85 + .8 \times .15)$. Each individual identification would be right with probability .8 (that is, given that the taxi is X, $p = .8$ that the witness will report X, whether X is Green or Blue). And the witness is also 80% right overall $(.96 \times .71) + (.41 \times .29) = .8$. For although the witness is right only 41% of the time when he reports Blue (as calculated earlier), that is offset by being right 96% of the time he reports Green. The arithmetic involved in all this is trivial. Nevertheless, illustrating the main point of the text, a reader who is not a professional statistician will almost certainly have to go over the numbers repeatedly to get a clear picture of why it is true that the witness is right 80% of the time whether the cab is Green or Blue, but is right only 41% if he reports the cab is Blue.

11. Bar Hillel (1980) and Kahneman & Tversky give examples of problems that elicit neglect of base rates (or apparent neglect) even though the particular difficulties discussed here are not involved. If the analysis here of the taxi problem is correct, then these other cases should yield to a similar analysis. And indeed that seems to be so. For Bar Hillel's examples, it often is sufficient to slightly reword the problem, making the language more natural. But some of the other examples—such as Ajzen's "exam" puzzle (Kahneman et al. 1982:155 ff.)—are more challenging, leading us to expect (on the analysis here) instructive scenario effects. These often interact with the $\text{probability}_b/\text{probability}_g$ issue to be discussed in the context of the Linda problem.

12. We could not expect a person to have good facility at dealing with a problem merely on the basis of being familiar with it in a secondhand way: we expect that what counts is actual experience, as a person readily learns how to get to a place by driving there a few times but often does not learn from riding as a passenger many times. Acting is what is most effective for learning, probably for the rather obvious reason that performing an act engages a much richer neurophysiology than merely watching.

Chapter 9

1. A good deal of work on toy versions of concept learning has followed Bruner et al.'s (1956) pioneering study. But such studies obviously cannot deal di-

rectly with learning that changes habits of mind that are important in the lives of the individuals involved. One promising avenue would be to exploit quasi-natural experiments, analyzing differences within and across groups exposed to different sorts of learning experience and taking advantage of differences among teachers and among learners. On the argument here, there also is something to be learned relevant to habits of mind from study of how people learn a sport, learn to compensate for injuries, and so on.

2. Obviously, from my remarks on p. 1, I don't think this was a tactical mistake. The cognitively interesting point is how a tactical move gets transmuted—even for people who would explicitly claim to be making only a tactical move—into a habit of mind difficult to eliminate when circumstances make it tactically inappropriate, or even tactically absurd.

3. In the short run, the account I will give runs strongly contrary to a rationalist view of science (judgment is easily dominated by habits of mind of which the individual is unaware and sometimes by tacit habits he would denounce as absurd). But over the longer run, rational argument dominates. In the account I will give, it makes little more sense to say that scientists "negotiate" the outcomes of their disputes than to say that football teams "negotiate" the final score of their games. And although human beings in something like Goodman's sense make their world, they do not make it any way they like, or that their political or other interests and preferences make appealing.

In terms of the belief matrix [5.8], what Kuhn's normal science is doing is taking advantage of opportunities to move from puzzles ($I^+ C^\circ$) to knowledge ($I^+ C^+$). The belief state of paradox ($I^+ C^-$) will most often yield to either puzzle or doubt ($I^+ C^\circ$ or $I^\circ C^-$). Both "puzzle" and "doubt" encourage search but do not mandate it. Once a person is seized with pursuing some puzzle or doubt—or more exactly, what generates that pursuit is that a person becomes sensitized to incompatible intuitions associated with the puzzle or doubt—that person now sees a paradox where others within the community only see puzzle or doubt. The characteristic resolution (when there is a successful resolution; naturally there are also many failures, where the matter is eventually just dropped without substantial progress toward a resolution) is in the form of new information and arguments that fit within the old view. Revolutionary science occurs when, on the contrary, the search is successful but only by leading to a new view within which the old view will be seen to fit in at least the sense that the new view can account for how the old view could have seemed believable.

"Paradox" here is used in the technical sense of the belief matrix. In ordinary language, we use the word in a more relaxed way, so that what we call a paradox is usually only a puzzle. On one of the short-cut arguments [7.9], we are in no doubt that the anomaly can be fixed; but there is a puzzle about just how to do it.

Chapter 10

1. Darwin himself was more cautious—by today's perspective—allowing room for Lamarckian evolution; from the perspective of his day, he was exceeding-

ly bold, always insisting that natural selection was the primary mechanism, not the sole one.

2. There has been a small epidemic of Darwin-debunking in the last decade, aimed at the sophisticated general reader, with the general message that while the claims of "scientific creationism" are overstated, they aren't as absurd as establishment biology claims. The claim, as might be expected, turns on accepting evolution but attacking natural selection.

Chapter 11

1. It is only in relatively recent years that the work of Price, Swerdlow, and Neugebauer has made a reasonable understanding of the technical details of Copernican astronomy widely available. The account here could never have been put together without the benefit of their work. But it will be obvious to knowledgeable readers that the cognitive argument leads me to a different view of the historical significance of those technical details.

Recent proposals to account for the discovery are Ravetz (1974), C. Wilson (1975), Swerdlow (1973:471–78), and Gingerich (1983). Each focuses on some technical issue that very plausibly would have played a role in the story. But on the account here, we would expect each of the effects they discuss to play its role in the consolidation and downhill phases of the discovery (Chap. 9), not in the uphill phase which climaxes with first seeing the heliocentric idea as believable.

Briefly: Ravetz argues that Copernicus was first led to belief in a rotating (but not orbiting) earth to account for precession. Having gone that far made it relatively easy (Ravetz argued) to go the further step to considering an orbiting as well as rotating earth. But his evidence that Copernicus was concerned with precession before reaching the heliocentric idea is weak (C. Wilson 1975). And, as discussed in the text, the idea of a rotating earth was conceived many times before Copernicus, and never led to the more radical notion.

Wilson and Swerdlow, on the other hand, each make an argument that turns critically on an explicit Copernican concern with the problem of constructing a nested-sphere account of the sort illustrated by figure 11.3. Although not at all the only argument against these versions of the path to the discovery, it is an important point that Copernicus shows remarkably little interest in nested-sphere models [13.6].

2. It is not clear how original Copernicus' work on the equant was (see Swerdlow & Neugebauer 1984:45 ff.). A good deal of his technical work repeats what was done by Arab astronomers several centuries earlier. But since, so far as known, the Arab work was not directly available to Copernicus, how far if at all he drew on it is uncertain. On the other hand, since Copernicus did little about the equant that had not been done centuries earlier by the Arabs, we have a further reason (subsidiary to the argument in the text) that there is nothing about working on the equant itself that leads to the heliocentric view; there is no hint of the heliocentric idea in the Arab work.

3. The first example seems to be Maestlin's analysis of the Comet of 1577 (see Westman 1975a).

4. The common claim that Copernicus was ignored in the sixteenth century (Koestler 1959; Cohen 1985; Hanson 1973) seems to be flatly wrong. To the extent there is evidence for such claims it consists almost entirely in noticing that elementary treatments almost always ignored Copernicus; but that is so even when they are written by people (Maestlin, Galileo) who we know were Copernicans. But the leading astronomers and their students were thoroughly familiar with Copernicus, as shown by the detailed annotations of his book and the many references to him in their work (Gingerich 1983).

5. In the relation between Copernican and Ptolemaic models, a common source of confusion is a failure to see why there is no contradiction between the nested-sphere (fig. 11.3) and plane diagram (fig. 11.1 or 11.2) representations of Ptolemaic astronomy. But they are two ways of envisioning the same thing. In a *different* sense, there is also no contradiction between a Ptolemaic diagram like figure 11.1 and the diagrams Copernicus uses in his book. As a geometrical exercise, you can show the latter pair are mathematically equivalent: the orbit of the earth can be translated into the epicycle of each outer planet and into the deferent of the inner planets, such that the angle of observation from the earth is unchanged. But there is *no* sense in which the Copernican world is equivalent to the Ptolemaic three-dimensional structure that goes with figure 11.3.

Confusion arises because it ordinarily is correct to say that figure 11.3 is equivalent to figure 11.1 or 11.2; *and* it ordinarily is correct to say that figure 11.1 is equivalent to the Copernican plane diagrams. But the phrase "equivalent to" doesn't mean the same thing in the two contexts, so that this is a case in which $a = b = c$ does not warrant $a = c$. The equivalence between figures 11.1 and 11.3 is in the strong sense of "physically equivalent"; but Ptolemaic/Copernican equivalence is only "mathematically equivalent with respect to naked eye observations." In terms of physics or cosmology the Copernican and Ptolemaic schemes are flatly incompatible; but until the telescope came along no observations could favor one over the other.

Logically, the minimum number of observations to specify the Copernican system *must* also be the minimum number for the Ptolemaic system, given the observational equivalence of the rival schemes. You could construct a Copernican model, and then transform it into a Ptolemaic equivalent. But anyone who knew enough to understand how that could be done would find it hard to suppose that the Ptolemaic model gives a better description of how the world really is. Once the observational equivalence came to be well understood, belief in a Ptolemaic world began to wane. Contrary to what has often been asserted, the choice for expert astronomers came to be mainly between the Copernican system and its Tychonic inversion, even *before* the telescope produced the first empirical evidence against Ptolemy. See the detailed discussion in Chapter 13.

In [12.2] I remark that Copernicus could have settled the question of whether the heliocentric retrogression would match observations without actually doing any calculations. For the geometrical translations (between the major Ptolemaic epicycle and the earth's heliocentric orbit) guarantee that the heliocentric calculations must come out right, even if the world were actually Ptolemaic not Copernican!

6. After Tycho's death, supporters of his system diverged, reflecting more than anything else whether they were motivated mainly by loyalty to Rome or to Scandinavia. For the first, the option of supposing the earth rotates was ruled out, since the Church's opposition to Copernican astronomy was tied to its reading of the biblical texts, which would be only trivially less in need of reinterpretation for a spinning earth as for an orbiting as well as spinning earth. But for the second, allowing the earth to spin was a common move.

7. Not much is known about how far the system we call "Ptolemaic" originated with Ptolemy. His work so completely supplanted what had gone before that almost nothing survives, with the significant exception of Aristarchus' little book on the earth/sun distance. The significance is that Aristarchus' book had a role to play in lending enhanced credibility to Ptolemy's calculation of this distance, which plays a crucial role in the story. Since Archimedes (in his "Sand-reckoner," which refers to Aristarchus' work) gives a value of 30 times the earth/moon distance, not the 18–20 given by the surviving copies of Aristarchus' book (see the discussion to follow in the text), we might doubt that the coincidence between Aristarchus' and Ptolemy's calculations is more than a corruption (or "correction") of Aristarchus' text.

8. Aristarchus is often said to have written a book on his heliocentric view, which of course suggests that he said much more than the few lines that Archimedes passed on. But there is no clear evidence that he did so; see Heath (1913), which is both the source (p. 302) of the claim that Aristarchus wrote a book on the matter and (p. 303) sufficient grounds for doubting that claim. Essentially, Archimedes uses language that might be read as a reference to a book but might also be a reference to nothing more than a sketch in the sand supported by oral argument.

Chapter 12

1. Another indication of this isolation is an exchange reported by Rheticus between Copernicus and his close friend Bishop Tiedeman Giese. Copernicus, urged by Giese to publish, proposed instead to make available tables based on his work. Giese counsels against that. In politics, he argues, it often pays to act first and only explain later after favorable results have been obtained, but in science (Giese argues) you shouldn't be doing that. Rheticus' anecdote does not specify who would receive the explanation. But since the only people who would make use of the tables Copernicus proposed would be astronomers, it was apparently his fellow astronomers that Copernicus was tempted to outmaneuver by giving them tables they could see were effective, and only afterward tell them that the tables were constructed from heliocentric models.

2. Osiander mentions the Venus anomaly as an illustration of why no astronomical theory should be taken seriously as a physical representation.

3. A particular link between the discovery of the New World and the Copernican challenge to Ptolemaic astronomy has been suggested before (Goldstein 1972). But on Goldstein's argument, Florentine scholars had broken free of the traditional commitments several decades before Columbus sailed. In this,

and in many other details, his account is quite different from and in fact incompatible with the account here.

It was through Goldstein's article that I learned of Rosen's 1943 paper which will be discussed shortly; in turn, I am indebted to Robert Westman of UCLA for alerting me to Goldstein's work when I sketched out the oecumene argument in conversation with him.

4. As the book was going to press I found a reference to a paper in Polish (Babicz 1973) on Copernicus and geography. From the English abstract, it is apparent that the author was skeptical of any substantial connection between Copernicus and the voyages of discovery. So, at the least, the information Babicz provides has not been biased by an argument of the sort I am making.

Chapter 13

1. The account here is about the circumstances under which a novel idea comes to take hold as social knowledge. "Justification" as a context (vs. discovery) is sometimes discussed in this same sense, especially when the writer is not much concerned with, or even wants to deny, any essential distinction between what normatively would justify a new idea and what socially and historically has in practice been seen as justifying it.

2. By the mid-sixteenth century especially, and to an important extent much earlier, it is a serious error to conflate Aristotelian and Ptolemaic views. In the 1570s, not only the then-young Tycho but also the future cardinal Ballarmine (Coyne & Baldini 1985) were pugnaciously critical of Aristotelianism, though like Galileo and us they respected Aristotle.

Among scholars who were not experts in astronomy, Aristotle was primary and Ptolemy merely a technician who worked out how to construct tables of planetary motions. Among such scholars, skepticism about the complex Ptolemaic technical machinery (epicycles, eccentrics, equants) was common. Nothing like that machinery is in Aristotle, or even compatible with Aristotle; and all of it would ordinarily be well beyond the comprehension of anyone but a dedicated astronomer. So an Aristotelian philosopher found it congenial to treat the details of Ptolemaic astronomy as a fussy business useful in making calculations but irrelevant to questions of cosmology. Ptolemaic astronomers, equally naturally, saw the same machinery as essential for understanding cosmology, and indeed it must have seemed absurd to such a person to hear a philosopher deriding the physical reality of epicycles, when the three-dimensional structure of the heavens made sense only if the epicycles were somehow really there. A Ptolemaic astronomer, then, in contrast to the Aristotelian philosopher, learned to be comfortable taking some things from Aristotle (the plenum, for example) and ignoring others (the 55 homocentric spheres, for example). During the sixteenth century, new information cast doubt on Aristotelian cosmology; but there was nothing that of itself could have given a Ptolemaic astronomer even mild qualms. On the contrary, for Ptolemaic cosmology (the epicycles and the rest) to be right, Aristotelian cosmology must be substantially wrong (the homocentric spheres, in particular). Observations commonly cited as undermining Ptolemy could easily be read as supporting Ptolemy. In particular, the nova of 1572 and the comet of 1577 did indeed undermine Aristotelian cosmology, as is often pointed out; but if the Copernican alternative were not on the stage, they would have been easily interpreted as supporting Ptolemaic cosmology by confirming the inadequacy of Aristotle's.

3. See, for example, the article on Brahe in the *Dictionary of Scientific Biography* (*DSB*).

4. I can only conjecture the details. But if you set the radii and synchronize the motion of the Ptolemaic epicycles of the outer planets equal to the earth/sun relation, and the same for the radius of the deferents of the inner planets, you get a "Copernican" version of the Ptolemaic system, in which the golden chain properties are imposed on a Ptolemaic framework.

5. A common counterargument here is that giving up the solid spheres was really a radical step since it removed the logical underpinning of the traditional commitment to uniform circular motions. Hence, astronomers were freed to consider alternatives (leading to Kepler's ellipses); and they were forced to look for some other way of accounting for what moved the planets. So on this sort of argument Tycho is also the starting point for Kepler's speculations about forces from the sun, and eventually for Newton's theory of gravity. But neither aspect of this argument, I think, holds up. Brahe himself, and most other astronomers down to the time of Newton, remained committed to the idea of uniform circular motion even after explicitly abandoning the solid spheres. On the other hand (as will be spelled out shortly), Tycho's argument against the solid spheres was in fact sufficiently weak that if we want a cognitively plausible account about why belief in solid spheres faded after around 1590, it will be as a *consequence* not as a cause of the contagion of the Copernican ideas. Astronomers wanted to account for astronomical details much more complicated than the main effects covered in the discussion of the Kepler diagram [12.3]. There were a variety of additional anomalies (departures from uniform motion), each of which required further machinery—usually yet more minor epicycles, leading to a very intricate clockwork of wheels within wheels within wheels. It was not—we can see from what astronomers of the period wrote—a hard thing to give up belief in the physical reality of this clockwork once an astronomer had lost the nested-sphere sense of the structure of the world, and the essential role in that world played by the major epicycles.

6. Boas & Hall (1942) translate Tycho's argument and give an interpretation of the sort I argue against here.

Chapter 14

1. More exactly, omniscience as well as omnipotence would be required to preempt politics. For if there were omnipotence only, an intricate politics would develop among rival advisors to the ruler.

2. It is hard to resist a left brain/right brain conjecture these days to account for this. But I have nothing to add that seems worth spelling out.

3. A possible fifth indicator is the frontispiece, which is a drawing of three figures engaged in discussion. Labels have been put on the hems of the gowns to tell us that they are Aristotle, Copernicus, and Ptolemy, though Aristotle

is shown as bald and homely, in contrast to other portrayals. Perhaps Renaissance publishing conventions made it unremarkable that the figures shown in intent discussion should be people being talked about, not the characters of the *Dialog* who do the talking. But if that is as odd as it sounds today, there is a curious and perhaps significant puzzle here: perhaps someone wanted to be sure that the three could not be taken to be Ptolemy, Copernicus, and Tycho.

4. There were two key documents: a version of the tides argument which Galileo had begun to circulate in manuscript in 1618, and a reply to an attack on Copernicanism (the "Letter to Ingoli"). Together, these drafts provided a sketch of what would eventually be transformed into the *Dialog*. That Galileo was quite openly circulating such material within a few years of the events of 1616, when memories were still fresh, adds powerfully to the (substantial on other grounds) doubt that Galileo had pledged himself in 1616 that he would never discuss the Copernican claims in any way. A document purporting to show that he had made such a promise played a large role in the trial. But neither that document nor any of the other explicit charges play any large role in the reports of the Florentine ambassador of his discussions with the pope, which give us our only glimmer of insight into whatever the pope really thought about the matter. It is entirely consistent with the account here that the pope would *not* rely on the formal charges in remarks that Galileo himself might have an opportunity to rebut once back in Florence. It is also, of course, consistent with the account here that neither the pope nor the Inquisition ever complained that Galileo failed to so much as mention the Tychonic system. "Galileo," the ambassador reports the pope as saying (September 18, 1632), "is still my friend . . . but he has stumbled into a great tangled mess [gran ginepreto] he does not understand" (my translation).

5. Wisan (1986), whose article drew my attention to the pope's remark in the previous note, provides the most recent elaboration of the main account of why the pope turned on Galileo, which depends on the point that Galileo slighted the closing argument dictated by the pope. Yet we know that the final pages were reviewed with special care by the pope's immediate staff. Although Urban told the Florentine ambassador that he never read the manuscript himself, given his explicit interest in how the argument was handled (Santillana 1955:317), *and* the obvious political and religious significance of the decision to let Galileo publish, *and* the pope's long-standing interest in Galileo's work, it is in fact very hard to believe that the pope did not personally read the crucial few paragraphs at some point during the months they were under review in the Vatican. That is like supposing that Nixon knew nothing of Watergate or Reagan nothing of the Iran/Contra affair.

Another recent account is Redondi (1987). The American edition (announced but not yet available as I write) follows previous publication in Italian and French. Redondi apparently provides a more detailed account of the political context of the trial than is otherwise available. But his theory of the trial exhibits, by the sheer desperation of its hypothesis, how far from a plausible account the voluminous prior literature has left the situation. Redondi's claim is that the fuss over Copernicanism was not the real issue but rather an obscure paragraph in Galileo's

1623 *Assayer,* which had been dedicated to the pope. To avoid a more direct link of heresy to himself, the pope let Galileo be condemned instead for the *Dialog.* I think this is like arguing that Hitler really invaded Russia because he loved borscht.

6. Fr. Grassi's argument in a 1618 polemic against Galileo is instructive here. He takes it as hardly worth a detailed argument that Ptolemy is untenable; and since the Church has barred advocacy of Copernicus, a good Catholic will favor Tycho (quoted in Drake & O'Malley 1960). Even earlier (1610), in the dedication of his report on his discoveries with the telescope—so he is saying this before receiving the post in Florence the dedication was intended to advance—Galileo himself refers in an offhand way, as if it mentioned nothing controversial, to Jupiter revolving around the sun ("the center of the universe," Drake trans. 1957:24). Later in the book, in its only other Copernican remark, Galileo frames an argument about the moons of Jupiter on a presumption that his reader is at least Tychonic (accepts the idea that the planets other than the Earth orbit the sun). The only detailed discussion in print is Schofield 1981. But the argument here is that things moved even further and faster than her account suggests. I have already cited equally unqualified remarks by Kepler and Maestlin years earlier [13.5]. So there is something bizarre in supposing that in 1630 Galileo was in the slightest doubt that his sophisticated critics would answer him by Tychonic, not Ptolemaic, arguments. I have asked a number of knowledgeable scholars to name a Ptolemaic astronomer active when the *Dialog* was published without learning of one.

Note that the Aristotle versus Ptolemy conflict discussed in n. 2 of Chap. 13 would not be an issue in 1630, since if there are no longer any Ptolemaic astronomers, the philosophers and theologians could interpret Ptolemy to suit themselves.

Literature Cited

Abu-Mustafa, Y. S., & Psaltis, D. 1987. Optical neural computers. *Scientific American* 256, no. 3:88–95.

Aiton, E. 1981. Introduction. In his trans. of Kepler 1596.

Ayer, A. J. 1956. *The Problem of Knowledge.* Penguin.

Babicz, J. 1973. Copernicus and geography. *Kwartalnik Historii Nauki i Techniki* 18:451–61.

Bar Hillel, M. 1980. The base-rate fallacy in probability judgments. *Acta Psychologica* 44:211–33.

Beer, A., & Strand, K. A. 1975. *Copernicus, Yesterday and Today.* Pergamon.

Bickerton, D. 1981. *Roots of Language.* Karoma.

Biskup, M., ed. 1973. *Regista Copernicana Calendar of Copernicus' Papers. Studia Copernicana VIII.* Polish Academy of Sciences.

Boas, M., & Hall, A. R. 1942. Tycho Brahe's system of the world. *Occasional Notes of the Royal Astronomical Society* 3:252–63.

Bonner, J. T. 1980. *Evolution of Culture in Animals.* Princeton University Press.

Bowler, P. J. 1984. *Evolution: the history of an idea.* University of California Press.

Braden, V. 1977. *Tennis for the Future.* Little.

Brahe, Tycho. 1578. *On the Comet of 1577.* Trans. in Christianson 1979.

______. 1588. *Recent Phenomena in the Aetherial World.* Chap 8. trans. in Boas & Hall 1959.

Bruner, J. S. 1957. Going beyond the information given. In H. Gulber et al., *Contemporary Approaches to Cognition.* Harvard University Press.

Bruner, J. S., et al. 1956. *A Study of Thinking.* Wiley.

Butterfield, H. 1965. *Origins of Modern Science.* Free Press.
Chomsky, N. 1975. *Reflections on Language.* Pantheon.
Christianson, J. R. 1979. Tycho Brahe's German treatise on the comet of 1577. *Isis* 17:110–40.
Cohen, I. B. 1985. *Revolution in Science.* Harvard University Press.
Cohen, L. J., et al. 1981. Can human irrationality be experimentally demonstrated? *Behavioral and Brain Sciences* 4:317–70.
Copernicus, N. 1543. *Of the revolutions of the heavenly spheres.* Trans. E. Rosen, Johns Hopkins University Press, 1978; A. M. Duncan, Barnes and Noble, 1976; C. G. Wallis, *Great Books of the Western World,* 1952.
Coyne, G. V., & Baldini, U. 1985. The young Bellarmine's thoughts on world systems. In Coyne et al., eds., *The Galileo Affair.* Specula Vaticana.
D'Andrade, R. G. 1979. Reasoning and the Wason problem. Unpublished manuscript.
Dennett, D. 1983. Why the law of effect will not go away. *Journal of the Theory of Social Behavior* 3:169–87.
Descartes, R. 1677. *The World.* Written c. 1630; publ. posthumously. Trans. M. S. Mahoney. Abaris, 1979.
Donne, J. 1611. "An anatomy of the world."
Drake, S. 1957. *Discoveries and Opinions of Galileo.* Doubleday, Anchor.
______. 1984. Galileo, Kepler and phases of Venus. *Journal for the History of Astronomy* 15:198–208.
Drake, S., & O'Malley, C. D. 1960. *The Controversy on the Comets of 1618.* University of Pennsylvania Press.
Dreyfus, H. L., & Dreyfus, S. E. 1986. *Mind over Machine.* Free Press.
Duhem, P. 1985. *Medieval Cosmology.* University of Chicago Press.
Eastwood, B. S. 1982. The "chaster path of Venus" in the astronomy of Martianus Capella. *Archive for History of the Exact Sciences* 32:145–58.
Edelman, G. M. 1985. Neural darwinism. In M. Shafto, *How We Know.* Harper & Row.
Evans, J. S. B. 1982. *The Psychology of Deductive Reasoning.* Routledge & Kegan Paul.
______. 1983. *Thinking and Reasoning.* Routledge & Kegan Paul.
Evans, J. S. B., & Wason, P. C. 1976. Rationalization in a reasoning task. *British Journal of Psychology* 67:479–86.
Evans, R. 1948. *An Introduction to Color.* Wiley.
Evans-Pritchard, E. E. 1976. *Witchcraft, Oracles and Magic among the Azande.* Oxford University Press.
Feller, W. 1968. *An Introduction to Probability Theory and Its Applications.* Vol. 1. Wiley.
Feyerabend, P. 1975. *Against Method.* Verso Editions.
Field, J. V. 1983. Kepler's rejection of solid celestial spheres. *Vistas in Astronomy* 23:207–11.
Finocchiaro, M. A. 1980. *Galileo and the Art of Reasoning.* Reidel.
Fodor, J. 1975. *The Language of Thought.* Crowell.

Frisby, J. P. 1979. *Seeing.* Oxford University Press.
Galilei, G. 1632. *Dialog on the Two Greatest Systems of the World.* Trans. Stillman Drake. California, 1953.
Geertz, C. 1973. Person, time and conduct in Bali. In Geertz, *Interpretation of Cultures.* Basic Books.
Gettier, E. L. 1963. Is justified true belief knowledge? *Analysis* 23:121–23.
Geymonat, L. 1965. *Galileo Galilei.* Trans. S. Drake. McGraw-Hill.
Gingerich, O. 1975a. Copernicus and the impact of printing. In Beer & Strand 1975.
______, ed. 1975b. *The Nature of Scientific Discovery.* Smithsonian Institution.
______. 1983. A fresh look at Copernican astronomy. In R. M. Hutchins et al., eds. *The Great Ideas Today 1983.*
Goldstein, T. 1972. The Renaissance concept of the earth in its influence upon Copernicus. *Terra Incognita* 4:19–51.
Gombrich, E. 1961. *Art and Illusion.* Princeton University Press.
______. 1982. *The Image and the Eye.* Cornell University Press.
Goody, J. 1977. *The Domestication of the Savage Mind.* Cambridge University Press.
Gordon, P., et al. 1984. One man's memory. *British Journal of Psychology* 75:1–14.
Gould, S. J. 1980. *The Panda's Thumb.* Norton.
Grant, E. 1984. In defense of the earth's centrality and immobility. *Transactions of the American Philosophical Society* 74.
Griggs, R. A. 1983. The role of problem content in the selection task and in the THOG problem. In Evans 1983.
Gruber, H. E. 1981a. *Darwin on Man.* University of Chicago Press.
______. 1981b. On the relation between "aha experiences" and the construction of ideas. *History of Science* 19:41–59.
Gruber, H. E., & Gruber, V. 1962. The eye of reason: Darwin's development during the Beagle voyage. *Isis* 53:186–200.
Hacking, I. 1975. *The Emergence of Probability.* Cambridge University Press.
Hadamard, J. 1945. *The Psychology of Invention in the Mathematical Field.* Dover (reprint).
Hanson, N. 1973. Copernicus. In *The Encyclopedia of Philosophy.*
Harman, G. 1973. *Thought.* Princeton University Press.
Harmon, L. 1973. The recognition of faces. *Scientific American* 229, no. 5:70–83.
Haugeland, J., ed. 1981. *Mind Design.* MIT Press.
Heath, T. 1913. *Aristarchus of Samos.* Oxford University Press.
Hebb, D. O. 1949. *The Organization of Behavior.* Wiley.
Hempel, C. G. 1973. Confirmation. In *The Encyclopedia of Philosophy.*
Henle, M. 1962. On the relation between logic and thinking. *Psychological Review* 69:366–78.
Herrnstein, R. J.; Loveland, D. H.; & Cable, C. 1976. Natural concepts in pigeons. *Journal of Experimental Psychology: Animal Behavior Processes* 2:285–311.
Hoch, S. J., & Tschirgi, J. E. 1985. Logical knowledge and cue redundancy in deductive reasoning. *Memory & Cognition* 13:453–62.

Hogarth, R. M. 1981. Beyond discrete biases. *Psychological Bulletin* 90:197–217.

Holland, J. H.; Holyoak, K. J.; Nisbett, R. E.; & Thagard, P. R. 1986. *Induction.* MIT Press.

Hull, D. 1973. Reduction in genetics: doing the impossible. In P. Suppes et al., eds., *Logic, Methodology and Philosophy of Science* 4:626.

Hull, D., et al. 1978. Planck's Principle. *Science* 202:717–23.

Jardine, N. 1984. *The Birth of History and Philosophy of Science.* Cambridge University Press.

Johnson-Laird, P. N. 1983. *Mental Models.* Harvard University Press.

Johnson-Laird, P. N., & Wason, P. C. 1970. Insight into a logical relation. *Quarterly Journal of Experimental Psychology* 22:49–61.

______. 1977. *Thinking: Readings in Cognitive Science.* Cambridge University Press.

Kahneman, D.; Slovic, P.; & Tversky, A. 1982. *Judgment Under Uncertainty: Heuristics and Biases.* Cambridge University Press.

Kahneman, D., & Tversky, A. 1982. Varieties of uncertainty. In Kahneman et al. 1982.

Kendall, M. G. 1956. Studies in the history of probability and statistics II. *Biometrica* 43:1–14.

Kepler, J. 1596. *The Cosmographic Mystery.* Trans. in Aiton 1981. Abaris.

______. 1601. *A Defense of Tycho against Ursus.* Trans. in Jardine 1984.

______. 1621. *Epitome of Copernican Astronomy.* Books IV and V. Trans. in *Great Books of the Western World.*

Kern, L. H., et al. 1983. Scientists' understanding of propositional logic. *Social Studies of Science* 13:131–46.

Koestler, A. 1959. *The Sleepwalkers.* Macmillan.

Koyre, A. 1973. *The Astronomical Revolution.* Cornell University Press.

Kuhn, T. 1957. *The Copernican Revolution.* Harvard University Press.

______. [1962] 1970. *The structure of scientific revolutions.* University of Chicago Press.

______. 1977. *The Essential Tension.* University of Chicago Press.

______. 1978. *Black Body Theory and the Quantum Discontinuity.* University of Chicago Press.

Lakoff, G. 1986. *Women, Fire, and Dangerous Things: What Categories Reveal about the Mind.* University of Chicago Press.

Libet, B. 1985. Unconscious cerebral initiative and the role of unconscious will on voluntary action. Published with critical commentary from reviewers and Libet's reply. *Behavioral and Brain Sciences* 8:529–67.

Magini, G. 1591. *New Theory of the Celestial Spheres, Congruent with the Observations of Copernicus.* Venice.

Margolis, H. 1982. *Selfishness, Altruism and Rationality.* Cambridge University Press. Reprint: University of Chicago Press, 1984.

______. 1987. Are preference reversals really reversals of preference? Unpublished manuscript.

Mayr, E. 1982. *The Growth of Biological Thought.* Harvard University Press.

Miller, G. A. 1956. The magical number 7, plus or minus 2. *Psychological Review* 63:81–97.
Minsky, M. 1975. A framework for representing knowledge. In Haugeland 1981.
Moesgard, K. P. 1972. Copernican influence on Tycho Brahe. *Studia Copernicana* 5:31–55.
Monod, J. 1972. *Chance and Necessity.* Vintage Books.
Neisser, U. 1967. *Cognitive Psychology.* Meredith.
Neugebauer, O. 1982. *A History of Ancient Mathematical Astronomy.* Springer-Verlag.
______. 1983. *Astronomy and History.* Springer-Verlag.
Newell, A., & Simon, H. A. 1972. *Human Problem Solving.* Prentice-Hall.
Nisbett, R. E., & Ross, L. 1980. *Human Inference: Strategies and Shortcomings of Social Judgment.* Prentice-Hall.
O'Gorman, E. 1961. *The Invention of America.* Greenwood.
Osborne, H. F. 1894. *From the Greeks to Darwin.* Ayer (reprint), 1975.
Pais, A. 1982. *Subtle Is the Lord.* Oxford University Press.
Poincaré, H. 1914. Mathematical discovery. Reprinted in Poincaré, *Science and Hypothesis.* Dover, 1952.
Polanyi, M. 1958. *Personal Knowledge.* University of Chicago Press.
Pollard, P., & Evans, J. 1983. The role of representativeness in statistical inference. In Evans 1983:107–35.
Popper, K. 1968. *Conjectures and Refutations.* Harper & Row.
______. 1972. *Objective Knowledge.* Oxford University Press.
Price, D. J. de S. 1959. Contra-Copernicus. In M. Clagett, *Critical Problem in the History of Science.* University of Wisconsin Press.
Pylyshyn, Z. 1984. *The Computational Theory of Mind.* MIT Press.
R: see Biskup 1973.
Randles, J. W. L. 1980. *De la Terre Plate au Globe Terrestre.* A. Colin.
Ravetz, J. R. 1974. Copernicus. *Journal of the British Astronomical Association* 84:257–71.
Redondi, E. 1987. *Galileo the Heretic.* Princeton University Press.
Rescher, N. 1982. *Empirical Inquiry.* Rowman & Littlefield.
Rheticus, J. 1540. *Narratio Prima.* Trans. in Rosen 1939.
Rock, I. 1985. *The Logic of Perception.* MIT Press.
Rosen, E. 1939. *Three Copernican Treatises.* Dover.
______. 1943. Copernicus and the discovery of America. *Hispanic-American Historical Review* (May):367–70.
Rumelhart, D. 1981. Schemata. In D. Norman, *Perspectives on Cognitive Science.* Erlbaum Associates.
Rumelhart, D., et al. 1986. *Parallel Distributed Processing: Explorations in the Microstructure of Cognition.* MIT Press.
Samuelson, P. A. 1952. Probability, utility, and the independence axiom. *Econometrica* 20:670–78.
Santillana, G. 1955. *The Crime of Galileo.* University of Chicago Press.
Savage, L. 1954. *Foundations of Statistics.* Wiley.

Schelling, T. C. 1978. *Micromotives and Macrobehavior.* Norton.
———. 1983. The intimate contest for self-control. In Schelling, *Choice and Consequence.* Harvard University Press.
Schofield, C. 1981. *Tychonic and Semi-Tychonic World Systems.* Ayer.
Schumpeter, J. A. 1954. *History of Economic Analysis.* Oxford University Press.
Selfridge, O., & Neisser, U. 1960. Pattern recognition by machine. *Scientific American* (August):60–68.
Seligman, M., & Hager, J. 1972. *Biological Boundaries of Behavior.* Appleton-Century-Croft.
Simon, H. 1955. A behavioral model of rational choice. *Quarterly Journal of Economics* 69:99–118.
———. 1969. *The Sciences of the Artificial.* MIT Press.
Steigler, G. J. 1966. The development of utility theory. In Steigler, *Essays in the History of Economics.* University of Chicago Press.
Stewart, I. 1987. Are mathematicians logical? *Nature* 325:386–87.
Swerdlow, N. 1973. The commentariolus of Copernicus. *Proceedings of the American Philosophical Society* 117:470.
Swerdlow, N., & Neugebauer, O. 1984. *Mathematical Astronomy in Copernicus' "de Revolutionibus."* Springer-Verlag.
Tversky, A. 1977. Features of similarity. *Psychological Review* 84:327–52.
Tversky, A., & Kahneman, D. 1974. Judgment under uncertainty. In Kahneman et al. 1982.
———. 1981. The framing of decisions and the psychology of choice. *Science* 211:453–58.
Van Helden, A. 1974. The telescope in the 17th Century. *Isis* 65:38–58.
———. 1985. *Measuring the Universe.* University of Chicago Press.
Walzer, M. 1985. *Exodus and Revolution.* Basic Books.
Wason, P. C. 1969. Regression in reasoning? *British Journal of Psychology* 60:471–80.
———. 1977. Self-contradiction. In Johnson-Laird & Wason 1977.
———. 1983. Realism and rationality in the selection task. In Evans 1983.
Wason, P. C., & Evans, J. S. B. 1975. Dual process in reasoning? *Cognition* 3:141–54.
Westfall, R. S. 1980. *Never at Rest.* Oxford.
Westman, R. S. 1975a. The Wittenberg interpretation of the Copernican theory. In Gingerich 1975b.
———. 1975b. Three responses to the Copernican theory. In Westman 1975c.
———. 1975c. *The Copernican Achievement.* University of California Press.
———. 1980. The astronomer's role in the sixteenth century. *History of Science* 18:105–47.
Wilson, C. 1975. Rheticus, Ravetz and the "necessity" of Copernicus' innovation. In Westman 1975c.
Wilson, E. O. 1975. *Sociobiology.* Harvard University Press.
Wisan, W. 1986. Galileo and God's creation. *Isis* 77:473–86.

Index